# Macromolecular Biomaterials


Editors

**Garth W. Hastings, D.Sc., Ph.D., C. Chem., F.R.S.C.**

Head
Biomedical Engineering Unit
North Staffordshire Polytechnic
and
Honorary Scientific Officer
North Staffordshire Health District
Medical Institute
Hartshill, Stoke-on-Trent
England

**Paul Ducheyne, Ph.D.**

Associate Professor of
Biomedical Engineering
University of Pennsylvania
Philadelphia, Pennsylvania


CRC Series in Structure-Property Relationship of Biomaterials


Editors-in-Chief
**Garth W. Hastings, D.Sc., Ph.D., C. Chem., F.R.S.C.**
**Paul Ducheyne, Ph.D.**



CRC Press
Taylor & Francis Group
Boca Raton London New York

CRC Press is an imprint of the
Taylor & Francis Group, an **informa** business

First published 1984 by CRC Press
Taylor & Francis Group
6000 Broken Sound Parkway NW, Suite 300
Boca Raton, FL 33487-2742

Reissued 2018 by CRC Press



**Library of Congress Cataloging in Publication Data**

Main entry under title:

Macromolecular biomaterials.

(CRC series in structure-property relationship of biomaterials)
Bibliography: p.
Includes index.
1. Polymers in medicine—Addresses, essays, lectures. I. Hastings, Garth W. II. Ducheyne, Paul. III. Series.
[DNLM: 1. Biocompatible materials. 2. Polymers. QT 34 M174]
R857.P6M33 1984 ‘610.28 83-7094
ISBN 0-8493-6263-6

A Library of Congress record exists under LC control number: 83007094

Publisher's Note
The publisher has gone to great lengths to ensure the quality of this reprint but points out that some imperfections in the original copies may be apparent.

Disclaimer
The publisher has made every effort to trace copyright holders and welcomes correspondence from those they have been unable to contact.

ISBN 13: 978-1-315-89512-3 (hbk)
ISBN 13: 978-1-351-07422-3 (ebk)

Visit the Taylor & Francis Web site at http://www.taylorandfrancis.com and the
CRC Press Web site at http://www.crcpress.com

## SERIES PREFACE

Biomaterials science is concerned with surgical implants and medical devices and their interaction with the tissues they contact. Their study, therefore, includes not only the properties of the materials from which they are made, but also those of the tissues which will accept them. Metals, ceramics, and macromolecules are the artifacts. Bone tendons, skin, nerves, and muscles are among the tissues studied. Prosthetic materials, implants, dental materials, dressings, extra corporeal devices, encapsulants, and orthoses are included among the applications.

It is not only the materials *per se* which interest the biomaterials scientist, but also the interactions in vivo, because it is at the interface between implant and tissues that the success of a procedure will be decided. This approach has led to the concept of a more aggressive role for biomaterials in the actual treatment of disease. Macromolecúlar drug delivery systems are receiving considerable attention, especially those with the capacity for targeting specific sites in the body. Sensing and control of body processes is a logical extension of this. There is much to be done before these newer developments become established.

The science of biomaterials has grown and developed over the last few years to become an accepted discipline of study. It is opportune, therefore, to systematize the study of biomaterials in order to improve their application in medical science, since that is the end point of all studies. That is the aim of this series of books on *Structure-Property Relationships in Biomaterials*. Knowledge of structure and the influence on properties is fundamental to any materials science study; it is a more complex problem to obtain the knowledge from tissue materials, as the living organism has a great capacity for change and adaptation in response to a stimulus. The stimulus may be chemical, electrical, or mechanical. The biomaterials scientist endeavors to identify and to use these stimuli and responses to improve the in vivo acceptability of the materials.

Many institutions and agencies have promoted the science of biomaterials. Societies now exist for this purpose. The Biological Engineering Society (U.K.) founded in 1960 formed a Biomaterials Group in 1974. In the same year the Society for Biomaterials was founded in the U.S. The European Society for Biomaterials (1976) was followed by Canadian and Japanese Societies (1979). All societies play a major role in disseminating knowledge through conferences and publications.

This series is complementary to these society activities. It is hoped that it will not only provide a basis of knowledge, but also its own stimulus for further progress. The series is inevitably selective. In part this is due to the editors' choice, in part to the availability of authors. The editors wish to thank those who fulfilled their agreements. Without them this series would not have been possible.

**G. W. Hastings**
**Series-Editor-in-Chief**

# PREFACE

The attraction of macromolecular materials lies in their diversity of types and, therefore, properties. They can be used as essentially inert materials (in joint replacement bearing surfaces or the silicone rubbers in plastic surgery) or they can be used in an interactive sense (drug release uses). Both uses depend on the properties of a macromolecular structure and this determines the nature of the material and hence its end use. Changes in bulk or mechanical properties are mainly changes in polymer backbone although the various groups pendant on the chain also have an effect. Interactive properties depend mainly on chemical groups attached to the backbone although the properties of the main chain will also influence their application, e.g., hydrophilicity, degradability.

The first part of this volume deals with some basic considerations of structure and its relation to function. After outlining some basic principles in the first chapter, including manufacturing effects, structural determination is discussed in detail. This is important since adequate characterization in terms of molecular size is necessary for a proper understanding and use. The theme of interrelationships between structure and properties is developed in the next chapters.

The second part of the volume deals with selected applications in which the development and use of macromolecular materials in specific applications is described. The selection of applications includes synthetic and natural polymers and the range of medical areas involved include orthopedics, dialysis, drug release (macromolecular pharmacology) blood contact, and plastic surgery. Since the list could not be completely comprehensive, these represent areas of special development or of continuing problems.

It is hoped that the volume will provide stimulus and information for established workers and for those new to biomaterials research. Future progress will continue to depend on the dedication of many individual researchers, many of whose names are in the extensive bibliographies. To them and to the chapter authors, the editors are grateful.

**G. W. Hastings and P. Ducheyne**
1983

## THE EDITORS

**Garth W. Hastings, D.Sc., Ph.D., C.Chem., F.R.S.C.,** is a graduate of the University of Birmingham, England with a B.Sc. in Chemistry (1953) and a Ph.D. (1956) for a thesis on ultrasonic degradation of polymers. After working for the Ministry of Aviation he became Senior Lecturer in Polymer Science at the University of New South Wales, Sydney, Australia (1961 to 1972). During this time he was Visiting Professor at Twente Technological University, Enschede, The Netherlands (1968-69), advising on their program in biomedical engineering. While in Australia, he became associated with Bernard Bloch, F.R.C.S., Orthopedic Surgeon, Sydney Hospital, and began a fruitful collaboration in the uses of plastics materials in surgery.

In 1972 he returned to England as Principal Lecturer in the Biomedical Engineering Unit of the North Staffordshire Polytechnic and the (now) North Staffordshire Health District with responsibility for research. With a particular interest in biomaterials research his own work has encompassed carbon fiber composites for surgical implants, adhesives, bioceramics, prosthesis performance in vivo, and electrical phenomena in bone. He is a member of British and International Standards Committees dealing with surgical implants and of other professional and scientific bodies, including Companion Fellow of the British Orthopaedic Association and Editor of the international Journal *Biomaterials*. He was elected President of the Biological Engineering Society in the U.K. (B.E.S.) in October, 1982. He was awarded a D.Sc. from the University of Birmingham in 1980 for a thesis in the field of biomedical applications of polymers. He has recently been appointed Acting Head of the department.

**Paul Ducheyne, Ph.D.** obtained the degree of metallurgical engineering from the Katholieke Universiteit Leuven, Belgium, in 1972. Subsequently he worked at the same university towards a Ph.D. on the thesis "Metallic Orthopaedic Implants with a Porous Coating" (1976). He stayed one year at the University of Florida as an International Postdoctoral N.I.H. Fellow and a CRB Honorary Fellow of the Belgian-American Educational Foundation. Thereafter he returned to the Katholieke Universiteit Leuven. There he was a lecturer and a research associate, affiliated with the National Foundation for Scientific Research of Belgium (NFWO). He recently joined the University of Pennsylvania, Philadelphia, as an Associate Professor of Biomedical Engineering and Orthopedic Surgery Research.

Dr. Ducheyne has published in major international journals on mechanical properties and design of prostheses, porous materials, bioglass, hydroxyapatite, and microstructural methods of analysis of biomedical materials. He is member of the editorial board of *Biomaterials, Journal of the Engineering Alumni of the University of Leuven, Journal Biomedical Materials Research*, and *Journal Biomechanics and Comtex System for Biomechanics and Bioengineering*.

He became active in various societies and institutions and has held or is holding the positions of Chairman-Founder of the "Biomedical Engineering and Health Care Group" of the Belgian Engineering Society, Secretary of the European Society for Biomaterials and member of the Board of Directors of Meditek (Belgian Institution to promote biomedical industrial activity).

## CONTRIBUTORS


**Francois Chabot, Ch. Eng.**
ERA CNRS No. 471
Université de Haute-Normandie
Mont-Saint-Aignan, France

**Pascal Christel, M.D.**
Laboratoire de Recherches Orthopédiques
Faculté de Médecine Lariboisière-St. Louis
Paris, France

**James M. Courtney, Ph.D.**
Bioengineering Unit
University of Strathclyde
Glasgow, U.K.

**Christine Fougnot, Ph.D.**
Laboratoire de Recherches sur le Macromolécules
Universitè Paris-Nord
Villetaneuse, France

**John D. S. Gaylor, Ph.D.**
Bioengineering Unit
University of Strathclyde
Glasgow, U.K.

**Neil B. Graham, Ph.D., F.R.S.C., F.P.R.I.**
Professor
Department of Pure and Applied Chemistry
University of Strathclyde
Glasgow, U.K.

**Garth W. Hastings, D.Sc., Ph.D., C. Chem., F.R.S.C.**
Head
Biomedical Engineering Unit
North Staffordshire Polytechnic
Honorary Scientific Officer
North Staffordshire Health District
Hartshill, Stoke-on-Trent, England

**G. Hoinkes, Ph.D.**
Institute of Minerology and Petrology
University of Innsbruck
Innsbruck, Austria

**Manfred Holtz, D.Sc.**
Clinic for Internal Medicine
Wilhelm Pieck University
Rostock, East Germany

**Jacqueline Jozefonvicz, Ph.D.**
Laboratoire de Recherches sur le Macromolécules
Université Paris-Nord
Villetaneuse, France

**Marcel Jozefowicz, Ph.D.**
Professor
Laboratoire de Recherches sur le Macromolécules
Université Paris-Nord
Villetaneuse, France

**Géza Kerényi, Ph.D.**
Director
Tisza Chemical Works
Budapest, Hungary

**Horst Klinkmann, M.D., D.Sc., Ph.D.**
Clinic for Internal Medicine
Wilhelm Pieck University
Rostock, East Germany

**Denis Labarre, Ph.D.**
Laboratoire de Recherches sur le Macromolécules
Université Paris-Nord
Villetaneuse, France

**Jean Leray, Ph.D.**
Unité INSERM No. 18
Hôpital Lariboisière
Paris, France

**G. Mikuz, M.D.**
Professor
Department of Pathology
University of Innsbruck
Innsbruck, Austria

**N. J. Mills, Ph.D.**
Lecturer
Department of Metallurgy and Materials
University of Birmingham
Birmingham, England

**Michael Parkany, Ph.D.**
ISO International Organization for Standardization
Geneva, Switzerland

**A. Propst, M.D.**
Professor and Head
Department of Pathology
University of Innsbruck
Innsbruck, Austria

**J. Schuijer, Ph.D.**
Professor
Polymer Materials Science
Department of Chemical Technology
Twente University of Technology
Enschede, The Netherlands

**J. W. A. van den Berg, Ph.D.**
Assistant Professor
Polymer Solution Thermodynamics
Department of Chemical Technology
Twente University of Technology
Enschede, The Netherlands

**Michel Vert, D.Sc.**
Research Professor
ERA CNRS No. 471
Université de Haute-Normandie
Mont-Saint-Aignan, France

**P. Wilflingseder, M.D.**
Professor and Head
Department of Plastic and Reconstructive Surgery
University of Innsbruck
Innsbruck, Austria

**David A. Wood, Ph.D.**
Department of Pure and Applied Chemistry
University of Strathclyde
Glasgow, U.K.

# TABLE OF CONTENTS

INTRODUCTION
Chapter 1
Structural Considerations of Plastic Materials ... 3
**G. W. Hastings**

PROPERTIES
Chapter 2
Molecular Structure Determination and Characterization ... 21
**J. W. A. van den Berg and J. Schuijer**

Chapter 3
The Relationship Between the Mechanical Properties of Polymers and their Structure ... 57
**N. J. Mills**

SELECTED APPLICATIONS
Chapter 4
Polymers of Natural Origin as Biomaterials. 1. Fibrin ... 91
**G. Kerényi**

Chapter 5
Polymers of Natural Origin as Biomaterials. 2. Collagen and Gelatin ... 111
**M. Parkany**

Chapter 6
Bioresorbable Plastic Materials for Bone Surgery ... 119
**M. Vert, P. Christel, F. Chabot, and J. Leray**

Chapter 7
Polymer Membranes ... 143
**J. M. Courtney, J. D. S. Gaylor, H. Klinkmann, and M. Holtz**

Chapter 8
Polymeric Inserts and Implants for the Controlled Release of Drugs ... 181
**N. B. Graham and D. A. Wood**

Chapter 9
Modifications to Polymer Surfaces to Improve Blood Compatibility ... 215
**C. Fougnot, D. Labarre, J. Jozefonvicz, and M. Josefowicz**

Chapter 10
Tissue Reactions with Silicone Rubber Implants (Morphological, Microchemical, and Clinical Investigations in Humans and Laboratory Animals) ... 239
**G. Mikuz, G. Hoinkes, A. Propst, and P. Wilflingseder**

Index ... 253

# *Introduction*

Chapter 1

# STRUCTURAL CONSIDERATIONS OF PLASTICS MATERIALS

**G. W. Hastings**

## TABLE OF CONTENTS

I. Introduction .... 4

II. Structure in Plastics Materials .... 5

III. Structure and Strength .... 6

IV. Structure and the Surface .... 7
- A. Surface Chemistry .... 7
- B. Surface Morphology .... 8
- C. Effect of Surface Properties in Biological Performance .... 8
  - 1. Chemical Effects .... 8
  - 2. Long-Range Effects .... 9
  - 3. Macroscopic Effects .... 10

V. Mechanical Properties and Biological Performance .... 10

VI. Effects of Manufacturing Processes .... 11
- A. General Considerations .... 11
- B. Polymerization .... 11
  - 1. Prepolymers and Intermediates .... 11
  - 2. Complete Polymerization .... 13
  - 3. Surface Grafting .... 14
- C. Fabrication of Plastics Materials .... 14
- D. The Physical Form of the Material .... 15

VII. Concluding Remarks .... 16

References .... 16

# I. INTRODUCTION

This volume in the series of *Structure-Property Relationships in Biomaterials* deals with plastics. The aim, as in the other volumes, will be to develop an understanding of the way in which properties are affected by structure and thereby to understand better the behavior of macromolecular materials presently in use. This will in turn provide a framework for the systematic study of new polymers for use as biomaterials.

In the period spanned by the two editions of the book by the author and surgical colleague,[1] it was noted that there had been few new developments, but rather a refining and elimination of clearly unsuitable plastics with experience being gained in the use of a relatively few products. In one sense, this process has continued over the last 10 years so that there are relatively few plastics materials in regular use as surgical implants (Table 1).

Each of these has contributed in a major way to the development of a particular branch of surgery. Joint replacement surgery has been dominated by the combined application of acrylic cement and ultra high molecular weight polyethylene (UHMWPE) to the problems of fixation and low friction/low wear articulation. Plastic surgery has seen over 2 decades of silicone elastomer use, with postmastectomy mammary prostheses being one important aspect. The use of polyester fiber in various forms as vascular prostheses made possible another advance in surgical methods. The treatment for hydrocephalus, ophthalmic disorders, and incontinence are some of the other significant developments closely related to availability of a suitable plastics material. Dramatic progress in microsurgery for nerve and blood vessel repair, limb reattachment, and grafting of vascularized soft tissue and bone relies on very fine sutures made from synthetic polymer fibers.

As well as this consolidation, there has been an increasing awareness of the limitations and actual disadvantages of these materials. Some of these are a consequence of the structure of the polymer as such, e.g., cold flow (creep) in polyethylene (typical of thermoplastics) and diffusion of silicone gel through an elastomer envelope. Others are a result of the addition of fillers, plasticizers, or stabilizers to the basic polymer to modify its properties. Abrasive particles may lead to adverse tissue reaction, and silica filler in silicone elastomer has been considered as a source of fibrosis around mammary implants.[2] The first polymer-related effect leads to material failure; the second additive-related effect leads to biocompatibility problems. Further consideration is given to these in Section IV.

The limitations in present materials have initiated research into alternatives either for improved mechanical and physical properties or for biocompatibility considerations. Attention has, for example, moved towards reinforced plastics (composite materials) which offer considerable advantages not only in terms of strength-to-weight considerations, but in terms of the anisotropic nature of their properties. Clinical uses have been reported for carbon fiber composites.[3]

However, it is not only for the purposes of improvement that materials are sought, but also for new applications, some of which have tended to rely on the polymer as a mainly inert substrate on or within which physical and chemical interactions can take place. Thus, polymer membranes are used to control diffusion of gases and ions and permeability is modified by application of chemical reactions to a mechanically suitable polymer film. The idea of an inert polymer as a substrate for other reactions has led to developments in chemotherapy. Controlled and sustained drug release is possible by encapsulation within or attachment to a polymer, and the targeting of specific disease sites has become feasible, e.g., in cancer treatment.[4,5]

Macromolecular biomaterials are thus well into another phase of application. The aim is not only replacement or support of tissues, but in some instances an active participation in therapy. This is either mechanically biased, e.g., in less rigid internal fixation of bone fractures using polymer matrices for the materials, or chemically biased, e.g., in drug release

**Table 1**
**POLYMERS IN CURRENT CLINICAL USE**

| | |
|---|---|
| Polyethylene (low density LDPE) | Reconstructive surgery |
| Polyethylene (UHMWPE) | Orthopedics |
| Silicone rubber | Plastic surgery, orthopedics |
| Polyacetals | Orthopedics |
| Epoxy resins (composites) | Orthopedics |
| Polyester (Dacron®) | Cardiovascular |
| Polyamides | Sutures |
| Fluoropolymers | General surgery, cardiovascular |
| Hydrogels | Ophthalmics |
| Poly(vinyl chloride) | Tubing |
| Polylactic acid | Sutures |
| Polyglycolic acid | Sutures |
| Rubbers, synthetic | Anesthetics |

systems, whereas blood compatible uses combine physical and chemical interactions. One area that should benefit from use of polymer-tissue chemical interactions is the development of adhesives for surgical use, but this has not progressed to any extent since the use of epoxy resins[6] and cyanoacrylates.[7] It may not yet be possible to achieve the objective proposed by Bloch[8] for an injectable glue for fracture fixation, but it may not seem so improbable with continuing progress in polymer chemistry. The continuing importance of polymers is the justification for this volume in the series and also emphasizes the need for an understanding of structure-property relationships.

## II. STRUCTURE IN PLASTICS MATERIALS

The essentials of the processes whereby polymers are made may be found in various works and in summary in the author's book.[1] Further definitions are given.[9]

All plastics and rubbers are chain macromolecules, i.e., molecules in which the component atoms or groups are joined by chemical bonds to form a long essentially linear molecule. This may be modified by coiling by side groups or branch chains which generate from the main chain or by short segments which join one chain to another (cross-linking).

Structure-property relationships may be considered at two levels. The first depends on molecular size, i.e., chain length and the statistical distribution of chain lengths within a particular sample, and is usually expressed as molecular weight. The usefulness of a polymer as a plastics material will depend markedly on the attainment of a minimum molecular weight during synthesis and its subsequent properties, upon the actual value of this, and its distribution. Hence, there is an emphasis in polymer characterization on measurement of molecular weight and related characteristics. The numerical value of molecular weight will depend upon the form of average used to express it and different experimental methods are used to obtain the various averages, as discussed later in this volume.

The second level of consideration deals with the way in which each molecule is spatially arranged and the nature of intermolecular and intramolecular associations. This level of organization determines the nature of the actual material, i.e., whether it is an elastomer or a fiber, whether it will crystallize or be amorphous. The type of component units in the chain structure have a profound effect upon these properties and Figure 1 shows the basic molecular structures, explained in the accompanying key. This provides a classification for polymers which directly relates the chain type and arrangement (or conformation) to the nature of the plastics material. It is thus a materials science classification and leads to an understanding of the related properties.

However, because the interatomic and inter- or intramolecular interactions are complex

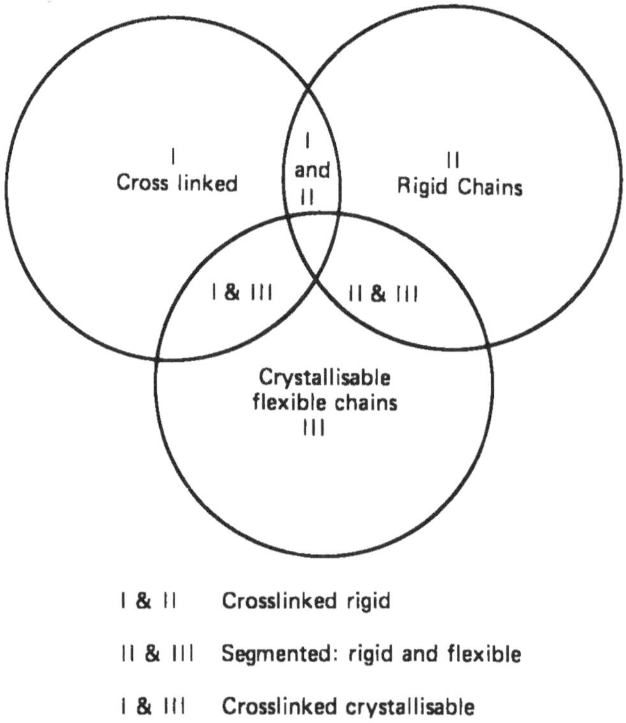


FIGURE 1. Basic polymer structures. See Reference 1 for full information.

in polymers, resulting from the diversity of atoms or groups present, a study of polymer structures must involve thermodynamic considerations, e.g., the free energy and entropy associated with chain coiling. Use has been made of this in a very practical way by Watson[10] in the development of endotracheal tubes; whereas natural rubber becomes stiffer on warming from room to body temperatures, poly(vinyl chloride) (PVC) softens, explainable in terms of thermodynamics of rubber elasticity.[11] PVC is therefore more comformable to the trachea and a softer inflatable cuff can be used to produce a seal. The advantage of this complexity is that it is possible to achieve wide variation in properties by small amounts of chemical modification. The combined effects of conformation and chemical reactivity are factors underlying the biological and physiological actions of macromolecules in living systems. Small changes in conformation greatly alter the capacity of macromolecular membranes for ion transport or for structural macromolecules to bind ions to control tissue mechanical properties or to control various processes.[12]

The properties of plastics or rubbers are directly relatable to the structural considerations arising from their component macromolecules and can be studied as such, but this gives only a partial understanding. The reason for this is that a pure polymer is seldom used, but is compounded with various additives which will modify the properties, often considerably.

## III. STRUCTURE AND STRENGTH

There are three main factors in polymer structure which contribute directly to strength related properties. These are

1. The groups which form part of the component units in the polymer chain
2. The nature of the inter- and intramolecular forces
3. Chain conformation effects

The relationships involved in 1 and 2 are closely related in their effects, since a group which promotes strong interchain forces will increase the strength of the resultant polymer.

Hydrogen bonds, i.e., electrostatic type secondary valence forces between a hydrogen atom and an electronegative atom located at another site are important strength contributors and the effect has been well-demonstrated in polyamides (nylons). These polymers are used for suture materials. Reduction in the number of potential hydrogen bond formers considerably decreases the tensile strength. The ridigity of unplasticized PVC results from hydrogen bond interactions between chlorine and hydrogen. Aromatic or other cyclic structures within the chain or as side groups are also strength imparting. Polystyrene is a good example in the way in which it gives a vulcanizing type of effect in thermoplastic SBS rubbers.

Segmented polyether polyurethanes are another good example of the incorporation of a hard segment into a softer elastomeric material to improve strength. A typical example appears to be based on a poly (tetra methylene ether glycol) for the elastomeric part and a methylene diisocyanate and diamine adduct for the hard segment. These polyurethanes are used in various cardiovascular implants.

Polyester fiber (Dacron®, Terylene®) used in vascular prostheses has a ring structure within the polymer backbone and this imparts rigidity to the molecule and tensile strength to the fibers, although regularity in structure which permits adjacent chain alignment is as important in fiber formation as the generation of interchain bonds. These interchain bonds need not only be H-bonds. Polar groups exert van der Waals' attractions and ionic groups are also used, e.g., in ionomers which utilize an ionized form of acrylic acid. Dimensionally stable polymers result from this structure.

Chain conformation effects determine the nature of the polymer as a plastics material and to a certain extent help to determine strength. Increase in crystallinity, for example, will increase stiffness, hardness, and tensile strength. The most important consequences of crystalinity are illustrated by consideration of the glass transition temperature, Tg. This is the temperature at which, on cooling, a macromolecule changes from a flexible to a brittle glass state. It is useful as a measure of the range of temperature over which the material may be exploited. For example, elastomeric material for use in the body should have a glass transition below room temperature. In molecular terms, it is the point at which available thermal energy is insufficient for chain mobility to occur and hence represents a discontinuity in mechanical properties.

## IV. STRUCTURE AND THE SURFACE

The nature of the surface of a plastics material is a consequence of molecular structure, manufacturing processes, and the subsequent history of the material. Each will exert its effect upon the biological performance of the material and careful control of each is therefore important. It is at the interface between the bulk plastic and the tissues that the interactions occur which determine in vivo behavior and these interactions will be a response to chemical and morphological features.

### A. Surface Chemistry

In natural macromolecules, the conformation of the chain controls the spatial arrangement of groups pendant to the main chain and any reaction which changes the primary conformation will vary the secondary chemistry or physical interaction of the molecule. Similarly, the surface chemistry of synthetic plastics materials is modified by chain structure.

It is not only the general chemistry, but the localized arrangement of groups which should be considered. Not only the presence, but also the arrangement of chemical moieties at a surface will be the determining factor in biological processes (Figure 2). Adhesion phenomena, ion attachment or transfer, macromolecule (protein or polysaccharide) deposition, ad-

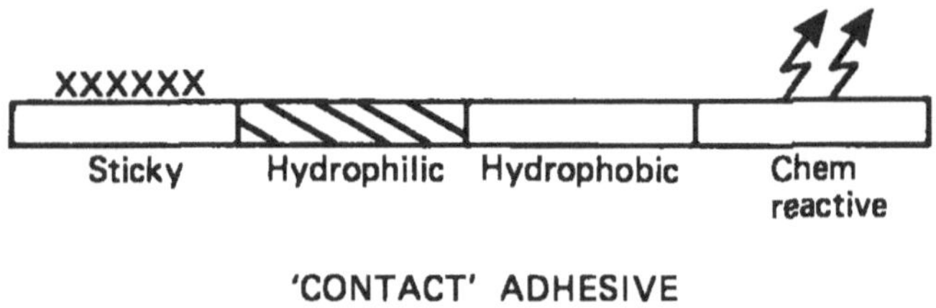


FIGURE 2. Structure and effect on surface chemistry, types of surface for adhesion. (Presented to Biomedical Polymers Conference, Durham, Biological Engineering Society, London, July 1982.

sorption and absorption of small molecules, diffusion, and processes associated with chemical activity (thermodynamic) are expected to be influenced by density of groups. The complex behavior of cell membranes in control processes occurring in cells illustrates the role of surface potentials and membrane structure. Plastics materials are not studied in terms of control for most of their uses as implants, but when used as membranes or in other systems in contact with blood, the control effect of the surface on the biological milieu is significant.

The crystal/amorphous ratio in a polymer will govern the availability and surface distribution of groups and these will be more ordered in the crystalline regions. In a fiber, this regulation of surface chemistry will be achieved by the amount of drawing, i.e., stretching, given to the fiber. Uncoiling and elongation of the polymer chains takes place on drawing and longitudinal alignment results with consequent changes in position of all groups reactive and nonreactive. Chemical variations can also be produced by copolymerization and this introduces the possibility of modifying the hydrophobic/hydrophilic nature of the surface which is now realized to be important for blood compatibility.

### B. Surface Morphology

The electron microscope study of collagen illustrates the effect of macromolecular conformation on morphology. Native collagen fibers have a very characteristic banded appearance with a 640-Å repeat. However, when the alignment of the tropocollagen macromolecules is varied, the banded appearance alters completely to give several different, though still characteristic, repeating structures. Studies on synthetic polymers have also revealed morphological changes when chain orientations are modified.[13] More information is needed as to the extent to which morphology will exert any specific effect on in vivo performance, but there must always be an awareness that a variation in molecular structure may exert an effect at a different level of interaction from that envisaged.

### C. Effect of Surface Properties in Biological Performance

In terms of interfacial effects, surface properties may be considered via a threefold classification:

1. Chemical effects are the most localized and operate at atomic group or molecular dimensions.
2. Long-range effects operate over a bigger span of surface dimensions and may, for example, be associated with dimensional changes.
3. Macroscopic effects are often mechanical in their effects, e.g., surface roughness.

#### *1. Chemical Effects*

Some of these are inherently related to the macromolecular system used, whereas others may be introduced deliberately. A polymer containing ionic groups may react with metallic ions in the tissues and form a new complex system. Calcification of implants could occur by this type of mechanism, but could also occur by chelate- and ligand-forming reactions which involve electron transfer.

Polar groups on the polymer surface will promote adsorption of tissue proteins. The interaction of protein and other tissue components is more important for cardiovascular and related implants than for those intended for musculo-skeletal applications. In fact, for these latter, mechanically related properties and factors associated with size and shape may have more effect on eventual success. In blood compatibility, protein adsorption is the critical factor since subsequent reactions will be in terms of that protein layer and its characteristics. For example, platelets adhere to this layer rather than to the actual implant surface. The surface chemistry response in adsorption is very rapid and soon removes the effect of that chemistry. It is observed that irreversible protein adsorption is obtained with hydrophobic surfaces, whereas hydrophilic surfaces will readily desorb proteins.[14] Most studies have been made with hydrophobic surfaces and it will be of interest to study hydrophilic surfaces and a hydrophilic/hydrophobic balance on one surface.

The production of negatively charged surfaces has been one attempt to reduce thrombogenicity by means of electrostatic repulsion since erythrocytes and platelets have net negative surface charges. Applied to metals, this approach has demonstrated that metals with negative electromotive potential form a positive interfacial potential with blood and have a high interaction with blood.

The position is, in fact, more complex because the protein adsorbed at a surface would be expected to be very different from unattached material. Any denaturation occurring may itself change the interaction with tissues and obscure the effect of the underlying material. The characterization of surface structure and the nature of charge distribution are generally ill-defined in any case and, therefore, the interactions at the interface are not easy to interpret.

It will be seen that surface chemistry is therefore of major importance for blood contact applications.[15] Joint replacements and other applications have different requirements.

*2. Long-Range Effects*

The consequences of a localized chemical reaction may be considered over a range of dimensions greater than those involved at the immediate reaction site. For example, although a protein may be adsorbed at specific sites, this process of adsorption may modify the macromolecular structure so that its biological effect is also modified. In some cases it will be deactivated; in others it will be activated to trigger further sequences of events. The adhesion of molecules at the interface is subject to the energy requirements expressed in terms of surface tensions of the species involved at the interface. The contact angle of the liquid phase on the solid surface is a measure of surface forces and this will prescribe to some extent the structure of the adhering molecules. When there are charges present on the surface, there will be additional constraints imposed on adhering molecules. Even electrically neutral molecules will change their form due to dipole interactions, whereas the subsequent conformation of charged molecules will depend on net repulsions of like charges and attraction of unlike. Ions will also be attracted to such surfaces and the ion flux will be a key factor in the surface interactions.

Another application of these effects is seen when a study is made of the wear processes taking place between the articulating surfaces of joint prostheses. In vitro studies of the wear of metals against UHMWPE show that when these are done in a dry (i.e., unlubricated) condition, a transfer film of polymer forms on the metal counterface. Subsequent wear is then a result of plastic/plastic contact and is much reduced from the initial phase in which metal to plastic contact prevails. The adhesion of the transfer film determines wear behavior. In "wet" conditions, the behavior is different and this author has not observed a transfer film on the surface of the retrieved hip or knee prostheses.

*3. Macroscopic Effects*

The fibrous capsule which develops around implants could be considered as a long-range effect, but mechanical (movement) factors are also involved. The motion at an interface can prevent direct tissue attachment. This has led to development of stepped stem prostheses and various types of porous or indented surfaces to encourage stabilization of orthopedic implants by tissue ingrowth. Acrylic cement injected into the medullary canal to fill interstices in the bone bed serves a similar purpose in limiting movement.

The various mechanical manufacturing artifacts exert a largely macroscopic effect. Wear is dependent on the asperities on the metal and on the roughness of the plastic, and there appears to be an optimum surface roughness. Surface properties exert a profound effect on biological performance at all levels, and there will be increasing need to specify surface condition as well as the properties of the bulk material.

## V. MECHANICAL PROPERTIES AND BIOLOGICAL PERFORMANCE

If a plastics material used for implant purposes deforms or breaks, it will have failed its intended purpose. It is not only the adequacy of mechanical properties at the time of implantation, but the change of these in vivo which counts. A design therefore should not only call for minimum values of yield stress or elastic modulus, but for the retention of this value under the influence of in vivo stresses and environment. Safety factors required in calculating minimum requirements may be surprisingly greater than expected. It is not only the effects of normal activity, but the effects of unexpected peak loadings due to excessive activity which must be considered, remembering also that operated patients may remain in a somewhat "shaky" condition. In fact, patient training is just as important as implant design.

The main design or materials parameters of elastic modulus, yield stress, and ultimate stress can be interpreted in terms of macromolecular structure and additives. Environmental conditions in vivo can affect all of them. Penetration of tissue components, e.g., lipids, will plasticize polymers and reduce elastic modulus, and the removal of added plasticizer will have the reverse effect. (An additional effect of plasticizer removal may be local or systemic toxic effects.) Crystalline polymers (UHMWPE) will resist ingress of fluids or low molecular weight materials more than certain elastomers (silicone rubbers take up lipids) or amorphous polymers (acrylic cement absorbs water). It should be noted that yield strength may in fact be raised by absorption of tissue molecules in inhomogeneous plastics.

Degradation reactions may also occur in vivo whereby chains are broken. Low molecular weight fragments may then be released, e.g., formaldehyde from cyanoacrylate by hydrolytic degradation and possibly from polyacetals by a combined mechanical and hydrolysis mechanism. Depending on the type of degradation, the strength may progressively decline. An adjunct to the in vivo degradation is the effect produced by sterilization which may be sufficient to compromise the material before it is used.[16] Elastic modulus and yield strength are reduced by degradation processes.

A particular manifestation of in vivo mechanical loads is creep. This is deformation which takes place at loads below the yield stress. Thermoplastic materials are subject to this effect and this includes UHMWPE. Dowling et al.[17] have reported creep as one factor in the long-term wear of acetabular cups. In most applications in which the UHMWPE component can be adequately supported by underlying bone or by a metal surround, the consequences of creep can be minimized. Reinforcement with carbon fiber has been investigated to try to reduce creep.

Fracture strengths which depend on the mode of crack propagation are generally environment sensitive. Water or ions at the site of crack growth may produce a stress-corrosion type of crack. Polymer surfaces are subject to cracking (crazing) and this occurs in the presence of nonsolvents, possibly by an effect on macromolecular dimensions on the surface. Acrylics craze following treatment with high energy radiation, e.g., gamma irradiation.

Fatigue failure must always be considered among design parameters. This is fracture at stress levels below the ultimate strength in a material undergoing cyclic deformations. The various design features which exacerbate this, sharp corners or notches, as well as surface damage due to careless handling or storage must be borne in mind. Acrylic cement failure may result mechanically from fatigue together with creep, although being a glassy amorphous polymer it is a brittle material.

The behavior of reinforced plastics (composites) will depend on the nature of the reinforcement material, fibrous or powder. Generally anything that weakens the bond between the reinforcing material and the polymer matrix will reduce the elastic modulus. Fatigue in composites may not be evidenced by failure, but by reduction in stiffness. This property change was used to investigate fatigue properties of carbon fiber reinforced epoxies in the form of bone plates. The reduction in stiffness at $10^6$ c in dependence on initial load enabled the determination of a fatigue limit. Some plates showed partial delamination for the highest loads used, but none fractured.[18]

The in vivo environment will markedly influence the design parameters for mechanical properties of plastics materials implants. Although a general understanding can be gained from macromolecular considerations, considerable modification may be observed. Both environment and mechanical factors should be used in the design specification.

## VI. EFFECTS OF MANUFACTURING PROCESSES

### A. General Considerations

The control of polymer properties during manufacture is effected at the time of polymerization and also in the subsequent compounding and fabrication processes during which additives are incorporated and the final form of the plastics article is made. The polymerization stage affords the opportunity to modify properties related to the physical chemistry of the material. These include solubility, melting temperature which carries its effect forward into processing characteristics, the character of the material as elastomeric, plastic fiber-forming, crystallizing, thermoplastic or thermosetting properties, wetting characteristics, adhesive nature, environmental resistance, and basic strength properties.

The other stages follow after the polymer is made. At this time, it is blended or compounded with pigments, fillers, stabilizers, lubricants, or plasticizers and then formed into an intermediate stage for ease of transport and handling or into the finished article. The control here is over materials properties, strength, toughness and impact resistance, flexibility and rigidity, resistance to degradation, flex and fatigue life, abrasion (wear) resistance, and environmental resistance. In addition to these end-use related properties, control of processability may be achieved.

There is a further stage in which the actual forming of the material will affect its properties, e.g., film-forming, fiber diameter, surface texture, or porosity. It is not intended to discuss in detail workshop processes, machining, drilling, cutting, although insofar as they affect the surface, they have an effect upon in vivo behavior.

### B. Polymerization

The aim is to convert liquid or solid monomers into an intermediate prepolymer which serves as a basis for further chemical reaction, polymerization or cross-linking, or into a fully polymerized material which will go straight to fabrication processes.

#### *1. Prepolymers and Intermediates*

Prepolymers and intermediates are usually prepared by step-growth types of polymeri-

**Table 2**
**TYPICAL DIOLS USED IN BIOMEDICAL POLYURETHANES**

| | |
|---|---|
| $HO(CH_2)_4\text{-}[O(CH_2)_4]_n\ OH$ | Poly(tetramethylene oxide) glycol |
| $HOCH_2CH_2\text{-}[OCH_2CH_2]_n\ OH$ | Poly(1,2-oxyethylene) glycol |
| $HO(CH_2)_4\text{-}[O\text{-}\overset{O}{\overset{\Vert}{C}}\text{-}(CH_2)_5]_n\ OH$ | Poly(caprolactone) glycol |
| $HO(CH_2)_4\text{-}[O\text{-}\overset{O}{\overset{\Vert}{C}}\text{-}(CH_2)_4\text{-}\overset{O}{\overset{\Vert}{C}}\text{-}O\text{-}(CH_2)_4]_n\ OH$ | Poly(tetramethylene adipate) glycol |

**Table 3**
**PROPERTIES OF TYPICAL SILICONE RUBBERS**

| Class[a] | Hardness Shore A | Plasticity | Ultimate tensile strength, min. psi, MPa | Ultimate elongation (% min) | Tear resistance Min. lb$_f$/in. (kN/m) | Specific gravity |
|---|---|---|---|---|---|---|
| PVMQ | Soft, 29 + 5 | 0.080 ± 0.015 | 600 (4.13) | 400 | 65 (11.4) | 1.12 ± 0.03 |
| VMQ | Medium, 50 ± 5 | 0.095 ± 0.015 | 750 (5.17) | 360 | 71 (12.5) | 1.14 ± 0.03 |
| VMQ | Hard, 70 ± 5 | 0.135 ± 0.025 | 700 (4.83) | 190 | 80 (14.0) | 1.24 ± 0.03 |

[a] PVMQ, silicone rubbers having both methyl and phenyl substituent groups on the polymer chain; VMQ, silicone rubbers having both methyl and vinyl substituent groups on the polymer chain.

zation, typical of this being epoxy resins and polyethers or polyesters intended for polyurethane preparation. The aim is to prepare a chain segment of sufficient length to impart the desired properties to the final molecule. Rigid or elastomeric segments need to be of sufficient length to enhance or direct the properties, otherwise the affect will be only that of a copolymer or polymer blend when the individual characteristics of each component monomer are masked.

Thermoplastic polyurethanes, for example, used in biomedical applications are usually made from macromolecular diols reacted with a diisocyanate and a glycol chain extender. The reaction produces hard and soft segments in the chain. The balance of hard to soft segments determines hardness, elastic modulus, and extensibility. Generally, polyethers are used as the long chain diol for biomedical polyurethanes since they are more hydrolysis resistant and biologically acceptable than polyesters. Another version uses a polyether with a polydimethylsiloxane which combines the mechanical properties of the polyurethane with the surface tissue properties of the silicone. The biomedical polyurethanes having this segmented structure show good flex life and, in particular, blood compatibility. The properties of these polyurethanes are achieved by intermolecular interactions which lead to a domain type of chain entanglement structure. During heat processing, the domains melt and permit flow, reforming on cooling. Some of the typical diols used are shown in Table 2.

Silicone elastomers are made by heat vulcanization (cross-linking) of a lower molecular weight polymer "gum-stock" and the chain length of this controls the density of cross-links in the vulcanized elastomer and the length of the flexible segments. Properties of some typical silicones are shown in Table 3. Room temperature vulcanizing stocks are also availalbe and are used where the rubber is to be formed at the time of surgery or for building up contoured articles. Most prostheses, e.g., mammary, hydrocephalus, are made from heat vulcanized stocks.

Another type of intermediate is acrylic cement for room temperature curing. This is different from those above for two reasons. First, it is based on a polymer mixed with a monomer. It is made at the time of surgery by mixing together a powder, comprising fully polymerized polymethyl methacrylate incorporating an initiator and the liquid monomer methyl methacrylate incorporating an amine activator. Polymerization takes place in and around swollen polymer beads. Second, it is a free radical polymerization in which polymers are formed more rapidly than the types described previously. Control over the product is exercised by particle size variation, initiator concentration, and incorporation of other monomers with methyl methacrylate into the liquid component. Butyl methacrylate imparts a more rubbery character to the product. Styrene may also be copolymerized in the solid phase. In the example of this very useful adjunct to total joint replacement surgery, properties are controlled by polymerization and by the physical form of the solid phase, and this latter factor is directly seen in the coherence of the final cured cement. A further variable is the addition of barium sulfate as a radio opacifier and this modifies crack behavior.[19,20]

*2. Complete Polymerization*

The majority of biomedical polymers will have been completely polymerized by the various industrial methods. In this context, it should be remembered that certain initiator or catalyst systems leave residues in the polymer which may be undesirable for biomaterials. Traces of catalyst substrates are other contaminants.

At present, only two polymer materials have been standardized internationally for medical use, UHMWPE and acrylic cement, and work is proceeding with silicone elastomers. The polyethylene standard specifies that pure ethylene monomer is to be used and gives limits for trace elements and contaminants.[21]

The principal aim of the manufacturer will be to achieve a molecular weight which satisfies his product specification. Quality control methods, often of an "in house" nature, will be used to ensure a consistent product, but some will be somewhat empirical and bear a tenuous relationship to the application, rather than to inherent properties of molecular weight. For example, a sand slurry test is specified in trade literature for polyethylene which does, in fact, appear to correlate with molecular weight ranges.[22] It should be further realized that some condensation polymers may be blends of various batches to achieve an average molecular weight, and although this may not be important for most uses as intermediates, for biomedical polymers where biological interaction may depend on segment lengths, this may be critical.

Batch to batch variation is to be expected, but the individual user will need to determine how important this is. Molecular weight and molecular weight distribution are a means for achieving the desired properties, and most plastics materials will be specified and characterized in terms of certain "secondary" properties, melt index, viscosity, tensile, or impact properties. Hence, primary molecular properties may change and be modified to meet any process variability.

Examples of materials prepared by bulk polymerization are the hydrophilic polymers (hydrogels). All hydrogels are characterized by a hydrophilic polymer network incorporating water or an aqueous solution, gelatin being of course, a well-known natural example used in surgery. The synthetic materials used for contact lenses, opthalmic hydrogels, are made from hydroxyethyl methacrylate cross-linked with a glycol dimethacrylate.[23] Variation in cross-linking leads to different equilibrium amounts of water in the gel and hence different extent of swelling.

Although brittle when dry, as typifies the glassy polymers, hydrogels are soft and flexible when swollen and are permeable to ions and to oxygen. They are also optically transparent. In manufacture, a prepolymer is polymerized in bulk and lenses are machined from the rods produced. Other hydrophilic methacrylate esters or vinyl pyrrolidone have been copolymerized with hydroxyethyl methacrylate. Graft copolymers onto a hydrophilic polymer backbone have also been made.

The aim with these lens materials is to produce a high water content with continual improvement in mechanical properties. Hydrophobic comonomers have been used as well as cross-link variation and use of copolymers of hydrophilic monomers. Methyl methacrylate and methacrylic acid (as the sodium salt) have both been used in various combinations with hydroxyethyl methacrylate, vinyl pyrrolidone, or other hydrophilic monomers with a variety of cross-linkers. A high rate of oxygen diffusion is an important property and if the lens is thin enough to give this, it is usually necessary to decrease water content to improve mechanical properties.

These are not the only polymers used for contact lenses. The traditional material has been poly(methylmethacrylate), the lens being machined from completely polymerized rods. Silicone rubbers have also been examined, but as in other sites of the body, they absorb lipids.[24]

*3. Surface Grafting*

Another polymerization process used to change surface properties is a type of copolymerization in which a monomer is caused to polymerize to active sites on the surface of another polymer. The substrate material will have been selected usually for its mechanical properties, e.g., strength or stiffness. The surface character, however, must be changed, most usually for blood compatibility reasons. The active sites at which the second polymerization takes place may be produced by:

1. Prior chemical treatment of the surface to introduce active groups which will couple with monomers, e.g., carboxyl, hydroxyl, amine groups
2. Incorporating groups or double bonds into the substrate polymer at the time of polymerization
3. Activating the surface by high energy radiation, mechanical action, heat treatment, electrochemical processes, electrical discharge, aggressive chemicals (e.g., sodium) (This may be done either with the second monomer present or this may be introduced after treatment of the surface.)

The reactive sites will therefore be functional groups, double bonds, free radicals, or ions, the usual centers for polymerization. In some cases, the polymerization reaction of the new second monomer will include the substrate by hydrogen abstraction reactions.

The second monomer does not graft exclusively, but also forms unbonded homopolymer which must be removed by solvent treatment at the conclusion of the reaction. If an excessive amount of homopolymer is produced, this can deplete the amount of monomer available for grafting or can make purification of the grafted surface very difficult.

Grafting has made available the possibility to use structurally weak polymers which are otherwise biologically acceptable and multicomponent surfaces can be made which are thereby multifunctional. Interest in surface grafting was stimulated by the need to develop more permanent surface properties than were obtained by the graphite-quarternary ammonium-heparin treatment of Gott et al.[25] used to promote nonthrombogenic behavior. Graft polymerization onto polyester arterial prosthesis fabrics has been tried, but requires full evaluation. Various hydrogel coatings have been applied.[26] Controlled drug release applications can utilize drugs bonded to polymers by biodegradable bonds.

## C. Fabrication of Plastics Materials

The macromolecular material usually contains additives or polymer blends to develop a range of properties either to assist and stabilize during processing or to improve end-use requirements and viability. Since biomedical applications are small scale, it is not usually possible to request special materials from the supplier. Furthermore, a wide range of additives will be suitable for use and which ones are present in a particular batch will depend upon availability at the time of compounding the base polymer with the additives. It is sometimes a fact that no assurance can be given to the user that composition will always be the same, nor if the polymer is bought from a stockist that it will be from the same manufacturer. The best that may be hoped for is that the additives will have some sort of approval for food-contact use, e.g., from the Food and Drug Administration (FDA) in the U.S. or the Federal German Government regulations. Polyacetal is an example of a material for which a very large number of complex organic chemical additives are FDA approved and for which no certainty can be given as to which will be used from batch to batch.

Many manufacturers give a disclaimer for medical use of their products. Therefore, toxicity testing is of prime importance for plastics materials, but the uncertainty of batch to batch variation would make this a continuing problem for some materials. However, if a good grade is the starting material, then additives required may be more closely controlled. The problems of toxicity testing and the inadequacy of protocols available together with the lack of international standards has been referred to by Vincent.[16]

Even acceptable additives may be extracted in vivo and change the properties of the material in consequence. For this as well as for toxicity reasons, attention has been given to alternative polymers used without additives or with only a few well-characterized additives and to the use of polymer blends. Plasticized PVC is widely used for tubing and containers for i.v. fluids, and since leachable plasticizers may lead to problems in very ill patients or in children,[27] polymeric plasticizers have been developed.[28]

In evaluating the suitability of additives and the relationship to in vivo performance or to service lifetime, it should be noted that this is difficult to evaluate for engineering applications. Four requirements are proposed for nonmedical uses.[29]

1. Selection of correct test data to enable choice of right material
2. Correlation with actual service
3. Correct component design
4. Correct manufacture of component

The need for design specification is emphasized and this is very necessary for biomedical uses, although the criteria are harder to define. In fabricating a plastics material, the necessity of sterilization must be remembered. This may impose another restriction on the use of certain additives depending on the method of sterilization used.[16,17] Implant and device design must be considered with attention to sterility requirements.

Although additives are employed primarily to modify mechanical properties, processing conditions, and durability in service, the effect on other properties should not be ignored for biomedical applications. Surface condition may be considerably changed, e.g., by protruding hard particles of reinforcer. Soft tissue may then be irritated into excessive fibrosis and blood cells may be damaged. Friction and wear properties will certainly be impaired. Silica particles reported in polyethylene[32] would harm the metal counterface in a joint prosthesis. The same comment applies to nonuniformity of material and galling-type wear has been observed associated with inhomogenous material.[30] Extensive studies have been made on the fibrosis caused by filler particles ($SiO_2$) in silicone rubber used as breast prostheses.[2]

### D. The Physical Form of the Material

This is also a manufacturing artifact and has an effect on in vivo performance. Solid materials are processed by molding, extrusion, or casting and each imposes structural consequences on the product. During extrusion, as the polymer melt is forced through a die, alignment of chain molecules occurs and there is a resultant anisotropy in the product. Tensile strength in the direction of the extrusion will be greater than that measured at 90°.

This is a similar effect to that observed with a unidirectional fiber laminate. A large molded block will have differences between the center and edges due to variation in rate of cooling across the specimen.[31] Considerable attention must be given to controlled cooling in complex molded shapes to avoid stress being frozen into the specimen and subsequent stress relief and distortion. Alignment effects are used to advantage in manufacture of polypropylene hinges. Anisotropy in polypropylene film is used to fibrillate it and produce fibers for subsequent cording into strings and ropes.

The tensile properties of a fiber as well as its physical dimensions are changed by the amount of tension (drawing) given to it in the spinning process. This is another process producing chain alignment. In addition to the obvious uses of fibers, microfibril-covered surfaces have found application in blood contacting systems. Less damage to blood cells is experienced.

Properties of films are controlled by the extensions given to them as well as by chemical treatments. The subject of films and membranes is covered in Chapter 7.

Finally, foams are another form which may be grouped with porous materials designed to encourage tissue growth. Polymer foams have not had a very successful history of use. The initial insertion is usually for soft tissue replacement and tissue ingrowth is accompanied by subsequent fibrosis or calcification with consequent hardening. Polyester polyurethane foams prepared *in situ* for osteosynthesis caused necrosis and were discontinued! The properties of the foam are controlled by the base polymers used and the type of blowing agent. Rigid to very soft are possible in polyurethanes, silicone elastomers, polyvinyl alcohol, and polyethylene being other examples.

## VII. CONCLUDING REMARKS

The plastics material is a result of many varied processes at each stage of which control of properties can be exercised. This gives the versatility which has led to widespread investigation and use as biomaterials. Close control over all variables is essential and attention needs to be given to this in standards proposals and biocompatibility investigations.

## REFERENCES

1. **Bloch, B. and Hastings, G. W.,** *Plastics in Surgery*, Charles C Thomas, Springfield, Ill., 1967; 2nd ed., Charles C Thomas, Springfield, Ill., 1972.
2. **Wilflingseder, P., Hoinkes, G., Mikuz, G., and Propst, A.,** Constructive fibrosis post-augmentation mammaplasty, in *Mechanical Properties of Biomaterials*, Hastings, G. W. and Williams, D. F., Eds., John Wiley & Sons, Chichester, 1980, chap. 45.
3. **Tayton, K. J., Johnson-Nurse, C., McKibbin, B., Bradley, J., and Hastings, G. W.,** The use of semi-rigid carbon fibre reinforced plastic plates for fixation of human fractures, *J. Bone Jt. Surg.*, 64b, 105, 1982.
4. **Heller, J.,** Controlled release of biologically active compounds from bioerodable polymers, *Biomaterials*, 1, 51, 1980.
5. **Duncan, R., Kopecek, J., and Lloyd, J. E.,** Design of soluble polymers as drug carriers, in *Proc. Conf. Biomedical Polymers*, Durham, Biological Engineering Society, Royal College of Surgeons, London, 1982.
6. **Bloch, B., Hastings, G. W., and Wallwork, G. R.,** Major prosthetic replacement of bone for chondrosarcoma, *Med. J. Aust.*, 1, 938, 1970.
7. **Claes, L., Burri, C., Helbing, G., and Lehner, E.,** Comparison of the strength of cartilage glued by fibrinogen and cyanoacrylate, in *Biomaterials 1980*, Winter, G. D., Gibbons, D. T., and Plenk, H., Eds., 1982 John Wiley & Sons, Chichester, 1982, 683.
8. **Bloch, B.,** *Bonding of Fractures by Plastic Adhesive*, Monogr., Surgical Research Centre, Sydney Hospital, Sydney, Australia, 1960.

9. British Standard BS3531 Surgical Implants, Surgical Implant Glossary, British Standards Institution, London, in press.
10. **Watson, W. F.**, Development of the pvc endotracheal tube, *Biomaterials*, 1, 41, 1980.
11. **Treloar, L. R. G.**, *Physics of Rubber Elasticity*, 3rd ed., Oxford Clarendon Press, Oxford, 1975.
12. **Williams, R. J. P.**, On first looking to Nature's chemistry, *Chem. Soc. Rev.*, 9(3), 281, 1980.
13. **Okano, T., Shimada, M., Shinohara, I., Kataoka, K., Akaike, T., and Sakurai, Y.**, Role of microphase separation structure in interaction between polymer and platelet, in *Biomaterials 1980*, Winter, G. D., Gibbons, D. F., and Plenke, H., Eds., John Wiley & Sons, Chichester, 1982, 445.
14. **Bagnall, R. D.**, Implant biocompatibility, *Biomaterials*, 1, 97, 1980.
15. **Ferrutti, P., Barbucci, R., Danzo, N., Torrizi, A., Puglisi, O., Pignataro, S., and Spartano, P.** Preparation and characterisation of poly (vinyl chloride) surface grafted with heparin complexing poly (amido-amine) chains, *Biomaterials*, 3, 33, 1982.
16. **Vincent, W. W.**, Plastics in medical tubing applications, *Biomaterials*, 2, 194, 1981.
17. **Dowling, J. H., Atkinson, J. R., Wright, V., and Charnley, J.**, The contribution of creep deformation to the penetration rate of UHMWPE sockets, in *Proc. Conf. Biomedical Polymers*, Durham, Biological Engineering Society, Royal College of Surgeons, London, 1982.
18. **Hastings, G. W. and Bell, P.**, Fatigue studies of carbon fibre reinforced epoxy bone plates, in preparation.
19. **Owen, A. B. and Beaumont, P. W. R.**, Fracture characteristics of surgical acrylic bone cements, in *Mechanical Properties of Biomaterials*, Hastings, G. W. and Williams, D. F., Eds., John Wiley & Sons, Chichester, 1980, chap. 21.
20. **Rudigier, J., Kirschner, P., Richter, I. E., and Schweikert, G. W.**, Influence of different x-ray contrast materials on structure and strength of bone cement, in *Mechanical Properties of Biomaterials*, Hastings, G. W. and Williams, D. F., Eds., John Wiley & Sons, Chichester, 1980, chap. 22.
21. International Standards Organization Draft International Standard, ISO DIS 5843, Implants for Surgery, Ultra High Molecular Weight Polyethylene, International Standards Organization, Geneva, 1980.
22. Hercules Chemical Co., "Hifax", Ultra High Molecular Weight Polyethylene, product literature, Wilmington, Del.
23. **Contact Lens Symposium**, *Proc. Conf. Biomedical Polymers*, Durham, Biological Engineering Society, Royal College of Surgeons, London, 1982.
24. **Keogh, P. L.**, Approaches to the development of improved soft contact lens materials, *Biomater. Med. Devices Artif. Organs*, 7, 307, 1979.
25. **Gott, V. L., Whiffen, J. D., and Dutton, R. C.**, Heparin bonding on colloidal graphite surfaces, *Science*, 142, 1297, 1963.
26. **Jansen, B.**, Radiation induced grafting of polyetherurethane films and tubes with hydrophilic monomers, in *Proc. Conf. Biomedical Polymers*, Durham, Biological Engineering Society, Royal College of Surgeons, London, 1982.
27. **Hastings, G. W. and Green, A. J.**, Problems in using plasticised poly vinyl chloride for transfusion sets with particular reference to new born infants, in *Proc. Conf. Plastics in Medicine and Surgery*, Enschede, The Plastics and Rubber Institute, London, 1979.
28. **Biggs, M. S. and Baldwin, J.**, A flexible pvc compound for long-term contact with human tissue, in *Proc. Conf. Plastics in Medicine and Surgery*, Enschede, The Plastics and Rubber Institute, London, 1979.
29. **Sims, D.**, Durability of polymers in service. Conference report, *Plast. Rubber Int.*, 6, 211, 1981.
30. **Hastings, G. W.**, In vivo wear studies of polyethylene used in the knee joint, *Wear*, 55, 1, 1979.
31. **Wright, K. W. J., Dobbs, H. S., and Scales, J. T.**, Wear studies on prosthetic materials using the pin-on-disc machine, *Biomaterials*, 3, 41, 1982.
32. **Scales, J. T.**, private communication.

# *Properties*

Chapter 2

# MOLECULAR STRUCTURE DETERMINATION AND CHARACTERIZATION

**J. W. A. van den Berg and J. Schuijer**

## TABLE OF CONTENTS

I. Introduction ........ 22

II. Solubility of Polymers; Nonideality of Polymer Solutions ........ 23

III. Inhomogeneity on Molecular Scale ........ 25

IV. Viscometry ........ 27
A. Theory ........ 27
B. Experimental ........ 30
C. Applicability ........ 30

V. Methods Based on Colligative Properties ........ 31
A. Osmometry ........ 32
1. Theory ........ 32
2. Experimental ........ 33
3. Applicability ........ 34
B. Other Methods for the Determination of $\overline{M}_n$ ........ 35
1. End Group Analysis ........ 35
2. Ebulliometry and Cryometry ........ 35
3. Vapor Pressure Osmometry ........ 36

VI. Light Scattering ........ 36
A. Theory ........ 37
B. Experimental ........ 40
C. Applicability ........ 41

VII. Ultracentrifugation ........ 41
A. Theory ........ 41
B. Experimental ........ 42
C. Applicability ........ 42

VIII. Size Exclusion Chromatography ........ 44
A. Theory ........ 44
B. Experimental ........ 44
C. Applicability ........ 46

IX. Fractionation ........ 46

X. Copolymers ........ 47
A. Chemical Composition ........ 47
B. Molecular Weight and m.w.d. ........ 48
1. Methods Based on Colligative Properties ........ 49

2. Light Scattering .......... 49
3. Ultracentrifugation .......... 50
4. Viscometry .......... 50
5. Size Exclusion Chromatography .......... 50
6. Fractionation .......... 51

XI. Polyelectrolytes .......... 51

References .......... 53


# I. INTRODUCTION

This chapter presents a survey of methods that are frequently encountered for structure determination and characterization of polymers, with special emphasis on size and shape of their molecules. Structure determination is part of research and development of materials as it may help to answer a question like: "What can we use this polymer for?" Characterization is a means for classification by routine measurements and therefore is also important for research and development of materials, because it may help to answer a question like: "What is this polymer like?" The distinction between these two terms is not very sharp and they are sometimes interchanged.

The experience of polymer materials science, grown in the last 50 years, enables us to predict something about the behavior of large amounts of material from small-scale measurements of characteristic properties like molecular weight or crystallinity. As an example, see Table 1.

This table illustrates not only the usefulness of characterization, but also the close relationship between characterization and structure determination, as well as the vagueness of any distinction between the two terms (is "crystallinity" a structural or a characteristic parameter?).

For each method described below, attention is paid to theory, experiment, applicability, and literature. The experimental and theoretical descriptions are as concise as we thought justified. More, and more precise, details can be found in the general references,[1-6] and in the specific references given abundantly in the coming subchapters. As general references, we recommend the books referred to in Reference 1a and 1b. These are very useful reference works, easily understandable for workers in the medical or biochemical field with a good background in analytical chemistry but a poor background in mathematics and physics. References 1c and 1d refer to excellent collections of reviews on subjects related to elementary polymer concepts; Reference 1d contains an in-depth extension of Table 1. Advanced and very recent methods, on an analytical as well as a mathematical or physical level, but still applicable to biomaterials, are given in Reference 1e.

In this chapter, we consider mainly molecular properties that can be deduced from measurements of solution properties. Such measurements are relatively quick and easy, and they require only small amounts of (often very expensive) material. Examples of molecular properties derived from solution experiments are intramolecular order, intramolecular interactions, size* and shape of isolated polymer molecules, as well as the flexibility of polymer chains.

* Chain length or volume, depending on the method that is used. For the concept of the volume of an isolated polymer molecule in solution, see the end of Section II.

**Table 1**
**INFLUENCE OF CRYSTALLINITY AND MOLECULAR WEIGHT ON SOME BULK PROPERTIES OF CRYSTALLINE POLYMERS**

| | Change of property with increasing | |
|---|---|---|
| **Property** | **Crystallinity** | **Molecular weight** |
| Young's modulus | Increases | Increases slightly |
| Stiffness | Increases | Increases slightly |
| Brittleness | Increases | Decreases |
| Hardness | Increases | Increases slightly |
| Tensile strength | Increases | Increases |
| Cold flow | Decreases | Decreases |
| Melting point | Increases | No effect |

Knowledge of intramolecular properties may be important for the understanding of surface or bulk structures of the polymer. Unfortunately, in the majority of the cases there is no direct and easily understandable relationship between (intra-)molecular properties and material structure, as the latter depends also on intermolecular order and interaction. It has been found for many amorphous polymers, however, that their molecular volumes and shapes in solution are closely alike to that in solid specimens of the same polymer.[2]

Even more important may be the knowledge of molecular properties for the understanding of, for example, blood or tissue compatibility of polymers used in surgery. In cases like these, the point of interest is the interaction between dissolved biological macromolecules and synthetic macromolecules, held in a "structure" at a surface.

Among the molecular properties, molecular weights and molecular weight distributions (denoted in this chapter by m.w.d.) are very important. Remember what has been said in connection with Table 1. Furthermore, in research on, for example, blood or tissue compatibility, determination of changes in molecular weight (distribution) may be quite important for the establishment of enzymatic degradation of the polymeric material.

An elementary introduction to the vast and special subject of characterization and structure determination of "biopolymers" (like proteins, enzymes, polysaccharides) is given in this chapter under the heading "Polyelectrolytes" (Section XI). The vast majority of such "biopolymers" are polyelectrolytes indeed. A useful general reference to this subject of polymers present in, or extracted from, living tissues is the book referred to in Reference 6.

## II. SOLUBILITY OF POLYMERS; NONIDEALITY OF POLYMER SOLUTIONS

As molecular properties are mainly determined on solutions, a discussion of measuring methods should be preceded by a short review on solubility and on the quite different behavior of polymer solutions, compared to that of solutions of low molecular weight solutes. Furthermore, some knowledge of solubility and solution behavior might be desirable for the study of interactions between tissues (or blood) and implants of polymer materials.

Polymers dissolve very slowly in their solvents, even if "by eye" a solution seems to be formed readily. The dissolution process occurs in two stages. First, solvent penetrates into the polymer coils, which for the present remain entangled and/or physically cross-linked (e.g., by hydrogen bridges or by crystallites consisting of subchains of several polymer molecules; we do not mean real chemical cross-links here, where covalent or ionic bonds are involved). The penetration process is time consuming; it is the slowest in the most crystalline parts of the material. This first stage is called *swelling*; polymers with chemical cross-links remain in this stage and will never dissolve (but they may swell to many times their original volume).

The second stage is the decomposition of the swollen polymer (parts) into molecules; it takes place parallel to the swelling stage and is completed after it. In some cases, during the swelling stage the polymer decomposes into tiny gel particles, which still consist of millions of tightly cross-linked polymer molecules, but which already possess properties similar to that of the final solution; they may have the same refractive index and the same density as the solution itself. A real solution is not formed, until all these gel particles are dissolved. Filtration of such apparent, gel-containing solutions through narrow pore ($<0.45\mu m$) filter sheets is practically impossible and is thus a means for discriminating between real and apparent polymer solutions. Light scattering provides another means.

For a vast amount of polymers, solvents are known and tabulated.[1] For amorphous polymers, it is sometimes possible to calculate whether a certain liquid is a solvent to a certain polymer. To this end, solubility parameters (or $\delta$-parameters), both for solvent and polymer, must be known or determined. If they match, the liquid in question mostly is a solvent to the polymer indeed. (For these calculations and for the criteria of matching, we refer to Reference 1, where $\delta$-parameters have been tabulated for both solvents and polymers.

In polymer science, the distinction between good solvents and bad solvents has a special meaning. In a bad solvent there is much less volume expansion on swelling than in a good one. Whether a polymer readily dissolves in a solvent or not has nothing to do with solvent quality.

There is a parameter, the $\chi$-parameter,[2] that characterizes the solvent quality in the sense just defined. If it is close to or higher than 0.5, the solvent is called "bad"; if it is lower than 0.5 or 0 or even negative, the solvent is said to be "good". The $\chi$-parameters are tabulated.[1]

The distinction between good and bad solvents is of use in two cases. First, there is a direct practical application of the solvent quality concept when we wish to determine molecular weights. Second, when we wish to calculate size and shape parameters of dissolved macromolecules, we should realize that they may depend on solvent quality. We will discuss these matters briefly.

Generally, polymer solutions show large deviations from ideal solution behavior, even at low (e.g., less than 1%) concentrations. A solution is said to be ideal when it obeys, for example, Raoult's law: the vapor pressure of the solution is equal to the product of the solute's mole fraction and the vapor pressure of the pure solvent. Nonideality is caused *inter alia* by differences in size and shape between the molecules of solvent and solute; this is certainly the case for solutions of macromolecules in a low molecular weight solvent. Dipolar or ionogenic forces may enhance nonideality considerably, too.

Nonideal solutions are characterized by the occurrence of "virial coefficients" in the equations which describe the concentration dependence of, for example, their vapor pressure or their osmotic pressure or their boiling point elevation. In order to eliminate them when an exact molecular weight has to be determined, one must always extrapolate the results of measurements on polymer solutions to zero concentration. In bad solvents, the virial coefficients are lower than in good ones, which makes straight line extrapolations allowable.

For the so-called "$\theta$-solutions", the virial coefficients are even zero and the solution behaves like an ideal one, which enables us to perform an exact molecular weight determination with only one solution. (A $\theta$-solution is a solution at the verge of liquid-liquid demixing; it should be maintained as precisely as possible at the so-called "$\theta$-temperature". Solutions for which the $\chi$-parameter is exactly 0.5 are $\theta$-solutions.) $\theta$-Solvents for polymers are tabulated.[1]

The majority of synthetic linear (included branched) polymers and part of the natural linear polymers have no definite structure, neither in the solid state nor in solution. In solution they will be randomly coiled. The majority of the polymer molecules behaves like particles with a volume that may be 50 to 100 times larger than the volume occupied by the segments of the polymer molecule themselves. In a good solvent, the polymer's solution volume may be twice as large as in a bad one.

This solution volume of one single polymer molecule may theoretically be considered as a sphere with an effective radius $R_e = \xi R_g$. The factor $\xi$ is a dimensionless parameter depending on measuring conditions (e.g., viscous flow, diffusion) and slightly on solvent quality.

$R_g$ is called the radius of gyration. In fact it is the root-mean-square distance of all mass elements of which the polymer molecule is built up, from its center of gravity:

$$R_g = \left(\frac{\overline{\sum_i m_i r_i^2}}{\sum_i m_i}\right)^{1/2}$$

Its absolute value can be determined only by means of light scattering. Viscosity measurements or determination of virial coefficients may give information about $R_g$, provided the influence of interfering factors like solvent quality is well-known. For further details see Tanford.[3]

Part of the natural polymers and similar synthetic polymers form definite structures in the solid state as well as in solution. Some proteins, for instance, occur in a helix form stabilized by intramolecular hydrogen bridges. A molecule with this so-called α-helix structure can be considered as a cylindrical rod.

As a consequence of the definition of the radius of gyration, $R_g$ can also be applied to these rods. Therefore, it is an important parameter for structure determination of polymer molecules in general.

## III. INHOMOGENEITY ON MOLECULAR SCALE

In a sample of a synthetic or a natural polymer material, there is usually only a very poor chance of finding two molecules of equal molecular weight. (Exceptions to this "rule" are, for example, most proteins and some polymers, prepared under very special conditions.) Still lower is the chance for finding two molecules with equal distribution of tacticity, or of monomer type (in random copolymers), or of electrically charged monomers (in polyelectrolytes).

Hence, there are three types of inhomogeneity leading to m.w.d., monomer sequence distribution, and electrical charge distribution. They even may occur together in one polymer sample. Research on heparin and heparin analogs, for instance, where postpolymer reactions frequently occur, has to account for this phenomenon. This is a consequence of the fact that reactive sites on a polymer (even if they are regularly dispersed over a molecular chain before a reaction begins) are not equally reactive and may be subject to all the side reactions that might occur when these sites formed part of a low molecular weight compound. Hence, the result of such postpolymer reactions on heparin or heparin analogs is mostly a random copolymer, which may be charged randomly.

In the following, only some aspects of the m.w.d. will be discussed. Aspects of monomer sequence distribution and of electrical charge distribution will be referred to in the sections dedicated to copolymers and polyelectrolytes (Sections X and XI).

The ultimate chain length of each polymer molecule depends on reactions leading to further growth. Consequently, the m.w.d. depends on polymerization kinetics. Polyethylene, for instance, may consist of molecules built up of different numbers of monomers, ranging

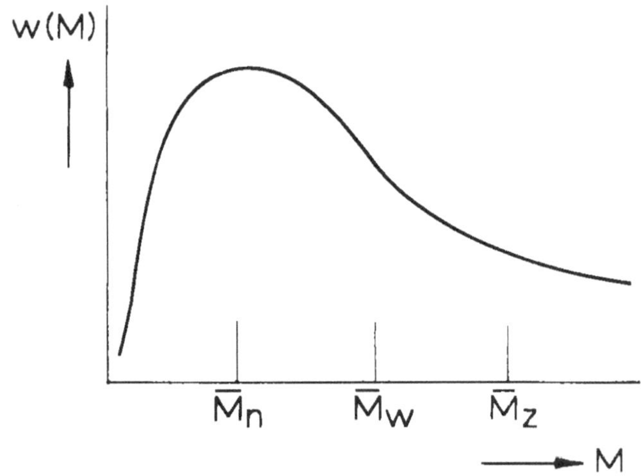


FIGURE 1. Example of a molecular weight distribution, with an indication of the positions of different molecular weight averages.

from about 100 to over 100,000. A polymer like this consists of more than 100,000 compounds, only differing in chain length, not mentioning the distribution of side chains, which causes the number of different compounds in the polymer to rise to astronomic figures.

The m.w.d. can be expressed in different ways, e.g., as the weight of $w_i$ of the fraction of polymer molecules with a molecular weight $M_i$ as a function of M. This is essentially a step function, but the steps given by the molecular weight of the monomer are small enough to replace it by a continuous function. Figure 1 gives an example.

Measurements of the molecular weight of a polymer with an m.w.d. always lead to an average value of M, depending on the method. The most simple way of averaging is to divide the sum of the molecular weights of all molecules by the number of molecules. This average is called the number average molecular weight, $\overline{M}_n$.

$$\overline{M}_n = \frac{\sum_i N_i M_i}{\sum_i N_i} = \frac{\sum_i w_i}{\sum_i w_i/M_i}$$

It will be measured by each method where the measured effect depends on the mole fraction of the polymer in solution, irrespective of molecular weight. When we know the weight concentration, we can easily calculate $\overline{M}_n$ from the measured mole fraction. These methods will be described in Section V.

$\overline{M}_n$ may be considered, mathematically, as the ratio of the first to the zeroth moment of the m.w.d.[1b] The next higher ratio, viz. of the second to the first moment of the m.w.d., gives the weight average molecular weight, $\overline{M}_w$.

$$\overline{M}_w = \frac{\sum_i N_i M_i^2}{\sum_i N_i M_i} = \frac{\sum_i w_i M_i}{\sum_i w_i}$$

where $w_i$ = weight of the fraction of molecules with a molecular weight $M_i$. It is determined by those methods where the measured effect depends on the size of the molecules (light scattering, ultracentrifuge).

For polydisperse polymers, $\overline{M}_w/\overline{M}_n > 1$ (Figure 1). This ratio is called the polydispersity ratio of the polymer. The larger this ratio is, the more inhomogeneous with respect to molecular weight the polymer is. It is often used to express the breadth of the m.w.d. For many polymers its value ranges from 1.5 to 2.0, but there are also polymers, e.g., polyolefins, with ratios from 20 to 50.

Also the next higher ratio of moments, viz. the third to the second, is important. It is called z-average molecular weight ($\overline{M}_z$) because it was first determined by the ultracentrifuge (German: Zentrifuge):

$$\overline{M}_z = \frac{\sum_i N_i M_i^3}{\sum_i N_i M_i^2}$$

$\overline{M}_z/\overline{M}_w > 1$ for polydisperse polymers (Figure 1) and is a measure of the asymmetry of the m.w.d.[1h]

As important polymer properties may be influenced by the high and low molecular weight tails of the m.w.d., measurement of all three averages might be necessary for a good characterization. The average molecular weights may be calculated from the distribution, but the reverse is not true. Hence, it is impossible to calculate an m.w.d. from measurements of $\overline{M}_n$ by osmometry, $\overline{M}_w$ by light scattering, and $\overline{M}_z$ by means of the ultracentrifuge; many distributions may be assumed to belong to one set of $\overline{M}_n$, $\overline{M}_w$, $\overline{M}_z$. Special methods are needed for the determination of the m.w.d., such as gel permeation chromatography (see Section VIII) or a special computer-assisted technique for the interpretation of data obtained from ultracentrifuge measurements.

The m.w.d. can be determined in different ways. These will be discussed in Sections VII to IX.

## IV. VISCOMETRY

By far the easiest way for polymer characterization is dilute solution viscometry, because of its simple experimental requirements and its well-reproducible results. In combination with light scattering, viscometry gives interesting information about the structure of polymers in solution, which is sometimes nearly equal to that found in the solid state.[2]

### A. Theory

The viscosity, $\eta$, of a liquid depends on intermolecular forces and collisions, and hence it depends also on number, kind, volume, and shape of the particles of which the liquid is composed. For a suspension of solid, spherical particles in a viscous liquid of viscosity $\eta_0$, for instance, Einstein derived[3]

$$\eta = \eta_0 (1 + 2.5\ nv)$$

Here n is the number of particles per unit volume, and v is the volume of one particle. The more particles there are, the higher the suspension's viscosity is.

Defining c as the concentration of a polymer with molecular weight M in, for example, grams per milliliter, we may write n as

$$n = \frac{c\ N_A}{M}$$

Now suppose the dissolved macromolecules are coiled and possess (on the average) the shape of a sphere with effective radius $R_e$. Then,

$$v = \frac{4\pi}{3} R_e^3$$

Insertion of these expressions for n and v in Einstein's equation gives:

$$\frac{\eta - \eta_0}{\eta_0 c} = \frac{10\pi\, N_A}{3M} R_e^3$$

For a solution of spherical macromolecules of molecular weight M and radius $R_e$, the quotient $(\eta - \eta_0)/\eta_0\, c$ should be a constant. However, in practice $R_e$ appears to depend on c. Therefore, this quotient is always extrapolated towards $c = 0$. If we define the extrapolated value as $[\eta]$,

$$[\eta] = \lim_{c \to 0} \frac{\eta - \eta_0}{\eta_0 c}$$

then,

$$[\eta] = \frac{10\pi\, N_A}{3M} R_e^3$$

$[\eta]$ is generally called the "intrinsic viscosity", although its dimension is not that of a viscosity.

Simha (in Tanford[3]) extended Einstein's theory to nonspherical particles. For particles which are ellipsoids of revolution, he found

$$\eta = \eta_0 (1 + \nu n v)$$

the shape factor $\nu$ depending on the axial ratio and differing for prolate ellipsoids (semiaxes a, b, b) and oblate ellipsoids (semiaxes a, a, b). In both cases, $\nu$ tends to 2.5 for axial ratios tending to 1 in agreement with Einstein. Consequently,

$$[\eta] = \nu \frac{N_A}{M} v$$

This equation enables us to calculate one of the parameters $\nu$, M, or $R_e$, provided the other two are known from independent measurements. Hence, viscometry may be helpful for the determination of shape or size of macromolecules in solution.

Theoretically, it can be derived that the polymer volume in solution is proportional to a power of its molecular weight, hence

$$v \sim M^b$$

Insertion in the equation for $[\eta]$ derived above leads to the well-known Mark-Houwink relationship:

$$[\eta] = K\, M^a$$

first suggested by these authors as an empirical equation. K values often range 0.5 to 5 $\times 10^{-2}$ when $[\eta]$ has the dimension $m\ell/g$, as is usual. The exponent "a" normally varies

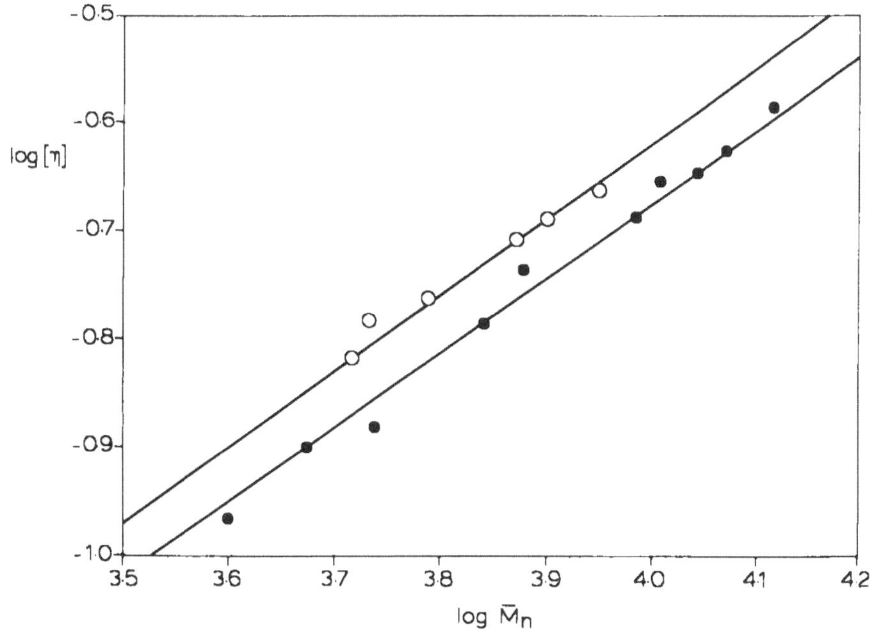


FIGURE 2. Log [η] vs. log $\overline{M}_n$ plot for unfractionated (○) and fractionated (●) samples of polyvinyl alcohol. (From Misra, G. S. and Mukherjee, *Colloid Polym. Sci.*, 258, 152, 1980. With permission.)

from 0.5 (the theoretical value of a randomly coiled polymer in a θ-solvent) to about 1.0, but may exceed this value in case of rod-shaped molecules. Then "a" may have a value of about 1.8.

The Mark-Houwink relationship is often used for molecular weight determination. So far, nothing has been said about the kind of molecular weight, $\overline{M}_n$ or $\overline{M}_w$, derived from viscometry. This depends on the calibration, i.e., on the kind of molecular weight, used for the determination of the Mark-Houwink equation. The constant K depends on the type of molecular weight:[7]

$$[\eta] = K_1(\overline{M}_n)^a = K_2(\overline{M}_w)^a$$

Only when $\overline{M}_w/\overline{M}_n = 1, K_1 = K_2$. Moreover $K_1$ and $K_2$ depend on the type of m.w.d. Hence, use of the Mark-Houwink relation gives only exact information about $\overline{M}_n$ or $\overline{M}_w$, if the m.w.d. of the sample to be determined does not differ essentially from that of the samples used for the calibration of this relationship.

Figure 2 gives an example of the influence of the m.w.d. on the K value in the Mark-Houwink relation measured by Misra and Mukherjee[7] on solutions of polyvinyl alcohol in doubly distilled water. $\overline{M}_n$ values were measured by end group titrations.

There exist also tabulations of K and a values in which the K values are considered independent of the m.w.d. They lead to a "viscosity-averaged molecular weight" $\overline{M}_v$:

$$[\eta] = K(\overline{M}_v)^a$$

where

$$\overline{M}_v = \left(\frac{\sum_i c_i M_i^a}{\sum c_i}\right)^{1/a}$$

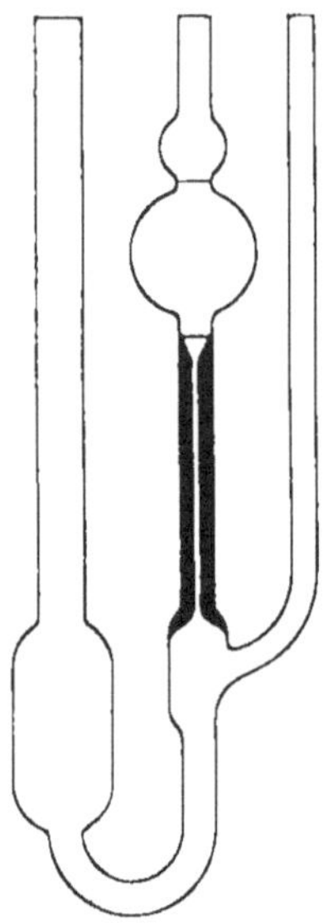

FIGURE 3. Ubbelohde capillary viscometer.

Now it is not K that depends on the m.w.d., but it is $\overline{M}_v$. Fortunately, $\overline{M}_v$ does not differ very much from $\overline{M}_w$ (but it is always somewhat lower).

### B. Experimental

Apparatus: thermostat-bath (temperature variation within ±0.002°C; ηs are very temperature dependent), stopwatch, viscometers (preferably of the Ubbelohde-type) (Figure 3); for polymer solution work mostly capillary lengths of 10 to 15 cm and capillary diameters of about 0.4 mm are applied. The solution has to be made dust free by filtration through a fritted glass filter.

Determinations of η by means of a capillary viscometer are based upon Poiseuille's law, which can be written as

$$\eta = A\rho t$$

A is the (geometrical) viscometer constant (independent of the kind of liquid), ρ is the specific mass of the liquid in the viscometer, and t is the time required for the liquid to pass from one mark to another. For low concentrations it is allowed to put $\rho = \rho_0$, where $\rho_0$ is the specific mass of the solvent. Hence,

$$\frac{\eta - \eta_0}{\eta_0} = \frac{t - t_0}{t_0}$$

This quantity has to be determined for 5 to 6 concentrations from 2 to about 0.5%. Dilutions should be made preferably in the viscometer. Now, a plot of $(t - t_0)/t_0c$ vs. c should give a straight line, with [η] as intercept. Tabulations[1] of K and a values allow $\overline{M}_n$, $\overline{M}_w$, or $\overline{M}_v$ to be calculated from the measured [η].

### C. Applicability

If $t > 1.5\ t_0$, well-reproducible values of [η] are obtained within 1 hr; corrections for $\rho \neq \rho_0$ may be necessary. For extremely precise work, viscometers of various capillary lengths and diameters should be used in order to check the validity of Poiseuille's law for the $\eta$ values under consideration. (This will mostly be the case for normal polymers and for sufficiently low concentrations.)

Anomalies in the plot of $(\eta - \eta_0)/\eta_0 c$ vs. c may be (1) it is slightly concave upwards or (2) it is concave downwards, with an extreme rise in $(\eta - \eta_0)/\eta_0 c$ for low c values. The first anomaly usually indicates (provided that experimental errors are low) that the highest c was too high; the second anomaly may indicate that either the capillary diameter was too small, dust particles were present, the polymer under consideration is a polyelectrolyte, or the solution did not contain enough salt (see Section XI).

In cases where no K and a values are known, the method is still useful when a rough classification of polymers according to their molecular weight is desired. Mostly, increasing $[\eta]$ values correspond to increasing $\overline{M}_w$ values.

Automatic capillary viscometer set-ups (with programs for dilution and repetition of measurements and with photoelectric time determination) are available. They save time considerably, but they are not more precise than nonautomatic devices.

Capillary viscometers should be distinguished from rotating disk viscometers; these are designed especially for work with high $\eta$s that may depend on the rate of shear. They should be used for the characterization (i.e., property determination) of highly concentrated (say 10%) polymer solutions or of melts, not for molecular weight determinations.

## V. METHODS BASED ON COLLIGATIVE PROPERTIES

Properties depending on the mole fraction of the solute are called colligative properties. They can be used for the determination of $\overline{M}_n$ as mentioned in Section III. The measurable effect rises with decreasing $\overline{M}_n$. Increasing the concentration may rise the measurable effect, but then difficulties may arise when extrapolation to zero concentration is carried out. It is advisable to determine $\overline{M}_n$ with those methods, which show the largest measurable effects in the $\overline{M}_n$ range which one expects:

| Method | Upper limit of applicability ($\overline{M}_n$) | Remarks |
|---|---|---|
| Membrane osmometry | 500,000 | A lower limit of approximately 25,000 is set by the nonsemipermeability of the membrane. |
| End group analysis | 50,000 | A chemical method; only applicable when there exist 1 or 2 end groups per macromolecule that are detectable by chemical (titration) or physical (IR) means |
| Ebulliometry | 10,000 | The polymer should not be degraded by the boiling of the solution. For large effects, the solvent should have a small heat of vaporization |
| Cryoscopy | 10,000 | The solvent should have a small heat of melting and a melting temperature in a range accessible to the apparatus available |
| Vapor pressure measurement | 1,000 | A very long time is needed to reach equilibrium; the temperature should be controlled within 0.0005°C |
| Vaporization rate (thermoelectric "vapor pressure osmometry") | 10,000 | Not a real colligative method; calibration is difficult, but always needed. |

All these methods but the last one lead to absolute values of $\overline{M}_n$ without calibration. Due to the large measurable effect, osmometry is the most widely used way for the determination of $\overline{M}_n$. Therefore, we will treat this method in more detail and mention only some aspects of other methods.

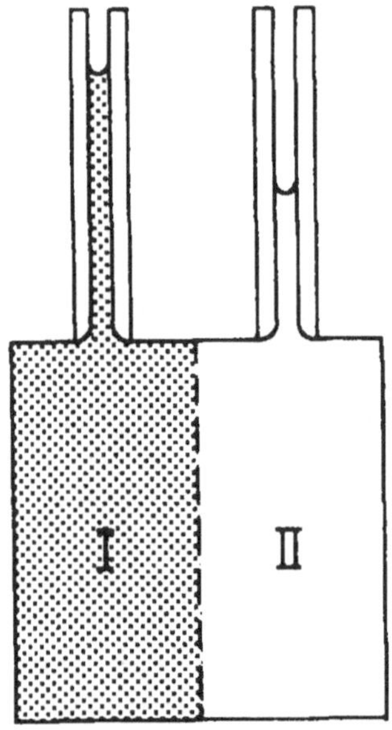


FIGURE 4. Osmometer. (I) Chamber with solution; (II) chamber with solvent.

**A. Osmometry**

*1. Theory*

A membrane osmometer essentially consists of two chambers, each provided with a capillary tube (Figure 4). The first contains the polymer solution; the second contains the pure solvent. The chambers are separated by a membrane, i.e., an ultrathin (but strong) film (e.g., made of cellulose acetate). It should swell, but never dissolve in the test solvent. Let us consider ideal semipermeable membranes first. These membranes are permeable to solvent only, not to the dissolved molecules.

At the beginning of the experiment, there is a disequilibrium situation. There exists a concentration gradient over the membrane. This gradient is "unacceptable" for the solvent, which tries to destroy it by penetrating into the solution chamber. The resulting net solvent flow in the direction of the solution causes a hydrostatic pressure difference between the two chambers. The net solvent flow stops and equilibrium is reached, as soon as this pressure difference is high enough to counterbalance the net solvent flow.

When the solution is an ideal one, then there exists a simple relationship between the equilibrium hydrostatic pressure, called the osmotic pressure, $\pi$, and the concentration, c (e.g., in grams per milliliter), of solute molecules in the solution chamber:

$$\Pi = \frac{RT}{M} c \qquad \text{(Van't Hoff's law)}$$

(This law is for ideal solutions the analog of Boyle's law for ideal gases.)

When the solute is polydisperse, the total osmotic pressure is the sum of the separate osmotic pressures caused by the components:

$$\Pi = \sum_i \Pi_i$$

For each component, Van't Hoff's law applies, so

$$\Pi = \sum_i \frac{RT}{M_i} c_i$$

and

$$\frac{\Pi}{c} = RT \frac{\sum_i c_i/M_i}{\sum_i c_i}$$

Hence, since $c_i$ is proportional to $N_iM_i$,

$$\frac{\Pi}{c} = \frac{RT}{\overline{M}_n}$$

When the solute molecules are so large that the solution is no longer an ideal one, extrapolation to zero concentration is necessary.

$$\lim_{c \to 0} \frac{\Pi}{c} = \frac{RT}{\overline{M}_n}$$

The equation for the best line connecting the data points is according to theory:

$$\frac{\Pi}{c} = RT \left( \frac{1}{\overline{M}_n} + Bc + Cc^2 + \ldots \right)$$

This is a virial series (the solution analog to the virial series for nonideal gases); the constants B,C, . . . depend on size and shape of the solute molecules and may give useful structural information.

From theory[2] it appears that often $C \approx 1/4\ B^2$. To take into account the contributions of the third virial coefficient C, we then may write:

$$\left(\frac{\Pi}{c}\right)^{1/2} = \left(\frac{RT}{\overline{M}_n}\right)^{1/2} \cdot \left(1 + \frac{1}{2} B\,\overline{M}_n c\right)$$

leading to straight lines when plotting $(\pi/c)^{1/2}$ against c. Figure 5 gives an example of measurements on monodisperse poly ($\alpha$-methyl) styrenes in toluene (samples A and C) and mixtures of these polymers (samples AC-1/7).[8]

Now let us consider the case of a nonideal semipermeable membrane. This membrane is also permeable to (not too large) solute molecules; the velocity of the solvent flow is higher, however, than that of the solute flow. Then the hydrostatic pressure rises at the beginning of the experiment, for the same reason as that mentioned for the case of the ideal semipermeable membrane. The difference is that after some time a transient maximum pressure is reached which decays towards zero or to an equilibrium value that is matched to the remaining concentration of those solute molecules that were too large for passage through the membrane.

At equilibrium, all the permeable solute molecules are evenly distributed over the two chambers and their concentration gradient over the membrane is zero. Experiments with nonideal semipermeable membranes are very difficult to interpret. Fortunately, natural and synthetic biopolymers mostly have no appreciable concentrations of low molecular fractions.

## 2. *Experimental*

There are hand-operated osmometers of several types (they differ mainly in the construction

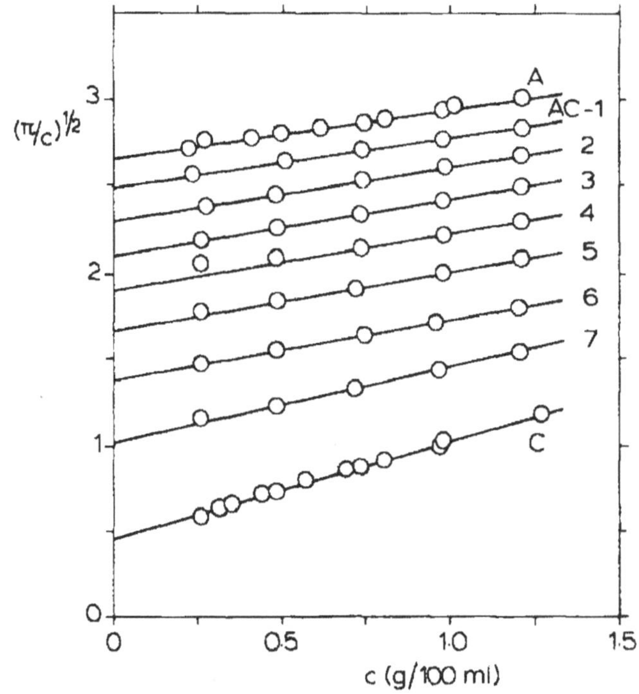


FIGURE 5. Plots of $(\pi/c)^{1/2}$ vs. c for samples A, C, and A-C mixtures. Sample numbers are indicated. (From Noda, I., Kitano, T., and Natasawa, M., *J. Polym. Sci. Polym. Phys. Ed.*, 15, 1129, 1977. With permission.)

of the membrane support) and automatically operated ones. For hand-operated experimental set-ups, a thermostat-bath (controlled within 0.001°C) is needed, as well as a cathetometer for the readings of the pressure, which are taken until equilibrium is reached.

The main disadvantage of the hand-operated osmometers is that it takes from several hours to several days before equilibrium is reached. The much more expensive automatic membrane osmometers help the establishment of equilibrium, e.g., by measuring the solvent flow and pressing the solvent gently to and from the direction of the solution chamber, as long as a net solvent flow is measured. Such devices shorten the time for reaching equilibrium appreciably (to within 1 hr). Hence, they enable us to discriminate between a true equilibrium osmotic pressure and a transient nonequilibrium one, within an acceptable experimentation time.

If one is sure that one has to deal with an ideal semipermeable membrane, measurements of the osmotic pressure for four to six concentrations should be performed and recorded in a graph, e.g., like the one given in Figure 5. Concentrations higher than 2 to 5% should be avoided, as higher terms in the virial series may be really present and bias a straight line, drawn through the data points. Special procedures are needed for the interpretation of experiments with nonideal semipermeable membranes.

*3. Applicability*

Most membrane osmometers allow a hydrostatic pressure height $\Delta h$ of 10 to 20 cm to be built up. It is advisable to estimate the maximum allowable concentration from an estimation of the molecular weight (e.g., based on viscosimetry) by means of the formula:

$$\Pi = \Delta h \, g\rho = \frac{RT}{\overline{M}_n} c$$

where g is the acceleration by the gravitational field of the earth and $\rho$ is the density of the solution. This formula demonstrates again that the measurable effect rises with decreasing $\overline{M}_n$.

The method is applicable to all polymers which dissolve in "suitable" solvents. Such a solvent should be not too expensive (rather large amounts are required for the experiments) and it should be fully inert with respect to membrane and apparatus. Remember what has been said in the table above about the limits of applicability.

## B. Other Methods for the Determination of $\overline{M}_n$

### *1. End Group Analysis*

In principle, the number average molecular weight can be measured if the polymer molecules possess reactive end groups. Polymers built up by condensation reactions, by their nature, have reactive end groups that can be determined, e.g., simply by titration.

Now, for example, suppose a polyamide with mass m (in grams), possessing $n_e$ end groups per molecule ($n_e$ is usually 1 or 2), is titrated with an acid and a base, respectively. If for these titrations $V_a$ cm$^3$ of the acid and $V_b$ cm$^3$ of the base are used and the normalities of the titrants are $t_a$ and $t_b$, respectively (in equivalents per liter), then

$$\overline{M}_n = \frac{1000\ n_e\ m}{V_a\ t_a + V_b\ t_b}$$

Unfortunately, this method is not always an absolute one, because assumptions often have to be made about the chemical nature of the end groups and/or their number per molecule. For instance, in case of polyaddition there are different termination reactions possible, leading to different end groups. If chain branching occurs, the number of end groups per molecule increases. Commercial polycondensates like polyamide fibers may possess end groups differing from the normal amino and carbonyl groups, e.g., acylated amino or amidated carbonyl end groups. In that case, use of intense dyes specifically bound to the end groups may be helpful.[9]

### *2. Ebulliometry and Cryometry*

Theoretically, ebulliometry and cryometry are very similar. When we perform ebulliometry, we measure the elevation of the boiling point of a solvent. This elevation is caused by the addition of a solute to the solvent.

Thermodynamics of polymer solutions leads to the expression:

$$\frac{\Delta T_b}{c} = \frac{R\ T_b^2\ M_0}{\rho_0\ \Delta H_v} \cdot \frac{1}{\overline{M}_n} + Bc + \ldots$$

where

$\Delta T_b$ = boiling point elevation (°K)
$T_b$ = boiling point (°K)
$M_0$ = molecular weight solvent
$\rho_0$ = density solvent
$\Delta H_v$ = molar heat of vaporization of the solvent

For cryometry, the depression $\Delta T_m$ of the melting point of a solvent is measured. This depression results also from the addition of a solute to the solvent. The above equation applies also to cryometry after replacing $T_b$ by the melting point $T_m$ and $\Delta H_v$ by the molar heat of melting $\Delta H_m$.

**Table 2**
**EBULLOSCOPIC AND CRYOSCOPIC CONSTANTS FOR VARIOUS SOLVENTS AT NORMAL PRESSURE**

| Solvent | $T_b$(°C) | $10^{-3}K_b$ (gK mol$^{-1}$) | $T_m$(°C) | $10^{-3}K_m$ (gK mol$^{-1}$) |
|---|---|---|---|---|
| Acetic acid | 118.0 | 3.07 | 16.6 | 3.9 |
| Benzene | 80.1 | 2.54 | 5.5 | 5.1 |
| Nitrobenzene | 210.8 | 5.24 | 5.7 | 7.0 |
| Phenol | 181.8 | 3.56 | 41.0 | 7.4 |
| Water | 100.0 | 0.51 | 0.0 | 1.8 |

In both methods, the temperature difference is measured mostly by using a pair of matched thermistors in a Wheatstone bridge circuit. It is customary then, to measure a galvanometer deflection which is proportional to $\Delta T$. Due to the nonideallity of most polymer-solvent systems, it is necessary to measure at about five concentrations and to extrapolate the values of $\Delta T/c$ towards $c = 0$.

The quantities

$$\frac{RT_b^2M_0}{\rho_0\Delta H_v}$$

and

$$\frac{RT_m^2M_0}{\rho_0\Delta H_m}$$

are known as ebullioscopic constant $K_b$ and cryoscopic constant $K_m$, respectively. Table 2 shows these constants for various solvents at normal pressure (760 mmHg).

From Table 2 it appears that cryometry will lead to higher $\Delta T/c$ values than ebulliometry using the same solvents. The usually occurring phenomenon of undercooling may cause errors in the determination of $\Delta T_m$. Hence, in spite of the larger measurable effects for cryometry, this method may be less precise than ebulliometry.

### 3. *"Vapor Pressure Osmometry"*

There exist commercially available apparatuses called "vapor pressure osmometers", which, in fact, are not osmometers at all. Moreover, they do not really measure vapor pressure — a colligative property — but a small temperature difference that results from the different rates of solvent vaporization from tiny droplets of solvent and solution. The vaporization takes place in a saturated solvent atmosphere.

It is assumed that the measured small temperature difference is a colligative property. This assumption should be checked every time that the "osmometer" is used for determinations; the results are heavily dependent on the polymers used for calibration. It is advised[10] not to calibrate with low molecular weight standards.

## VI. LIGHT SCATTERING

This method is a typical method for polymers: the measurable effect rises when the molecular weight rises. We will show why $\overline{M}_w$ is determined and how the structurally important radius of gyration can be derived from light scattering data.

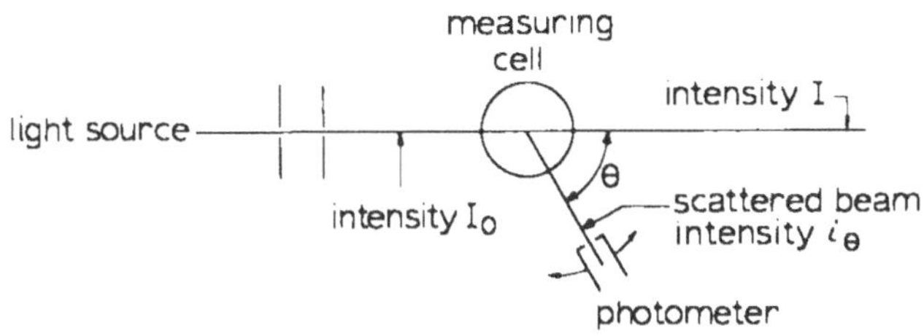


FIGURE 6. Schematic diagram of light scattering measurement.

## A. Theory

When a beam of light passes through a solution, a certain amount is absorbed, a certain amount is transmitted, and a certain amount is scattered. The scattering may be elastic ("Raleigh scattering") or inelastic ("Raman scattering" or "Brillouin scattering"). Absorption and inelastic scattering deal with transitions between quantum states, hence they depend on the wavelength or alter it. Structurally important parameters (like the "rotational diffusion coefficient") may be derived from Brillouin scattering and stereochemical parameters from Raman scattering or (IR) absorption.

The theory of elastic light scattering, which we will discuss in this section, has originally been developed for diluted gases by Lord Rayleigh (1871). The oscillating electric field of the light wave induces oscillating dipoles in the molecules. The dipole moments are proportional to the polarizability of the molecules. According to the Maxwell theory, these oscillating dipoles act as sources of secondary light waves, scattered to all directions. The intensity, i, of the scattered light is proportional to the intensity, $I_0$, of the primary beam: $i = \tau I_0$, where $\tau$ is called turbidity, because it is a measure of the diminution of the transmitted light by scattering (Figure 6).

A liquid also scatters light. This phenomenon has been explained[3] by Smoluchowski and by Einstein, who considered the liquid to consist of domains, small compared to the wavelength (diameter $\lambda/20$). The density of the domains and hence their polarizability fluctuates as a consequence of the Brownian movement. These domains may act as independent scattering sources like the molecules of the diluted gas. The intensity of the scattered light now depends on the fluctuation of the polarizability.

In solutions, there are local fluctuations of the concentration. The probability of the occurrence of a concentration fluctuation will depend on the variation of the partial free enthalpy $\overline{G}_0$ of the solvent with varying concentration (Debije) and hence, on $\partial\Pi/\partial c$, where $\Pi$ is the osmotic pressure.[3] This finally leads to

$$\tau = \frac{32\pi^3 n^2}{3\lambda^4 N_A}\left(\frac{dn}{dc}\right)^2 \frac{c}{\frac{1}{M} + 2Bc + \dots}$$

where n = refractive index; $\lambda$ = wavelength; B = second virial coefficient of osmotic pressure; and M = molecular weight.

The value of $\tau$ and hence the intensity of the scattered light is so low that the transmitted light beam has nearly the same intensity as had the incident beam. Therefore, it is not the diminution of the intensity of the incident beam that is measured in the light scattering experiment (this would be as imprecise as the determination of the captain's weight by weighing the ship with and without the captain), but the intensity of the light, scattered at an angle $\theta$ with respect to the incident light:

$$i_\theta = \frac{3I_0(1 + \cos^2\theta)}{16\pi r^2}\,\tau$$

where r = distance from the scattering center to the observer. The factor $(1 + \cos^2\theta)$ applies to unpolarized light. The ratio $i_\theta r^2/I_0$ is independent of the distance r and is known as the Rayleigh ratio. The final ratio to be determined is

$$R_\theta = \frac{i_\theta r^2}{I_0(1 + \cos^2\theta)}$$

Using the expressions for $i_\theta$ and $\tau$, respectively, and replacing $i_\theta$ by $\Delta I = i_\theta - (i_\theta)_0$ (the difference between solution and solvent scattering), we obtain

$$\frac{Kc}{R_\theta} = \frac{1}{M} + 2Bc + \ldots$$

with

$$K = \frac{2\pi^2 n^2}{\lambda^4 N_A}\left(\frac{dn}{dc}\right)^2$$

If the solute is heterodisperse, $R_\theta$ will be the sum of the contributions from all dissolved species and the limiting value as c tends to zero will be

$$\lim_{c \to 0} R_\theta = K\sum_i c_i M_i$$

or

$$\lim_{c \to 0} \frac{Kc}{R_\theta} = \frac{\sum_i c_i}{\sum_i c_i M} = \frac{\sum_i N_i M_i}{\sum_i N_i M_i^2} = \frac{1}{M_w}$$

Thus, light scattering measurements lead to weight average molecular weights.

In the above theory, the domains were considered to have dimensions $\lambda/20$. Most macromolecules have a much larger extension. Random coiling polymer molecules of, for example, $M = 10^5$ to $10^7$, have diameters of 10 to 100 nm. Hence, these molecules do not act as simple sources of scattered light. Radiation by induced dipoles in widely separated segments of the polymer molecule will have an appreciable phase difference, which leads to a diminution of the scattered intensity, by interference.

The effect of the large size of polymer molecules is generally described by a function $P(\theta)$, known as the particle scattering factor:

$$P(\theta) = \frac{\text{scattered intensity for large particles}}{\text{scattered intensity without interference}}$$

$P(\theta)$ may be much less than unity when $\theta$ is relatively large. It increases with decreasing $\theta$ and becomes unity when $\theta = 0°$. This particle scattering factor is important for structure determination. At $\theta > 0°$, it depends on the geometric shape. $P(\theta)$ has been evaluated for different models, e.g., for long thin rods,[11] flexible coils,[12,13] and stiff chains.[14] Figure 7 shows the measured value of $P^{-1}(\theta)$ for a sample of DNA,[15] together with the calculated values for a coil and a rod (dotted lines), from which it can be concluded that this sample resembles a random coil more closely than a rod.

For $\theta$ approaching zero, $P(\theta)$ is related to the radius of gyration $R_g$ by the equation:

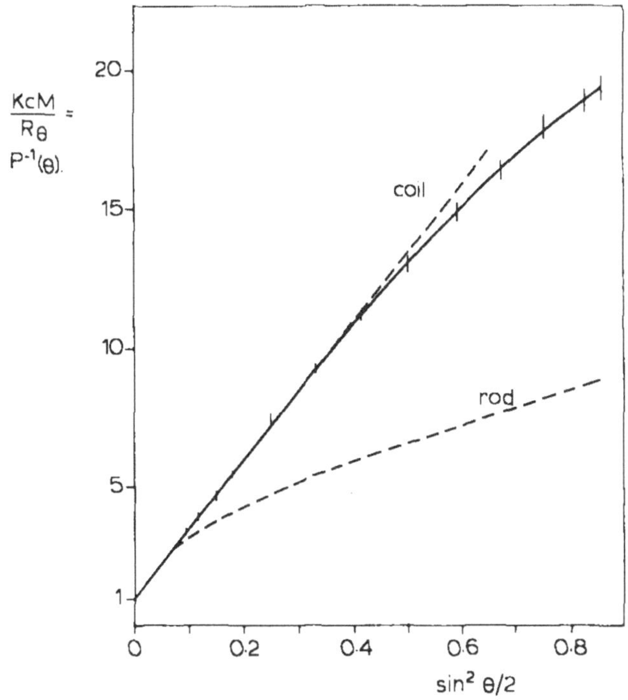


FIGURE 7. Reciprocal particle scattering factor of a sample of DNA, extrapolated to zero concentration. (From Doty, P. and Bunce, B. H., *J. Am. Chem. Soc.*, 74, 5029, 1952. With permission.)

$$\lim_{\theta \to 0} P(\theta)^{-1} = 1 + \frac{16\pi^2}{3\lambda^2} R_g^2 \sin^2 \frac{\theta}{2}$$

Consequently,

$$\lim_{\substack{c \to 0 \\ \theta \to 0}} \frac{Kc}{R_\theta} = \frac{1}{\overline{M}_w P(\theta)} = \frac{1}{\overline{M}_w}\left(1 + \frac{16\pi^2}{3\lambda^2} R_g^2 \sin^2 \frac{\theta}{2}\right)$$

N.B.: For polydisperse polymers, the measured value of $R_g{}^2$ is a z-average:

$$\langle R_g^2 \rangle_z = \frac{\sum_i N_i M_i^2 (R_g^2)_i}{\sum_i N_i M_i^2}$$

Thus, $\overline{M}_w$ is found by extrapolation of $Kc/r_\theta$ to zero concentration and zero angle as well. This can be achieved by a method, introduced by Zimm,[16] according to which $Kc/R_\theta$ is plotted vs. $\sin^2 \theta/2 + kc$, k being an arbitrary constant in order to provide a convenient spread of the data on a graph. This is demonstrated in Figure 8 where the measuring results for a sample of DNA[15] have been plotted. Extrapolation to $\theta = 0$ gives a plot of $Kc/R_\theta$ vs. Kc, which gives $1/\overline{M}_w$ as intercept and 2B as limiting slope. From extrapolation to $c = 0$ a plot is obtained of $kc/R_\theta$ vs. $\sin^2 \theta/2$ from which $R_g$ can be obtained.

For very high molecular weights, the light scattering may vary with θ in a complicated way, leading to curvatures of the lines obtained at constant c. Extrapolation to zero angle

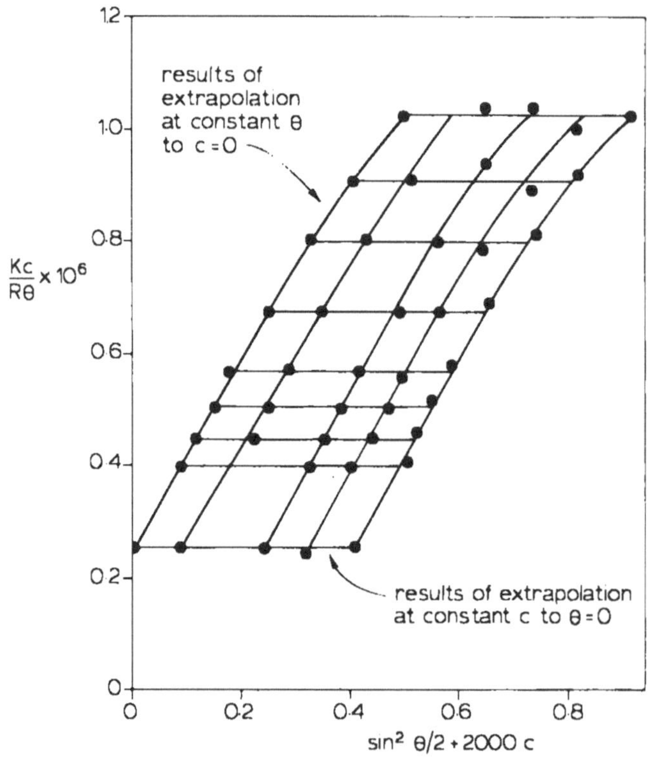


FIGURE 8. Zimm plot for extrapolation of light scattering data for a sample of DNA. (From Doty, P. and Bunce, B. H., *J. Am. Chem. Soc.*, 74, 5029, 1952. With permission.)

may then be difficult and measurements at very low angles (<25°) will be necessary. This has been shown by Harpst et al.,[17] who measured samples of DNA at angles of 10 to 25°. They found that $\overline{M}_w$ and $R_g$ measured at angles ≥30° are considerably lower than those measured at angles $10 < \theta < 25°$. The latter were in good agreement with the values calculated from intrinsic viscosity according to a method developed by Crothers and Zimm.[18]

Several methods have been developed for extrapolation by using numerical methods.[19-21] These methods have been compared and discussed by Llorente and Horta.[22]

## B. Experimental

There are several types of light scattering meters. The main difference is between the "absolute" meters and the other ones. Absolute meters measure the scattered intensities, $i_\theta$, relative to the intensity of the unscattered beam; the other ones measure them relative to the scattered intensity at scattering angle 90° of a calibration liquid for which $R_{90}$ is known (from data, obtained on absolute meters of well-known geometry and corrected for width and heighth of the incident beam, etc.). All meters allow unbiased measurements of intensities scattered at angles from 30 to 150° by means of sensitive photomultipliers that can be rotated around the measuring cell.

It depends on the design of the apparatus and on the skill of the experimenter in making his solutions, etc. "dust-free", whether measurements below 30° (but above 15°) can be performed reliably. A simple means to test the performance of the meter as well as the preparation of the calibration liquid and the solvent is to calculate the product $i_\theta \sin\theta$ for intensities, scattered from a vertically polarized incident beam. For all angles, this product should (for pure liquids, not for the solution) be $1.00 \pm 0.01$.

There exist laser light scattering meters, still more expensive than the "normal" meters, that allow measurements below 15°; some designs go down to 2°. These laser instruments are in some cases able to handle very small amounts of liquids, too (in contrast to the classic instruments, which require 10 mℓ or more for filling the measuring cell).

It is necessary to start light scattering work by the preparation of dust-free calibration liquid (toluene, benzene), pure solvent, and solutions (4 to 5; usually not above 5%; the higher the expected $\overline{M}_w$, the lower the concentrations can be). Measuring cells should be cleaned carefully, preferably in the specially designed apparatus of Schipmann and Farber.[23a] In order to prepare dust-free liquids, they can[23b] be filtered through inert (insoluble, low-swelling) filter sheets (pore diameter less than 0.45 μm), directly into the cleaned and dried measuring cell. Alternatively, the liquids can be centrifuged, preferably in specially designed measuring cells.[24] (Caution: not only "dust" can be removed in this way from the solutions, but the macromolecules themselves may be removed, too.[24])

From tabulations or from extra experiments, values should be known of the refractive indexes of solvent or calibration liquid, the refractive index increment of the solutions, and (when calibration is necessary) the Rayleigh ratio at 90° for the calibration liquid. Kratochvíl et al.[25] give a useful review of calibration liquids with their $R_{90}$ values.

The scattered intensities for the solution, from which those for the pure solvent are subtracted, are first corrected for depolarization.[3] The results can then be plotted according to the method of Zimm (Figure 7; sometimes it may be profitable to plot the results vs. $c_{1/2}$ instead of vs. c).

## C. Applicability

The light scattering method is applicable for all solute molecules which are large enough (but not too large; a useful rule is the radius of gyration should be in the range of roughly $\lambda'/20 < R_g < \lambda'/2$, where $\lambda'$ is the wavelength in the solution), provided that the intensity of the incident beam is not diminished by other effects than Rayleigh scattering, neither by the solvent nor by the solute. Thus, absorption or fluorescence should be absent. Corrections for such effects are unreliable and should be avoided (e.g., by means of lightfilters or measurement at other wavelengths).

A serious experimental limitation is that the refractive index increment dn/dc should exceed 0.100 mℓ/g. Otherwise, the $i_\theta$ values for the solution may be too close to the $(i_\theta)_0$ values for the solvent, and hence large errors in $\Delta I$ may result.

The method is not absolute, i.e., it does not give thermodynamically exact values of $\overline{M}_w$ and B. The reason is that the extrapolation to zero angle depends on the assumption to be made about the shape of the polymer molecules in the solution. This is not a serious disadvantage of the method; correction is possible[26] and only desirable for extremely precise determinations of $\overline{M}_w$, B, or especially $R_g$.

# VII. ULTRACENTRIFUGATION

The analytical ultracentrifuge, though as a tool in polymer materials science is somewhat outdated, is still widely used in biochemistry laboratories. Hence we present a brief survey of its possibilities, also because in some respects the analytical ultracentrifuge is unsurpassed by other methods: it is the only method which may give absolute (i.e., thermodynamically exact) m.w.d., and it can handle small amounts (less than 1 mℓ) of solutions, if necessary, in nasty solvents like formic acid.

## A. Theory

First, we should discriminate between the equilibrium method and the velocity method. The first method gives absolute values for $\overline{M}_w$ and $\overline{M}_z$; the rotation speed is relatively low

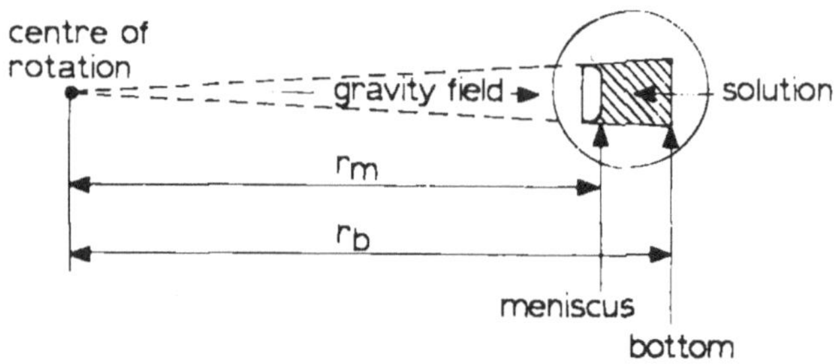


FIGURE 9. Measuring cell of the ultracentrifuge, looked upon from above. Direction of gravity field as indicated.

(less than 10,000 r/min) and the polymer molecules are not precipitated (no ''sediment'' is formed), but redistributed. In this case the centrifugal forces are counterbalanced by the diffusional forces which tend to abolish the generated concentration differences. This leads to a Boltzmann distribution of solute molecules in the cell, fully comparable, e.g., with the decrease of $CO_2$ concentration in air as function of increasing altitude.

The second method needs calibration and gives ''sedimentation averaged'' molecular weights (c.f., viscometry). The rotor is operated at speeds up to 64,000 r/min (which causes accelerations up to 250,000 times that by gravity) and the polymer molecules are precipitated (sedimentated). This is comparable with, for example, a metal ball falling down in a viscous liquid with a velocity that increases as (1) the difference in density between the metal and the liquid increases and (2) the viscosity of the liquid decreases. (Of course, if the solute molecules possess a lower density than the solvent, flotation occurs instead of sedimentation.)

The following diagrams show how the gravity field is directed with respect to the cell, containing the liquid, and how it influences the (re-)distribution or the sedimentation of the dissolved macromolecules (Figures 9 to 11).

Analytical ultracentrifuges contain various optical systems which enable us, after reading out photographic plates or recorder-outputs, to calculate the course of the concentration with r, the distance from a point in the solution to the center of rotation, and (for velocity experiments) with time t. From these graphs, molecular weights or m.w.d. may be calculated. There are so many ways for doing this, all with their own advantages and limitations, that we are obliged to refer here to the widely used review of Creeth and Pain,[27] to which we add the reviews of Adams et al. and Wan.[28,29] Structural information may be obtained from the sedimentation coefficient as a function of the concentration and/or the molecular weight.[3]

## B. Experimental

Analytical ultracentrifuges are mechanically and optically very sophisticated (hence, very expensive) instruments. For experimental details (which are so various and numerous that they fall outside the scope of this book), we refer to the excellent manufacturer's handbooks, supplied with each instrument.

## C. Applicability

High molecular weight polymers require low rotor speeds (both for equilibrium and velocity method); the reverse is true for low molecular weight polymers. Limits are set to the rotor speed for two reasons. The first is rotary instability at speeds lower than 3000 r/min. Then most rotors show a precessional movement, which causes errors in the c-r or the c(t)-r curves recorded by any optical system. The upper limit in speed is set by the gravity forces which a rotor can withstand. Old rotors, that have been used many times, may get tiny fissures. This may lead to a serious explosion and to damage of the instrument if such an old rotor is operated above an allowable speed. (The manufacturers deliver tables with

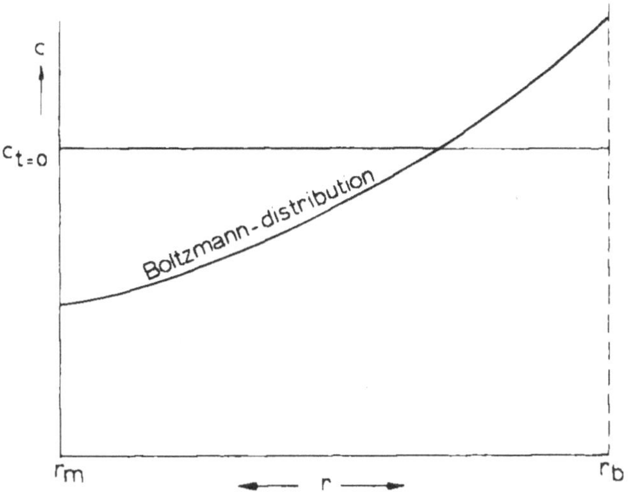


FIGURE 10. Equilibrium method leading to a Boltzmann distribution, independent of time and occurring at relatively low speeds after some time (~24 hr, c = concentration).

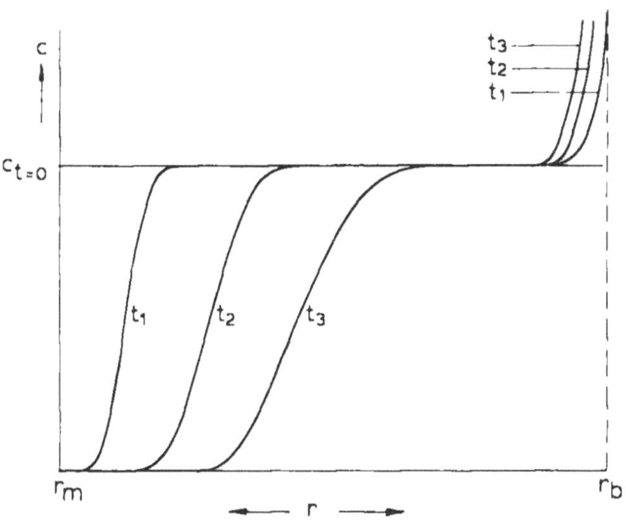


FIGURE 11. Velocity method leading to sedimentation coefficient; c as a function of time t (c = concentration).

rotor speeds as a function of "age" and rotor type.) The limitations to the rotor speed imply that the range of molecular weights covered is between $\overline{M}_w$ = 1,000 and $\overline{M}_w$=2,000,000.

The optical system has its own requirements. The refractive index increment (necessary to be known when the Schlieren optical system is used) should not be too small (preferably higher than 0.100 mℓ/g;) the solvent should not absorb light when an absorption optical system is used. The cell walls and packings should be entirely resistant to the solvent, chosen for optical or other reasons (the cell material should neither dissolve nor swell in the solvent).

Finally, theory sets some limitations. When the equilibrium method is used, molecular weights, especially m.w.d., should be determined in θ solvents (see Section II). Otherwise, tedious and approximate calculations may become necessary; determination of the m.w.d. will even become impossible. The velocity method is restricted to polymers with polydispersity ratio's preferably not exceeding 2.0.

# VIII. SIZE EXCLUSION CHROMATOGRAPHY

Attempts have been made to adapt thin layer and column chromatography, well-known to workers in biochemical and biophysical research, to molecular weight determinations of polymers. In some cases good results were obtained, but as a rule we may say that these methods are not apt for polymers.[30] Gel permeation chromatography (GPC), in its low pressure and in its high pressure version (HPLC), however, have been proven to be very successful instrumental methods for the characterization of polymers.

## A. Theory

The columns are packed with swollen gel particles, which form, together with the solvent enclosed by the particles, the stationary phase. The solvent, flowing through the interstitional volume (i.e., the volume between the gel particles), may be regarded as the mobile phase. Small polymer molecules, injected in the mobile phase stream, may be retarded by the stationary phase, because they are able to pass the pores of the gel particles and thus permeate into them. Hence, the largest polymer molecules leave the column system first; the smaller ones come later.

By using (e.g., four) different columns, each packed with gel particles of a certain mean pore size and a well-defined pore size distribution, a separation of the polymer solution into its components may be obtained. It is generally accepted nowadays that the mechanism of separation rests upon the hydrodynamic volume, i.e., the "hydrodynamic size" of the individual polymer molecules; this "size" is the product of the (individual) intrinsic viscosity, $[\eta]$, and the (individual) molecular weight, M, hence the name "size exclusion chromatography".

The hydrodynamic volume concept has been introduced by Benoît et al.[31,32] They showed that a variety of chemically and structurally different polymers yield a common calibration curve when $[\eta]M$ is plotted vs. the elution volume $V_e$:

$$\text{Log}\,([\eta]M) = f(V_e)$$

Coll and Prusinowski[33] developed a slightly different relationship, based on a viscosity relation of Ptitsyn and Eisner:

$$\text{Log}\,\frac{[\eta]M}{\Psi\,(a)} = g(V_e)$$

where $\Psi(a)$ is a function of the exponent a in the Mark-Houwink relation (Section IV).

Spatorico and Coulter,[34] among others, showed that calibration with polystyrene standards of very narrow m.w.d. leads to determinations of the molecular weight with an accuracy of 5 to 10%, whereas Samay et al.[35] affirmed this validity even for random copolymers (Figure 12; see also Section X).

The molecular weight derived from the elution volume depends on the shape of the molecules.[36]

## B. Experimental

An experimental set-up consists of a solvent container, a pump, a solution injector, an "oven" (thermostated within 0.1°C) containing the columns, a detector, and a recorder. The differences between the normal pressure and HPLC versions with respect to the instrumentation are that for HPLC the pump is an expensive precision instrument and that for the normal pressure GPC the columns are 3 to 4 times longer (total length about 6 m). In practical respects, the HPLC measurement is about 10 times faster (the result of an injection may

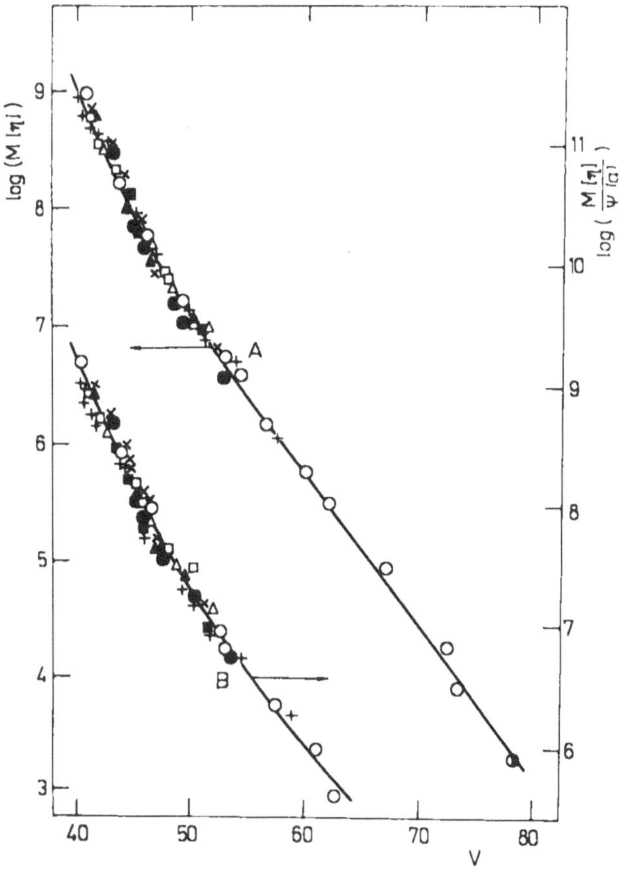


FIGURE 12. Universal calibrations according to Benoît[31,32] (A) and Coll and Prusinowski[33] (B), collected by Samay et al.[35] Symbols: O, polystyrene; □, poly(ethyl methacrylate); △, polyvinylchloride; ●, poly(*n*-butyl methacrylate; ■, poly(*n*-octyl methacrylate); +, poly(ethyl methacrylate-co-styrene); x, poly(butyl methacrylate-co-styrene); ▲, poly(*n*-octyl methacrylate-co-styrene); ◑, squalane. (From Samay, G., Kubín, M., and Podešva, J., *Angew. Makromol. Chem.*, 72, 185, 1978. With permission.)

appear on the recorder already after 10 min), but HPLC measurements are at present not as precise and accurate as low pressure GPC measurements. Developments in this field, however, are evolving rapidly.

The most frequently used detectors are differential refractive index increment detectors. Figure 13 shows an example of the recorder output.

The refractive index increment Δn, which may be safely assumed to be independent of molecular weight, provided that it is not "too low", is directly proportional to the weight of polymer in a volume element of the eluted liquid, but there is not such a simple relationship between the time-axis and molecular weight (the only thing which we may be sure of is the longer the time, the lower the molecular weight).

The time-axis needs calibration by means of samples of known $\overline{M}_w$ and known [η]; then their mean hydrodynamic volumes [η] $\overline{M}_w$ are plotted against the elution volume $V_e$ (calculated from time and flow velocity) that belongs to the maximum in the recorder output (see Figure 11). Different kinds of polymers should be used for this calibration curve, which should be independent of flow velocity, concentration of the injected solution, and kind of polymer, but may depend on the instrumentation and will hence differ from one laboratory to another.

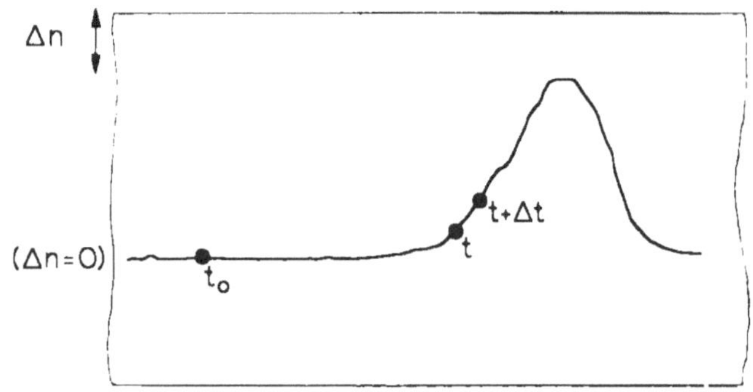


FIGURE 13. Recorder output of a GPC experiment, $t_0$ being time of injection of the polymer sample. Δ n = difference of refractive index between solution and solvent.

Once a calibration curve is available, average molecular weights and the m.w.d. may be calculated for any polymer that is soluble in one of the solvents, for which a calibration curve is prepared. For certain groups of solvents, the calibration curve is independent of the solvent.[37]

The calculated molecular weights and the calculated m.w.d. mostly need correction for several biasing factors; the most important is the correction for the "eddy-diffusion" (also well-known for gas chromatography). There exist several means to do this; Evans[38] gives a rather complete review of what one has to do before a recorder output can be finally called m.w.d.

### C. Applicability

The method is applicable to every polymer that is soluble in a suitable solvent, i.e., a solvent which does not damage the gel particles or the instrument (column material, packings of columns, or pump, etc.).

The calibration curve remains a weak point, as there are still polymers (e.g., polyelectrolytes, certain copolymers) which are not covered. This field, too, however, is rapidly in progress. A new development is the low angle laser light scattering detector (LALLS), which gives M as a function of the elution volume directly, without resort to a calibration curve.[39]

Even without calibration the instrument may be very useful, e.g., for qualitative checking of the progress of polymerization or depolymerization reactions or of the results of a post-polymer reaction (i.e., a reaction in which reactive groups on the polymer molecule are involved; often chain-break occurs at such reactions). Degradation of polymers by enzymes may also be followed in a qualitative way.

## IX. FRACTIONATION

Essentially, the size exclusion chromatographic techniques described above are fractionation techniques, by which fractions of different molecular weight are obtained. In an earlier stage, several other methods for fractionation have been developed, based on phase separation phenomena. For the theory and the experimental techniques, we refer to the recent book of Tung.[4]

Today, polymer fractionation based on phase separation as a tool for the determination of the m.w.d. has been superseded almost fully by size exclusion chromatographic techniques and, to a lesser extent, by ultracentrifugal analysis. It is still of interest for preparative purposes in order to obtain samples of different molecular weight. This may be important for the investigation of the influence of molecular weight and m.w.d. on the polymer properties. It is also important for the characterization of copolymers (see Section X).

Fractionation is a very simple technique and does not require the use of elaborate or special equipment. On the other hand, it is time consuming and the interpretation of the experimental data in terms of m.w.d. is much more complicated than one often has been aware of. This has been thoroughly discussed by Koningsveld.[40]

Fractionation is normally carried out by precipitation or by extraction, but the best way to obtain fractions with a narrow m.w.d. is to isolate the fraction remaining in dilute solution at a high degree of precipitation.[41] This is equivalent to extraction fractionation except that equilibrium is easier to achieve. It can be regarded as a process of distribution of each of the components over two liquid phases. As all components are present in both phases, a separation of "pure" components is impossible. Fractionation of a polymer with one maximum in its m.w.d. may even yield fractions with two maxima in their m.w.d. (never the reverse occurs). Moreover, the polydispersity ratio $\overline{M}_w/\overline{M}_n$ of a fraction is not necessarily lower than that of the parent sample, but may even be higher.[42]

Crystalline polymers can be fractionated by selective crystallization. Koningsveld and Pennings[43] reported the possibility of even sharp fractionation of polyethylene in this way. This has been confirmed later by Pennings[44] and Sadler;[45] it must be emphasized that chain irregularities like chain branching, stereoirregularity, and sequence distribution (in copolymers) have a bigger influence on crystallization temperature than have differences in molecular weight. In that case, fractionation may serve as an analytical tool for those irregularities, but not for the determination of the m.w.d.

## X. COPOLYMERS

In order to meet special requirements, polymeric biomaterials often consist of copolymers. The determination of the molecular structure of copolymers is more complicated than that of "homopolymers" as new parameters are involved. In particular we will be interested not only in the distribution of overall molecular weights, but also in the chemical composition, i.e., the weight fractions of the comonomers in the whole polymer as well as in the sequence distribution (numbers and lengths of sequences or branches per molecule). Moreover, the determination of volume and shape of the molecules generally appears to be more complicated than in the case of homopolymers. In this section we have to confine ourselves to only a brief introduction.

### A. Chemical Composition

A lot of chemical and physical methods are available for the analysis of chemical composition of polymers. These are reviewed by Mitchell and Chiu.[46] For copolymers, a selection of appropriate methods has to be made.

The chemical composition of condensation copolymers can be determined by analysis of the different constituents obtained after hydrolyzing. However, spectroscopic techniques may also be very helpful. For example, Kircheldorf and Hull[47] determined the chemical composition of ternary copolyamides with the 90.5-MHz $^{13}$C-NMR spectroscope from the shift of CO signals.

Short sequences of $(CH_2)_n$ in, for example, ethylenevinyl acetate copolymers, can be detected with IR spectroscopy because the rocking vibration shifts from 815 cm$^{-1}$ for n = 1 to 722 cm$^{-1}$ for n $\geq$ 5.[48]

Elemental analysis can be used if one of the comonomers contains an element not occurring in the other(s). For example, copolymers of vinylchloride and vinylacetate can be characterized simply by analysis of the chlorine or the oxygen content.

Harwood and Ritchey[49] introduced the run number R in order to interpret the results of chemical[50] and/or physical[51] analysis in terms of sequence distribution. To this end the percentage of A-B and B-A linkages in a copolymer AB has to be determined as well as the molar percentages of the monomers A and B (% A and % B).

R is the average number of monomer sequences (runs) occurring in a copolymer chain per 100 monomer units, which is equal to the percentage of A-B plus B-A linkages. Then % A-A linkages = % A − R/2 and % B-B linkages = % B − R/2.

The number average number of monomers A in a sequence is

$$\langle n_A \rangle_n = \frac{\%A}{R/2}$$

The probability of the occurrence of a sequence $A_n$ is

$$P_{A_n} = \frac{(\%A - R/2)^n}{(\%A)^n}$$

The same applies to the monomer B. In this way the fraction of A monomers occurring in the center of triads A-A-B or B-A-A can be expressed by:

$$f_{AAB} = \frac{R\,(\%A - R/2)}{(\%A)^2}$$

The fractions of A monomers occurring in the center of triads A-A-A or B-A-B are

$$f_{AAA} = \frac{(\%A - R/2)^2}{(\%A)^2}$$

and

$$f_{BAB} = \frac{R^2}{4(\%A)^2}$$

Consequently, the fraction of A monomers occurring in the center of A-B-A-B-A pentads is

$$f_{ABABA} = \frac{R^4}{16(\%A)^2\,(\%B)^2}$$

**B. Molecular Weight and M.W.D.**

The molecular weights of copolymers can be determined by the same instrumental techniques as are applied for homopolymers, but for theoretical reasons deviating measuring methods may be necessary. Moreover, the evaluation of the results may be more complicated and less precise.

For each of the species A and B in a copolymer AB number average molecular weights can be defined as:

$$\overline{M}_n^A = \frac{\sum_i n_i M_i^A}{\sum_i n_i}, \quad \overline{M}_n^B = \frac{\sum_i n_i M_i^B}{\sum_i n_i}$$

where $n_i$ is the number of moles of species i. Then,

$$\overline{M}_n = \overline{M}_n^A + \overline{M}_n^B$$

Weight average molecular weights are not additive. If

$$\overline{M}_w^A = \frac{\sum_i w_i^A M_i^A}{\sum_i w_i^A}, \quad \overline{M}_w^B = \frac{\sum_i w_i^B M_i^B}{\sum_i w_i^B}$$

where

$$w_i^A = n_i M_i^A \text{ and } w_i^B = n_i M_i^B$$

then it can be shown[53] that

$$\overline{M}_w < \overline{M}_w^A + \overline{M}_w^B$$

We will now discuss very briefly the applicability of the techniques, described in the preceding sections, to copolymers.

*1. Methods Based on Colligative Properties*

Number average molecular weights of copolymers can be determined in exactly the same way as described for homopolymers in Section V, because the measuring effects are not dependent on chain length, shape, or chemical composition of the solute molecules. This also implies that $\overline{M}_n^A$ and $\overline{M}_n^B$ cannot be determined by these methods.

The interpretation of the second and higher virial coefficients is much more complicated as they are related to at least three kinds of interactions between monomer sequences A, B, . . . and solvent molecules.

*2. Light Scattering*

Bushuk and Benoît[53] showed that the molecular weight, measured by light scattering in the way discussed in Section VI, has a value that strongly depends on the refractive index increment, which, according to these authors, is denoted by $\nu$. This increment varies with varying monomer-solvent combinations. Consequently, the local electric field strength of the light producing light scattering will also vary. This leads to a decrease of destructive interference and hence to an increase of scattered intensity. Consequently, the measured molecular weight is too high. This apparent molecular weight, $M_{app}$, is according to Bushuk and Benoît:

$$M_{app} = \overline{M}_w + 2\,P\,\frac{\nu_A - \nu_B}{\nu_0} + Q\left(\frac{\nu_A - \nu_B}{\nu_0}\right)^2$$

where P and Q are functions of $\overline{M}_w, \overline{M}_w^A$, $\overline{M}_w^B$ and the molar fraction $x$ of A in the copolymer AB, while $\nu_A$ and $\nu_B$ are the refractive index increments of the homopolymers in the same solvent and $\nu_0$ is the average refractive index increment that is obtained by measuring the copolymer.

Now, by plotting $M_{app}$ vs. $\nu_A - \nu_B/\nu_0$ a parabolic curve is obtained from which $\overline{M}_w$ can be obtained by putting $\nu_A - \nu_B/\nu_0 = 0$. Measurements in at least three solvents are necessary for the determination of $M_w$, leading also to values of $\overline{M}_w^A$ and $\overline{M}_w^B$ or P and Q, respectively. In practice it is recommended to use four to six solvents with $\nu_A$ and $\nu_B$ values differing as much as possible in order to reduce the experimental errors.[54] According to Kratochvíl et al.,[55,56] the maximal accuracy is obtained under the following conditions:

1st solvent: $\nu_A$ and $\nu_B$ are high and have the same sign
2nd solvent: $\nu_A$ is large; $\nu_B = 0$
3rd solvent: $\nu_A = 0$; $\nu_B$ is large

Spatorico[57] found that the method discussed above is insensitive for low molecular homopolymer in a blend with copolymer, whereas Chau and Rudin[58] stated that this method is less successful for random copolymers than for block or graft polymers. This is probably due to a failure of the additivity rule for $\nu$, which is the basis of the Bushuk-Benoît method, as the sequences are very short. When the solvent quality with respect to one of the comonomers decreases, $\overline{M}_w$ may increase as a result of the forming of micelles.[59,60]

### 3. *Ultracentrifugation*

For low molecular weight copolymers, the ultracentrifuge technique seems to be more appropriate,[61] especially when solvent mixtures with a density gradient are applied.[62] In fact, the ultracentrifuge technique is more sensitive to differences in density than to differences in molecular weight. Copolymers with chemical composition polydispersity and thus differences in solvation grade will possess differences in density. This leads to a broadening of the particle distribution in the measuring cell compared with homopolymers. This broadening reflects the chemical distribution.[62,63] As for a density gradient a mixture of solvents is needed, preferential adsorption of one of the solvents may occur leading to extra differences in partial specific volumes and hence to an extra influence on the position of the molecules in the measuring cell.[63]

### 4. *Viscometry*

Homopolymers with different chemical composition are normally incompatible, also in solution. Random copolymers in solution give rise to much more (repulsing) contacts between the different monomer units than block and graft polymers do. Consequently, random copolymers will show bigger hydrodynamic volumes than block and graft polymers of the same overall composition and the same molecular weight. Hence, these random copolymers will exhibit higher intrinsic viscosity values than the comparable block and graft polymers.[64] Hetero contact interaction is strongly solvent dependent. The better the solvent, the more compatible the solutes are and the smaller the difference in $[\eta]$ between random and block or graft polymers of the same monomer composition.[65]

Random copolymers and block or graft polymers of the same overall composition and the same molecular weight will show different θ-temperatures. Froehlich and Benoît[66] have shown that random and block copolymers of the same overall composition and the same molecular weight exhibit different θ-temperatures. At these θ-temperatures their $[\eta]$ values are about equal. They vary linearly with temperature, but their slopes are different. Consequently, viscometry may be helpful for differentiating random and block copolymers.

The Mark-Houwink relation (Section IV) should only be applied to copolymers of the same type as the standard materials used for deriving the constants K and a. Wang[67] proposed a method for the determination of the molecular weight of block copolymers by using $[\eta]$ and either the sedimentation coefficient or the translational diffusion coefficient.

### 5. *Size Exclusion Chromatography*

Determination of the molecular weight and the m.w.d. by means of size exclusion chromatography is based on the relationship between the hydrodynamic volume $[\eta]M$ and the elution volume (Section VIII). This method may give useful information also for copolymers,[4,35] provided that the polymer used for calibration has the same structure and chemical composition as the sample.

Chen and Blanchard[68] showed that a given species within an unfractionated sample has a longer residence time in the GPC columns than the same species within a narrow distribution fraction. Cantow et al.[69] and Runyon et al.[70] suggested use of dual detectors (e.g., differential refractometer and IR or UV spectrometer in series), in order to determine distribution of size and composition simultaneously. This may be necessary as the fractions often differ in both molecular weight and chemical composition.

*6. Fractionation*

In a copolymer-solvent system where $\chi_A = \chi_B$, fractionation is not affected by chemical position and thus will principally enable us to determine the m.w.d. by the usual techniques. In other cases the technique of "cross-fractionation" has to be applied, which is a combination of several successive fractionation procedures, using at least two solvents or solvent-nonsolvent systems. For a discussion of this method we refer to Riess and Callot (in Tung[4]).

## XI. POLYELECTROLYTES

Polyelectrolytes are synthetic or biological polymers, containing side groups (e.g., -COOH) or backbone groups (e.g., -NH-), that may be charged electrically when the polymers are dissolved in protolytic solvents like water (examples: polyacrylic acid, polyethyleneimine, polylysine, serum albumine, heparin). The characterization of these polymers consists of the determination of one or more molecular weight averages or of the amount and the kind of the charges or chargeable groups. Other desirable parameters may be ion binding capacities or equilibrium constants for ion binding. For some purposes the distribution of chargeable groups over the polymer chain may be of interest. This subject is related to the sequence distribution which has been dealt with in the preceding section; for an application of a spectroscopic method to heparin, see Reference 71. In this section, we will deal only with molecular weights and with charges, irrespective of their distribution.

The trivial fact, that the charges on a polymer can not move away from each other (as they are covalently bound to a chain that is held together itself by covalent forces), leads to a very strong cylindrically symmetric electric field along the chain of a dissolved polyelectrolyte molecule. Hence, a part of the counterions (i.e., particles of a size, small compared to that of the polymer molecule, e.g., like $Na^+$, $Cl^-$, $(CH_3)_4N^+$, and charged oppositely to the polymer charged groups) may be trapped in a region of several angstroms thickness along the polymer chain.[72] There they are free to move along the chain, but nevertheless are held so strongly that they give no contribution to the electrical conductivity of the solution. The other counterions are present in the volume, occupied by the polymer as a whole (characterized by the radius of gyration). In the space between the polymer molecules there are no counterions. This phenomenon differs fully from the homogeneous distribution of ions of opposite charge over the whole solution volume, e.g., in an aqueous solution of NaCl.

When a low molecular weight salt like NaCl, henceforward to be denoted shortly by "salt", is added to a dilute solution of a polyelectrolyte, the ions charged equally to the polymer charged groups (the coions) are expelled from the polyelectrolyte molecule domains. These expelled coions carry with them their own counterions (in order to maintain electroneutrality). The final result, again, is a nonhomogeneous distribution of ions over the solution volume; there are very tiny regions of polyelectrolyte molecules with their counterions and a minor amount of the added salt, while the major amount of salt is to be found in the vast regions between the polymer molecules.[73]

So the problem is raised: what is the blank solvent, against which a series of polyelectrolyte solutions of different concentrations should be measured in order to determine molecular weights? The answer is that blank solutions should be used of a composition equal to that of the regions between the polyelectrolyte molecules. When a polyelectrolyte solution is dialyzed exhaustively against a salt solution, a distribution of the salt through the dialysis membrane takes place, such that the composition of the dialyzing solution becomes equal to that of the regions between the polyelectrolyte molecules in the dialyzed solution.[74] A series of polyelectrolyte solutions ready for measurement may therefore be prepared either by diluting a dialyzed stock solution with its equilibrium salt solution or by dialyzing a series of polyelectrolyte solutions against salt solutions maintained at equal composition during the dialysis process.

Then, all methods mentioned in the foregoing sections for the determination of molecular weights may be applied,[73,74] except size exclusion chromatography. One should never try to determine molecular weights from measurements on solutions to which no extra salt is added or from measurements against water as the blank then the measured effects are a consequence of the electric field along the polymer chain and not a consequence of molecular weight. Addition of sufficient salt (a useful rule is given below) suppresses the electrical field effect.

When one desires to obtain evidence that a certain polymer is a polyelectrolyte, however, one should perform measurements without salt. For example, if a plot of $(\eta - \eta_0)/\eta_0 c$ vs. c shows the anomalous behavior that it strongly rises with decreasing c, one has an indication that the polymer in question might be a polyelectrolyte.[3,75]

Special precautions are necessary for the application of size exclusion chromatography, the main reason being the adsorption of much polyelectrolyte to the column material. This disadvantage is turned into an advantage by these special precautions; for details we refer to the reviews on GPC of proteins and polypeptides.[76] For salt effects in the size exclusion chromatography of polyelectrolytes, we refer to some recent papers.[77,78]

The molecular weight average is determined for the polyelectrolyte in the state wherein it was dissolved. If we try to dissolve a weighed amount, e.g., of a high molecular weight polyacrylic acid (a powder), dissolution takes place readily when NaOH is added. Then measurements, after addition of salt and after dialysis, lead to the molecular weight of wholly or partially neutralized polyacrylic acid.

When this is regarded as an undesirable ambiguity, one may determine the degree of polymerization, P, without ambiguity. To this end, one has only to plot the results of measurements against the monomer molar concentration, $c_m$, instead of the polymer weight concentration c. The relation between these concentration definitions is

$$c_m = \frac{\text{weight of polymer per unit volume}}{\text{monomer molecular weight}}$$

Thus,

$$c_m = \frac{c}{m}$$

As

$$\frac{c}{M} = \frac{c}{Pm} = \frac{c_m}{P}$$

plots as mentioned in the preceding sections should have 1/P ($1/\overline{P}_n$, $1/\overline{P}_w$, etc.) as their intercept.

The definition of $c_m$ allows us to formulate the following "rule" for the addition of salt to a polyelectrolyte solution. The dialyzing solution, as well as the polyelectrolyte solution before dialysis, should contain salt of a molar concentration $c_s$ (moles/liter) of at least ten times $c_m$. Then molecular weights may be obtained unambiguously, independent of the salt concentration. Omission of the dialysis procedure is sometimes allowable, especially when the polyelectrolyte bears not too many charges. It is advisable, then, to perform measurements at more than one $c_s$ (always $>10\ c_m$).

Equilibrium constants, e.g., for equilibria like

$$\wr\!-COOH + H_2O \rightleftharpoons \wr\!-COO^- + H_3O^+$$

or

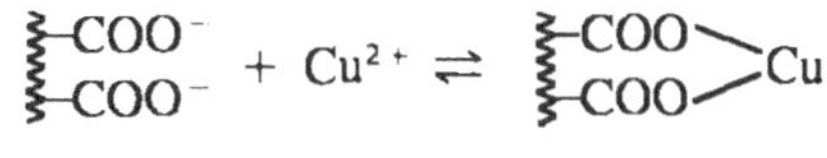


or for the complex association between polyelectrolyte molecules of different kinds, depend always on the amount of added salt. They may be determined by a great variety of techniques.[3,79]

It is possible to determine "intrinsic equilibrium constants", corrected for the electric field effect and thus independent of the amount of added salt.[3] These corrections may be carried out either by assuming the equality of the electric field strength under different circumstances[80,81] or by calculating it from a theoretical polyelectrolyte model[3] or by measuring it (the electric field around a polyelectrolyte is closely related to the ζ-potential, well-known from colloid chemistry; it is measurable, e.g., by means of electrophoresis[3]). Some ambiguity remains, however, as an assumption or a theoretical model is always involved (also necessary for the interpretation of electrophoresis measurements) in the calculation of intrinsic constants.

The type of binding, discussed here, has to do with the interaction of charged groups on polyelectrolyte molecules with certain low molecular weight ions (mainly metal ions of valency ⩾2) or with molecules or with (charged) groups on other polyelectrolyte molecules. This type of interaction — mostly denoted by "site binding" — should be distinguished from the preferential adsorption — or "domain binding" — of counterions, as discussed above. Site binding leads to the formation of lower (or zero) charged "complexes".

It is possible to distinguish between these two types of binding experimentally[82-85] and to study their effects on biological properties.[84-86] Biologically important systems are also "polyelectrolyte-complexes"; (solutions of) oppositely charged polypeptides, for instance.[87]

## REFERENCES

1a. **Brandrup, J. and Immergut, E. H., Eds.,** *Polymer Handbook*, 2nd ed., John Wiley & Sons, New York, 1975.
1b. **Rabeck, J. F.,** *Experimental Methods in Polymer Chemistry*, John Wiley & Sons, Chichester, 1980.
1c. *Polymer Chemistry, a State-of-the-Art Symposium*, published as a special issue of the *J. Chem. Ed.*, 58(11), 836, 1981.
1d. **Nunes, R. W., Martin, J. R., and Johnson, J. F.,** *Polym. Eng. Sci.*, 22, 205, 1982.
1e. *International Symposium on Advances in Polymer Characterization, and Microsymposium on Macromolecules Characterization of Structure by NMR methods*, published as special issues of Pure Appl. Chem., 54(2), 247, and 54(3), 559, 1982.
2. **Flory, P. J.,** *Principles of Polymer Chemistry*, Cornell University Press, Ithaca, N.Y., 1953.
3. **Tanford, C.,** *Physical Chemistry of Macromolecules*, John Wiley & Sons, New York, 1967.
4. **Tung, L. H., Ed.,** *Fractionation of Synthetic Polymers*, Marcel Dekker, New York, 1977.
5. **Rudin, A.,** *J. Chem. Ed.*, 46, 595, 1969.
6. **Walton, A. G. and Blackwell, J.,** *Biopolymers*, Academic Press, New York, 1973.
7. **Misra, G. S. and Mukherjee, P. K.,** *Colloid Polym. Sci.*, 258, 152, 1980.
8. **Noda, I., Kitano, T., and Nagasawa, M.,** *J. Polym. Sci. Polym. Phys. Ed.*, 15, 1129, 1977.
9. **Knott, J. and Rossbach, V.,** *Angew. Makromol. Chem.*, 86, 203, 1980.
10. **Morris, C. E. M.,** *J. Appl. Polym. Sci.*, 21, 435, 1977.
11. **Neugebauer, T.,** *Ann. Phys.*, 42(5), 509, 1943.
12. **Debije, P.,** *J. Appl. Phys.*, 15, 338, 1944; *J. Phys. Chem.*, 51, 18, 1947.
13. **Benoît, H.,** *J. Polym. Sci.*, 11, 507, 1953.
14. **Peterlin, A.,** *Makromol. Chem.*, 9, 244, 1953; *J. Polym. Sci.*, 10, 425, 1953.
15. **Doty, P. and Bunce, B. H.,** *J. Am. Chem. Soc.*, 74, 5029, 1952.

16. **Zimm, B. H.,** *J. Chem. Phys.*, 16, 1099, 1948.
17. **Harpst, J. A., Krasna, A. I., and Zimm, B. H.,** *Biopolymers*, 6, 595, 1968.
18. **Crothers, D. M. and Zimm, B. H.,** *J. Mol. Biol.*, 12, 525, 1965.
19. **Brice, W. A. J.,** *Polymer*, 10, 804, 1969.
20. **Fujita, H.,** *Polym. J.*, 1, 537, 1970.
21. **Miller, W. and Stepto, R. F. T.,** *Eur. Polym. J.*, 7, 65, 1971.
22. **Llorente, M. A. and Horta, A.,** *Biopolymers*, 15, 2287, 1976.

23a. **Schipmann, R. H. and Farber, E.,** *J. Polym. Sci. B*, 1, 65, 1963.

23b. **Kulicke, W. M. and Knieuwske, R.,** *Makromol. Chem. Rapid Commun.*, 1, 719, 1980.

24. **Tabor, B. E.,** in *Light Scattering from Polymer Solutions*, Huglin, M. B., Ed., Academic Press, London, 1972, 13.
25. **Kratochvíl, J. P., Deželić, G., Kerker, M., and Matijević, E.,** *J. Polym. Sci.*, 57, 59, 1962.
26. **Mijnlieff, P. F. and Coumou, D. J.,** *J. Colloid Interface Sci.*, 27, 553, 1968.
27. **Creeth, J. M. and Pain, R. H.,** in *Progress in Biophysics and Molecular Biology*, Vol. 17, Pergamon Press, Oxford, 1967, 217.
28. **Adams, E. R., Ferguson, W. E., Wan, P. J., Sarquis, J. L., and Escott, B. M.,** *Sep. Sci.*, 10, 175, 1975.
29. **Wan, P. J. and Adams, E. T.,** *Biophys. Chem.*, 5, 207, 1976.
30. **Inagaki, J.,** in *Fractionation of Synthetic Polymers*, Tung, L. H., Ed., Marcel Dekker, New York, 1977, 649.
31. **Benoît, H., Grubisic, Z., Rempp, P., Decker, D., and Zilliox, J. G.,** *J. Chem. Phys.*, 63, 1507, 1966.
32. **Grubisic, Z., Rempp, P., and Benoît, H.,** *J. Polym. Sci. B*, 5, 753, 1967.
33. **Coll, H. and Prusinowski, L. R.,** *J. Polym. Sci. B*, 5, 1153, 1967.
34. **Spatorico, A. L. and Coulter, B.,** *J. Polym. Sci. Polym. Phys. Ed.*, 11, 1139, 1973.
35. **Samay, G., Kubín, M., and Podešva, J.,** *Angew. Makromol. Chem.*, 72, 185, 1978.
36. **Elias, H. G.,** *Macromolecules*, John Wiley & Sons, London, 1977, 341.
37. **Kato, Y. and Hashimoto, T.,** *J. Appl. Polym. Sci.*, 18, 1239, 1974.
38. **Evans, J. M.,** *Polym. Eng. Sci.*, 13, 401, 1973.
39. **Ouano, A. C. and Kaye, W.,** *J. Polym. Sci. Polym. Chem. Ed.*, 12, 1151, 1974.
40. **Koningsveld, R.,** *Adv. Polym. Sci.*, 7, 1, 1970.
41. **Koningsveld, R. and Staverman, A. J.,** *J. Polym. Sci. A2*, 6, 305, 1968.
42. **Koningsveld, R. and Staverman, A. J.,** *J. Polym. Sci. A2*, 6, 367, 383, 1968.
43. **Koningsveld, R. and Pennings, A. J.,** *Rec. Trav. Chim.*, 83, 552, 1964.
44. **Pennings, A. J.,** *J. Polym. Sci. C.*, 16, 1799, 1967.
45. **Sadler, D. M.,** *J. Polym. Sci. A2*, 9, 779, 1971.
46. **Mitchell, J. and Chiu, J.,** *Anal. Chem.*, 45(5), 273r, 1973.
47. **Kircheldorf, H. R. and Hull, W. E.,** *J. Macromol. Sci. Chem. A*, 11(12), 2281, 1977.
48. **Lesch, W.,** *Kolloid Z. Z. Polym.*, 240, 737, 1970.
49. **Harwood, H. J. and Ritchey, W. M.,** *J. Polym. Sci. B*, 2, 601, 1964.
50. **Harwood, H. J.,** *Angew. Chem.*, 77, 405, 1965.
51. **Harwood, H. J.,** *Angew. Chem.*, 77, 1124, 1965.
52. **Benoît, H.,** *Ber. Bunsenges. Phys. Chem.*, 70, 286, 1966.
53. **Bushuk, W. and Benoît, H.,** *Can. J. Chem.*, 36, 1616, 1958.
54. **Kambe, Y.,** *Angew. Makromol. Chem.*, 54, 71, 1976.
55. **Tuzar, Z., Kratochvíl, P., and Straková, D.,** *Eur. Polym. J.*, 6, 1113, 1970.
56. **Kratochvíl, P. and Tuzar, Z.,** *Chem. Zvesti*, 25(3), 190, 1971.
57. **Spatorico, A. L.,** *J. Appl. Polym. Sci.*, 18, 1793, 1974.
58. **Chau, T. C. and Rudin, A.,** *Polymer*, 15, 593, 1974.
59. **Tuzar, Z. and Kratochvíl, P.,** *Makromol. Chem.*, 160, 301, 1972.
60. **Tuzar, Z., Petrus, V., and Kratochvíl, P.,** *Makromol. Chem.*, 175, 3181, 1974.
61. **Kotaka, T., Donkai, N., Ohnuma, H., and Inagaki, H.,** *J. Polym. Sci. A2*, 6, 1803, 1968.
62. **Hermans, J. J. and Ende, H. A.,** *J. Polym. Sci. C*, 1, 161, 1963; 4, 519, 1964.
63. **Hermans, J. J.,** *Ber. Bunsenges. Phys. Chem.*, 70, 280, 1966.
64. **Dondos, A., Froehlich, D., Rempp, P., and Benoît, H.,** *J. Chem. Phys.*, 64, 1012, 1967.
65. **Dondos, A., Rempp, P., and Benoît, H.,** *Eur. Polym. J.*, 3, 657, 1967.
66. **Froehlich, D. and Benoît, H.,** *Makromol. Chem.*, 92, 224, 1966.
67. **Wang, F. W.,** *Macromolecules*, 9, 97, 1976.
68. **Chen, H. R. and Blanchard, L. P.,** *J. Appl. Polym. Sci.*, 16, 603, 1972.
69. **Cantow, H. J., Probst, J., and Stajanov, C.,** *Kautsch. Gummi Kunstst.*, 21, 609, 1968.
70. **Runyon, J. R., Barnes, D. E., Rudd, J. F., and Tung, L. H.,** *J. Appl. Polym. Sci.*, 13, 2359, 1969.
71. **Perlin, A. S. and Hamer, G. K.,** *Polym. Prepr.*, 19, 242, 1978.

72. **Manning, G. S.,** *J. Chem. Phys.*, 51, 924, 1969.
73. **Casassa, E. F. and Eisenberg, H.,** in *Advances in Protein Chemistry,* Vol. 19, Academic Press, New York, 1964, 287.
74. **Eisenberg, H.,** *J. Chem. Phys.*, 36, 1837, 1962.
75. **van der Does, L., Hofman, J., and van Utteren, T. E. C.,** *J. Polym. Sci. Polym. Lett. Ed.*, 11, 169, 1973.
76. **Kluh, I. and Prusík, Z.,** in *Liquid Column Chromatography,* Elsevier, Amsterdam, 1975, 741.
77. **Sieburg, W., Lundberg, R. D., and Lenz, R. W.,** *Macromolecules,* 13, 1013, 1980.
78. **Rochas, C., Domard, A., and Rinaudo, M.,** *Eur. Polym. J.*, 16, 135, 1980.
79. **Steinhardt, J. and Reynolds, J. A.,** *Multiple Equilibria in Proteins,* Academic Press, New York, 1969.
80. **Kono, N. and Ikegami, A.,** *Biopolymers,* 4, 823, 1966.
81. **Gregor, H. P., Luttinger, L. B., and Loeb, E. M.,** *J. Chem. Phys.*, 59, 34, 1955.
82. **Boyd, G. E. and Wilson, D. P.,** *Macromolecules,* 15, 78, 1982.
83. **Kowblansky, M. and Zema, P.,** *Macromolecules,* 14, 1448, 1451, 1981.
84. **Cleland, R. L.,** *Macromolecules,* 15, 382, 1982.
85. **Strand, K. A., Bøe, A., Dalberg, P. S., Sikkelund, T., and Smidsrød, O.,** *Macromolecules,* 15, 570, 1982.
86. **Diakun, G. P., Edwards, H. E., Wedlock, D. J., Allen, J. C., and Phillips, G. O.,** *Macromolecules,* 11, 1110, 1978.
87. **Domard, A. and Rinaudo, M.,** *Macromolecules,* 13, 898, 1980.

Chapter 3

# THE RELATIONSHIP BETWEEN THE MECHANICAL PROPERTIES OF POLYMERS AND THEIR STRUCTURE

**N. J. Mills**

## TABLE OF CONTENTS


I. Introduction .... 58

II. Molecular Structure .... 58
- A. Interatomic Bonding .... 58
- B. Rotational Isomers and the Shape of Polymer Molecules .... 60
- C. Chain Regularity .... 62
- D. Molecular Weight Distribution .... 64
- E. Network Molecules and Cross-Linking .... 65

III. Microstructure .... 65
- A. Introduction .... 65
- B. Rubber .... 67
- C. Melt State .... 69
- D. Glassy State .... 71
- E. Spherulitic Semicrystalline State .... 73
- F. Fibrous Semicrystalline State .... 75

IV. Mechanical Properties .... 78
- A. Elastic Moduli and Their Variation with Temperature .... 78
- B. Viscoelastic Behavior .... 80
- C. Yield and Crazing .... 82
- D. Fracture .... 86

# I. INTRODUCTION

In order to keep this chapter brief, only the polymers that are commonly used as biomaterials will be considered. The main features of the mechanical behavior of selected examples of such polymers will be described, and the extent to which this behavior can be explained in terms of the molecular structure will be examined. Table 1 categorizes a number of polymers in terms of the state in which they occur at room temperature.

Detailed properties of each of these polymers are best obtained from manufacturer's literature, as each is sold in many grades to suit different processing methods and applications. Here we will concentrate on the properties of a typical member from each category and investigate the relationship between the molecular structure and the properties of that exemplar. The examples chosen and examples of their use as biomaterials are

| | |
|---|---|
| Polyethylene | As a bearing material in replacement joints |
| Polymethylmethacrylate | Teeth and dentures |
| Polyethyleneterephthalate | In fiber form, woven into fabric for reinforcing tubes |
| Polydimethylsiloxane | Rubber valve parts, tubing, etc. |

It is important to realize that structure-property relationships cannot be considered on their own; the processing method interacts strongly with both (Figure 1). With metal products, heat treatment can be used, after fabrication to the required shape, to give the required microstructure. However, with polymers this is not possible so the melt processing method has a permanent effect on the product microstructure. The flow requirements of the process influence the average molecular weight of the polymer that can be used, which in turn affects the mechanical properties of the product. These interrelations will be considered further in Sections III and IV.

# II. MOLECULAR STRUCTURE

There are many facets to the molecular structure of polymers. Since these can only be introduced one at a time, a balanced view of the structure only develops slowly. The various aspects of the structure that will be used later to explain mechanical properties will be introduced starting with those on the smallest scale and working upwards in size.

## A. Interatomic Bonding

Figure 2 shows the molecular structures of the four typical polymers mentioned in Section I. This figure shows the structure in the form of a graph with the vertexes representing the atoms and the edges representing the covalent bonds between the atoms; no information is given about the bond lengths or the interbond angles. The "backbone" of chain molecules usually consists of C–C, C–O, or C–N bonds and inorganic bonds such as Si–O sometimes occur. It will be necessary later to consider whether such bonds can be broken by macroscopic mechanical forces. A measure of the bond strength is given by the bond dissociation energy, the energy needed per mole of bonds (1 mol contains Avogadro's number or $6.02 \times 10^{23}$ bonds) to separate the two atoms to an infinite distance. Some bond dissociation energies and the effective range of the interbonds forces are given in Table 2.

It should be noted how much stronger the covalent in-chain bonds are than the van der Waals' or hydrogen bonds between chains. Hydrogen bonds only exist between specific chemical groups, such as between the amide groups illustrated, but van der Waals' bonds exist between all noncovalently bonded atoms. In order to appreciate the magnitudes of these energies, they should be compared with the thermal energy that all atoms possess. Making the assumption that each atom in a chain-molecule has two degrees of freedom in its kinetic motion, then the kinetic energy per mole of atoms is RT, where R is the gas constant 8.3

**Table 1**
**POLYMERS COMMONLY USED AS BIOMATERIALS**

| Name | Abbreviation | Category |
|---|---|---|
| Polyethylene | PE | Semicrystalline |
| Polyethyleneterephthalate | PET | |
| Polyoxymethylene | POM | |
| Polypropylene | PP | |
| Polytetrafluoroethylene | PTFE | |
| Polyethersulfone | PES | Glassy |
| Polymethylmethacrylate | PMMA | |
| Polystyrene | PS | |
| Polyvinylchloride | PVC | |
| Polydimethylsiloxane | PDMS | Rubber |
| Polyurethane | PUR | |

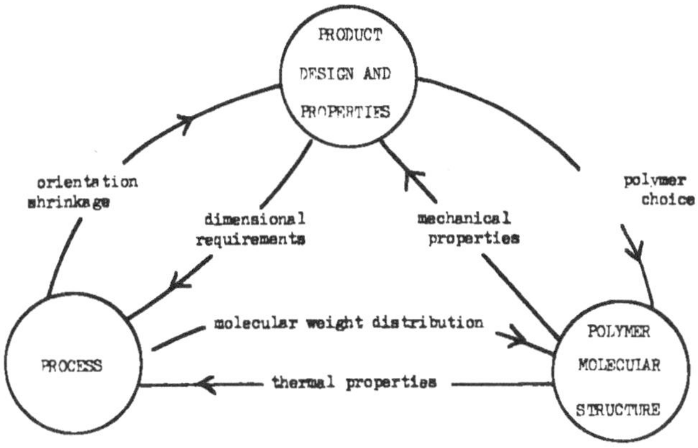


FIGURE 1. The interactions that exist between the process, the molecular structure of the polymer, and the product properties.

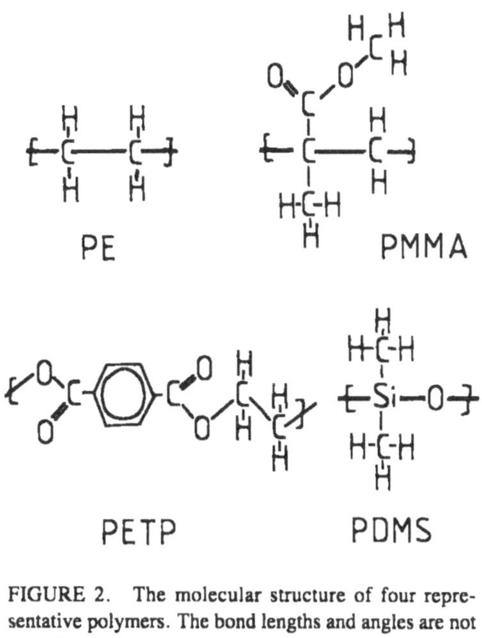


FIGURE 2. The molecular structure of four representative polymers. The bond lengths and angles are not to scale.

**Table 2**
**BOND DISSOCIATION ENERGIES**

| Bond | | Dissociation energy (kJ/mol) | Approximate range of action (nm) of force |
|---|---|---|---|
| R—$CH_2$— | —$CH_2$—R | 337 | 0.15 |
| —(C)—$CH_2$— | —O—(C) | 335 | 0.15 |
| R—C(=O)—$C_6H_4$—C(=O)— | —N(H)—$C_6H_4$—N(H)— | 224 | – |
| Hydrogen bonds | H—N(—)—C=O . . . H—N(—)—C=O | 20 | 0.28 |
| van der Waals' bonds | HCH . . . HCH (each bonded to HCH) | 8 | 0.4 |

J/mol K and T is the absolute temperature. Therefore at T = 293 K, the kinetic energy per mole is 2.4 kJ/mol which is of the same order as the dissociation energy of the van der Waals' bonds.

### B. Rotational Isomers and the Shape of Polymer Molecules

Figure 2 is not a complete representation of the structure of a polymer. The interatomic distances in polymer molecules have fixed values, i.e., the C–C distance is 0.154 nm. The covalent interatomic bonds consist of shared electrons; the orbitals of these electrons are directional, imparting a directional nature to the covalent bonds. The bonding of a carbon atom is explained in terms of the hybridization of electron orbitals; four $sp^3$ hybrids exist giving rise to four single bonds pointing to the corners of a tetrahedron. The angle between these bonds is 109.5°, so any plane drawing of them must necessarily be a projection.

At first it might appear that each C–C bond is potentially free to rotate about its axis. However, as will rapidly become apparent if ball-and-stick molecular models are used, the rotation of successive C–C bonds can cause parts of the chain to "knock" into each other. Figure 3a shows the situation of four successive carbon atoms in a polymer chain. The positions of atoms $C_1C_2$ and $C_3$ are used to define reference axes by which the motion of atom $C_4$ is measured; this is shown more clearly in the projection in Figure 3b. The rotation angle $\phi$ of the bond $C_3$–$C_4$ can take any angle in the range $-180$ to 180°. However, if $\phi = 0$, some of the other atoms bonded to $C_1$ and $C_4$ atoms come into very close contact and the interatomic repulsion of these atoms increases the potential energy of that conformation (shape). If the substituted atoms on $C_1$ to $C_4$ are all H atoms, as in polyethylene, then Figure 4 shows how the potential energy of the rotation of bond $C_2$–$C_3$ varies with $\phi$. There is a minimum energy at $\phi = 180°$, when the four bonds $C_1C_2C_3C_4$ lie in a plane; this is known as the "*trans*" (*t*) conformation, which occurs in the crystalline state of polyethylene. At an energy $\Delta E$ of 2 kJ/mol compared with the *trans* state, there are "*gauche*" ($g+$ and $g-$) positions of $\phi = \pm 60°$ where subminima exist. In the rotational isomer approximation, the bond $C_2$–$C_3$ can only exist in the *trans* or *gauche* forms.

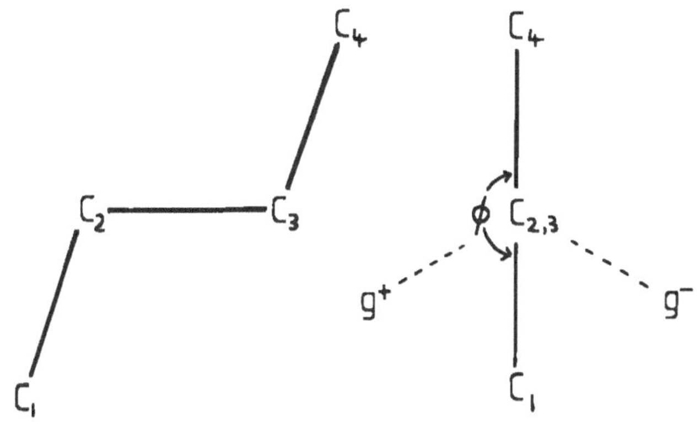


FIGURE 3. Projected views of four carbon atoms in a polymer chain. The bond $C_2$–$C_3$ is drawn in the *trans* rotational position.

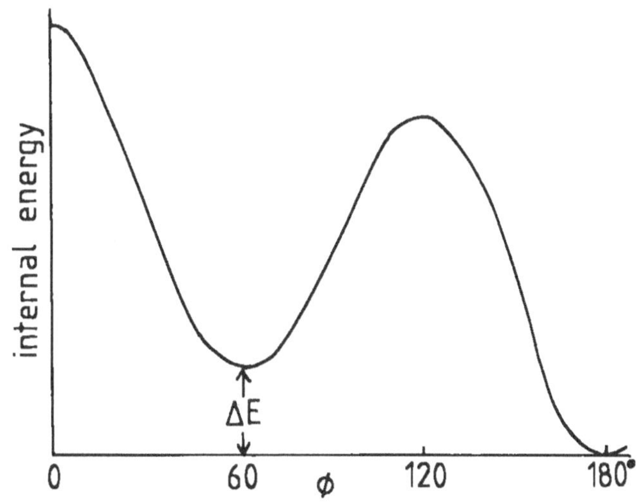


FIGURE 4. Typical values of the internal energy of the bond $C_2$–$C_3$ shown in Figure 3. The minimum at the *gauche* position $\phi = \pm 60°$ is an energy ΔE above the *trans* minimum at $\phi = 180°$.

These ideas can be extended to longer sections of polymer chain. If the chain contains n carbon atoms in its backbone, then there will $3^{n-3}$ conformational isomeric forms of the chain. Exactly one of these will be a planar zigzag containing all *trans* rotations (Figure 5). Other regular helices can be formed by using regularly recurring sequences of bond rotations — for instance, the sequence $tg+tg+tg+tg+\ldots$ occurs in one crystalline form of polypropylene. However, the overwhelming majority of irregular forms occur only in the glassy or liquid states.

In the liquid state the molecule is free to transform from one rotational isomer to another; the lifetime of a rotational isomer is about $10^{-10}$ sec. It is possible, knowing the potential energy difference ΔE in Figure 4, to estimate the relative numbers $n(g^+)$ and $n(t)$ of C–C bonds that have $\phi = 60$ or 180°.

Boltzmann statistics can be applied so that

$$\frac{n(g^+)}{n(t)} = \exp(-\Delta E/RT) \tag{1}$$

Therefore at T = 410 K, $n(g^+) \simeq 0.38\, n(t)$ and the proportions of bonds in the 3 isomeric

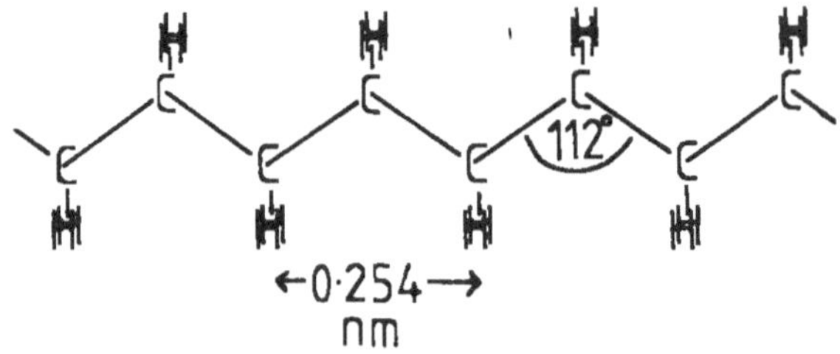


FIGURE 5. A projection of the crystalline form of polyethylene showing the planar all-*trans* conformation of the C–C bonds.

forms are roughly 2:2:5. This is close to the situation in which the three isomeric forms are equally populated. If this were the case then, as the number of conformations of the polymer molecule $W = 3^{n-3}$, the statistical entropy of the molecule is

$$S = k \ln W = (n - 3) k \ln 3 \tag{2}$$

where k is Boltzmann's constant. Later on we will be concerned with the entropy difference ΔS between a polymer melt and a polymer crystal and between a stretched and an unstretched network chain in a rubber. We can make a quick calculation of the magnitude of ΔS as follows. The all-*trans* form of the polyethylene chain has one conformation so S = 0. In the limit of n >> 1 for long chains the entropy difference per C–C bond of the liquid form is

$$\Delta S \simeq k \ln 3 \ (\equiv R \ln 3 \text{ per mole}) \tag{3}$$

so $\Delta S = 9.1\ JK^{-1}\ mole^{-1}$. This can be compared with an experimental value of $\Delta S_m = 10\ JK^{-1}\ mole^{-1}$, calculated from the enthalpy change $\Delta H_m$ on melting of 4.2 KJ/mol and melting temperature $T_m = 419$ K, using $\Delta H_m = T_m \Delta S_m$. Thus, the approximate calculation gives an answer within 10% of the correct value.

### C. Chain Regularity

Figure 2 is misleading in another way. It implies that the molecular structure of polymers is perfectly regular. The two most important ways in which irregularities occur are (1) copolymerization and (2) steroisomerism. Figure 6 shows two forms of copolymer that occur. With polyethylene we are largely concerned with the small amount (<5%) of co-monomers that are used to form random copolymers to control the overall level of crystallinity (see Section III.E). With the earlier method of polymerization of ethylene, it was impossible to avoid having ~1% of nonethylene units in the chain. Table 3 indicates the main varieties of polyethylene manufactured.

The original high pressure (2000 atm) ICI (Imperial Chemical Industries) process for making low density polyethylene (LDPE) produced a polymer that also contained long chain branches (Figure 6c), whereas the newer linear low density polyethylene (LLDPE) process produces linear molecules. Copolymerization is also used with the polydimethylsiloxane (PDMS) silicone polymers to introduce suitable sites for subsequent cross-linking of the polymer molecules (see Section II.E).

Stereoisomerism is a factor that must be considered when the monomer units do not have symmetry about the C–C backbone and the polymer is made by addition polymerization (when monomer units are added sequentially to the end of a growing polymer chain). Polypropylene (PP), polymethylmethacrylate (PMMA), polystyrene (PS), and polyvinylchloride (PVC) of Table 1 are such polymers. It is best to explain the possibilities of stereo-isomers in terms of triplets or "triads" of monomer units in a chain, because this is the

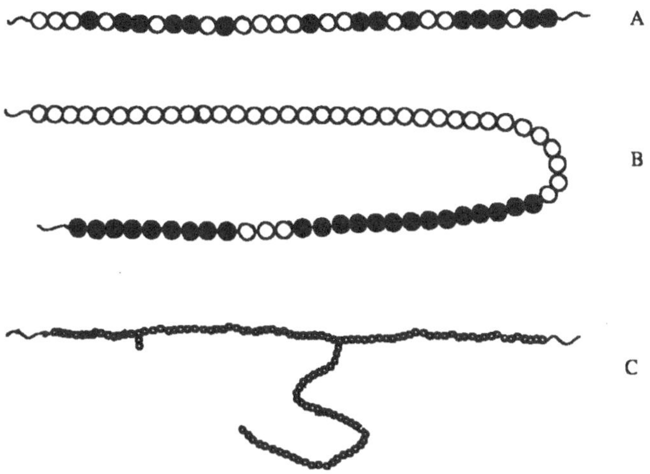


FIGURE 6. Copolymers of two monomers shown as ○ and ●. (A) Part of a random copolymer containing 60% ○; (B) part of a block copolymer in which there is a 90% probability that a monomer is joined to another of the same kind; (C) long chain branching in a low density polyethylene.

**Table 3**
**COMMERCIAL POLYETHYLENES**

| | Abbreviation | Density range ($kg/m^3$) | Counit (%) |
|---|---|---|---|
| High density | HDPE | 950—970 | 0.5 |
| Medium density | MDPE | 930—950 | — |
| Low density | LDPE | 910—930 | 2—3 |
| Linear low density | LLDPE | 910—930 | 2—3 |

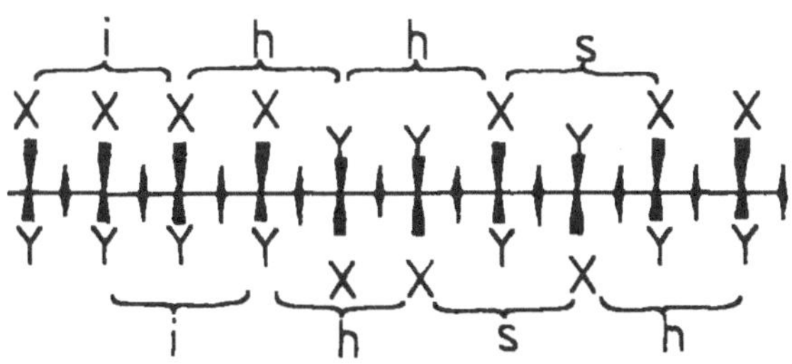


FIGURE 7. Examples of isotactic (i), syndiotactic (s), and heterotactic (h) triads in a section of atactic PMMA chain.

way in which the structure is determined (by nuclear magnetic resonance (NMR), for example). Figure 7 shows examples of three kinds of triads. Note that the monomer units have all added onto the chain head-to-tail and in the planar projection thicker bonds represent those that project towards the viewer. Because there is only one way in which the C–C backbone can be placed in the all-*trans* form shown, it is not possible for the chain to rotate to exchange the positions of the X and Y substituent groups. The monomer unit on the left establishes the reference frame and the subsequent monomer units are either added with X on the left- (L) or right-hand (R) side of the chain. There are eight possible triads and these are classified below in terms of whether they belong to idealized structures.

| Triads | Group | Ideal structure |
|---|---|---|
| LLL, RRR | Isotactic (I) | LLLL . . . or RRRR . . . |
| LRL, RLR | Syndiotactic (S) | LRLRLR . . . |
| LLR, LRR, RRL, RLL | Heterotactic (H) | |

In an ideally atactic structure, in which each of the eight possible triads occurred at random with equal probability, the ratio I:S:H should be 25:25:50. When a so-called atactic PMMA was examined by NMR, the ratio was found to be 7:55:38. However, this is still a sufficiently disordered structure for crystallization to be impossible. It is necessary for there to be long sequences of consecutive isotactic (or syndiotactic) triads for the chain section to be sufficiently regular to fit in a crystal lattice; for example, commercial "isotactic" polypropylene of the high degree of crystallinity, has >90% isotactic triads in its structure.

**D. Molecular Weight Distribution**

Having considered the structure of parts of a polymer molecule, we now consider the whole molecule. Commercial polymers fall into two main categories. The first is that of polydisperse linear polymers, considered here, and the second is network molecules, considered in the next section. "Linear" means not having long branches (Figure 6) and not that the shape of the molecule is constrained to be straight. It is impossible to manufacture polymer molecules of exactly the same length; there is a statistical distribution of the number of monomer units in a chains, and we term the material polydisperse. Molecular weights (strictly masses) are measured on a scale with carbon = 12 units, hydrogen = 1 unit, etc. So for addition polymers,

$$\text{Polymer molecule molecular weight} \simeq (\text{monomer molecular weight}) \times (\text{number of monomers in chain})$$

the approximation occurring because the ends of the chain may differ from the other repeat units. If the bulk polymer contains a number $N_i$ molecules of molecular weight $W_i$, then two commonly defined averages are the number average molecular weight

$$\overline{M_N} = \Sigma N_i M_i / \Sigma N_i \tag{4}$$

and the weight average molecular weight

$$\overline{M_W} = \Sigma N_i M_i^2 / \Sigma N_i M_i \tag{5}$$

$\overline{M_N}$ is the "mean" molecular weight in statistical usage and it gives an inverse measure of the number of chain ends in the polymer. Since each chain has two ends,

$$\text{Number of chain ends} = 2 \times \frac{\text{mass of polymer in atomic mass units}}{\overline{M_N}} \tag{6}$$

Chain ends are potential weak points, either for chemical attack that may cause depolymerization or in that they interrupt the transfer of stress through the microstructure, $\overline{M_W}$, which averages the molecular length according to the weight fraction of each species and controls properties such as the melt viscosity at low flow rates. Therefore a single molecular weight average cannot be used to correlate with all the polymer properties; knowledge of the molecular weight distribution is required. A common instrument for determining this is the gel permeation chromatograph (GPC), which produces data of the type shown in Figure 8.

For symmetric traces of the shape shown, $\overline{M_W}$ occurs near the peak deflection and $\overline{M_N}$ on

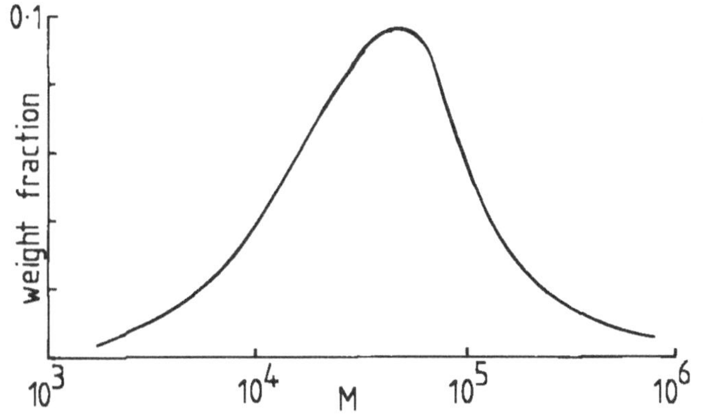


FIGURE 8. The molecular weight distribution of a polyethylene, determined using a GPC.

the lower molecular weight tail. If a given polymer always has the same shape of molecular weight distribution, the ratio $\overline{M}_W/\overline{M}_N$ can be used to characterize its width. Many polyethylenes have a high value of $\overline{M}_W/\overline{M}_n$ in the range 5 to 15, whereas other polymers such as polyethylene terephthalate (PET) are less polydisperse, with $\overline{M}_W/\overline{M}_N$ in the range 2 to 4.

### E. Network Molecules and Cross-Linking

If, after the shape of a polymer product is fixed, a chemical reaction is carried out to link all the molecules into a three-dimensional network, the resulting material is called a thermoset or a rubber, depending on its elastic modulus at room temperature. Figure 9 is an imperfect representation of such a network. Three- or four-functional cross-links are used to covalently bond together the initially linear molecules, into an effectively infinite molecule. Figure 9 shows a planar projection of only a very small part of the network, which extends to the boundaries of the macroscopic block of rubber/thermoset. The network is characterized either by the number N of network chains per cubic meter or by the number average network chain molecular weight $M_C$; a network chain is the length of polymer between neighboring cross-links. In typical thermosets, which are glassy at room temperature, the value of $M_C$ is usually much smaller than in typical rubbers. The formation of cross-links is usually an irreversible chemical reaction, so that subsequent heating of the network does not destroy it, and no further reprocessing of the material is possible. In contrast, the next section shows that the microstructure of thermoplastics forms each time the polymer is cooled from the melt and the material can be reprocessed indefinitely, so long as degradation of the chain length does not occur to too large an extent.

## III. MICROSTRUCTURE

### A. Introduction

The microstructure of the four polymer types will be described briefly in turn, concentrating on the details that are relevant to a subsequent discussion of mechanical properties and on the microstructures found in commercial products. In contrast with the situation for many other crystalline materials, there is still considerable ignorance of the details of the microstructure, and debate continues about the model that best explains the incomplete evidence. In describing the microstructure we are concerned with:

1. The way in which the molecules are packed together (especially if there is any crystal lattice)

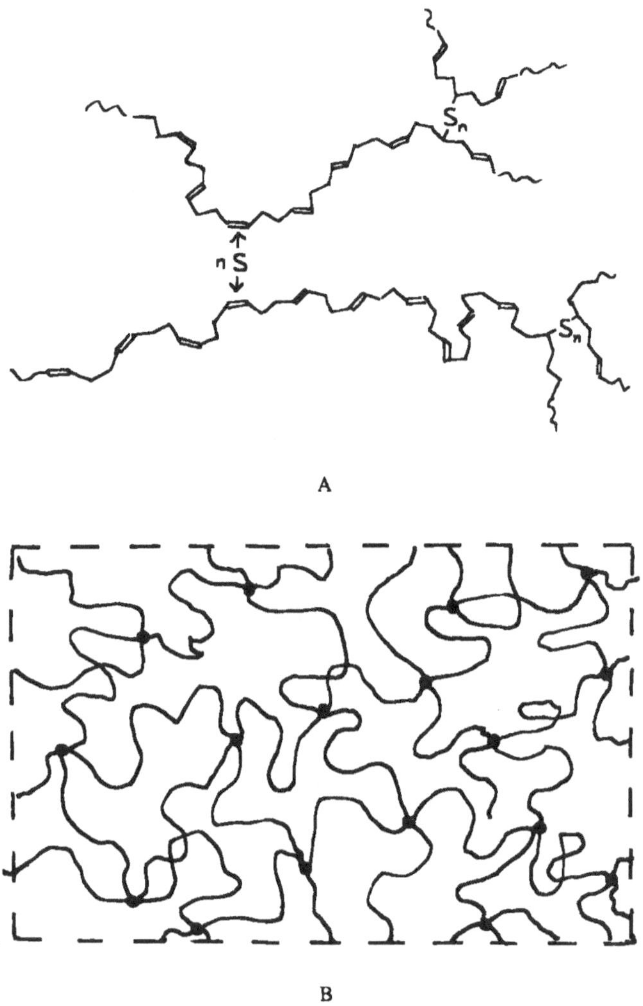


FIGURE 9. Rubber networks. In (A) a cross-linking agent reacts with C=C bonds on neighboring network chains and forms two new network chains. A projection of a small part of a rubber network is shown in (B).

2. Whether the microstructure is homogeneous on a scale greater than 20 nm
3. Whether the microstructure is anisotropic on a microscopic and on a macroscopic scale

Since we will then be concerned with explaining the mechanical properties of these polymers, we will want to know:

1. How the loads transferred fron one part of the microstructure to another
2. How the microstructure changes under elastic and plastic deformation
3. In what ways a crack could propagate through the microstructure

Only a small selection of the mass of available microstructural information will be presented, with the aim of answering the above questions.

The microstructure of polymers changes with temperature, the most important transition temperatures being $T_g$, the glass transition temperature, and $T_m$, the crystal melting point. Figure 10 shows the states in which the four polymers exist at different temperatures. Note

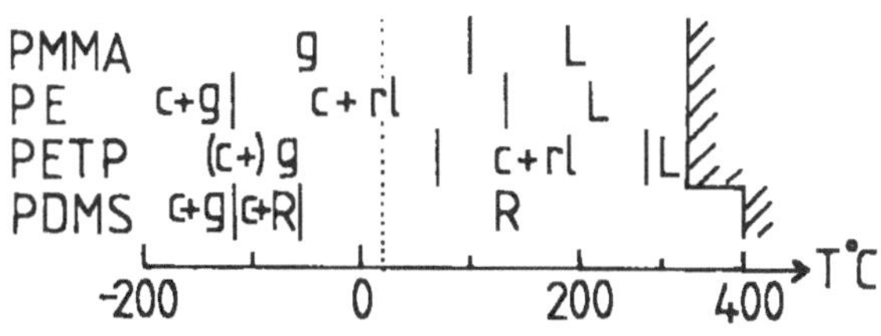


FIGURE 10. A chart indicating the states in which four representative polymers exist at different temperatures; g, glassy phase; c, crystalline phase; L, liquid, R, rubber, rl, rubbery liquid with the crystals acting as physical cross-links.

how at room temperature the four polymers are in different states. No attempt will be made here to explain the values of $T_g$ and $T_m$ in terms of the molecular structure; it is possible to detect some trends in transition temperatures between groups of similar polymers, but no comprehensive theory exists. The states will be described in an order which allows widely used concepts, such as that of rubber elasticity, to be introduced early.

### B. Rubber

This state differs from all the others chemically in that chemical cross-links occur and the rubber consists of a single three-dimensional network molecule, rather than huge numbers of individual polymer chains. We define a network chain as the length of the network joining two neighboring cross-links. Figure 9 shows that if one typical cross-link attacks two –C=C– bonds on neighboring network chains, two new network chains are formed. It is difficult to make realistic two-dimensional drawings of a rubber network that emphasize the "infinite" three-dimensional nature of the network and the lack of gaps between the network chains. When a block of rubber is stretched, the forces are transmitted through the network and via the cross-links to every network chain. It is assumed that if the rubber is stretched, say threefold in one direction, so is every network chain. It is essential that the cross-links do not break and reform under stress, otherwise the rubber will not return to its original shape (permanent set occurs).

The elastic properties of rubber are explained in terms of the behavior of each network chain behaving as an entropic spring (Figure 11). If a force f causes the end to end vector r of the chain to increase reversibly by an amount dr then, by the laws of thermodynamics, the internal energy change dE is given by

$$dE = TdS + fdr \tag{7}$$

where T is the absolute temperature and dS is the entropy change. The network chain molecule can be modeled by a chain of freely jointed links (Figure 11) by replacing about three C–C bonds by a link of fixed length $\ell$, having a spherical ball joint at each end. As the model cannot store internal energy, dE = 0 and the retractive force f is given

$$f = -T\frac{dS}{dr} \tag{8}$$

A qualitative argument, using the concepts of Section II.B, shows that when r is small there are a large number of conformations so the entropy S is large, whereas for the fully extended straight chain there is one conformation and S = 0. Therefore dS/dr is negative and by Equation (8) the retractive force f is positive. A quantitative calculation for a chain of n links gives the result

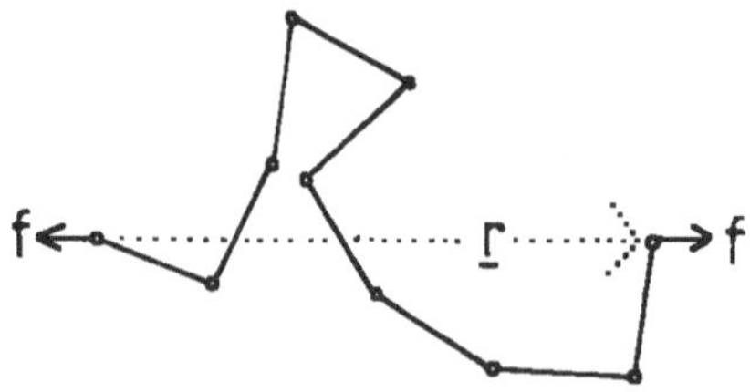


A

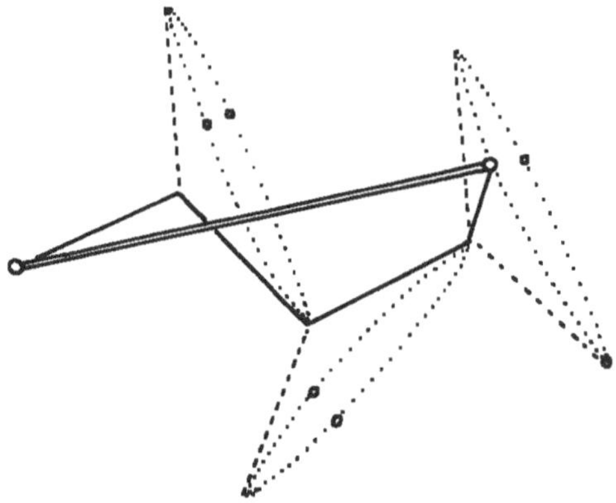

B

FIGURE 11. Modeling of rubber elasticity. (A) Shows the forces f acting on the ends of a network chain to keep them a distance r apart; (B) shows how one freely jointed link replaces several C–C bonds having restricted mobility.

$$f = \frac{k\,T\,r}{n\,\ell^2} \quad (9)$$

The kT term shows that the kinetic motion of the chain is essential for rubber elasticity and the 1/n term shows that longer chains are softer. In a rubber network the density of network chains (N chains/$m^{-3}$) is inversely proportional to their length, since the density of monomer units is constant. The spring constant f/r of the individual network chain controls the shear modulus G of the rubber and the equivalent to Equation (9) for a rubber network can be shown to be

$$G = NkT \quad (10)$$

Thus, the density of network chains (or equivalently of the chemical cross-links added) and not the chemical nature of the monomer units determines the elastic modulus of the rubber. There are limits on this because the chain must contain more than ten links for the statistical entropy arguments to be valid. Consequently rubbers can be thought of as akin to liquid polymers in their structure, with the network structure preventing flow and providing a low shear modulus between 0.3 and 3 $MNm^{-2}$.

The extensibility of a rubber network can be explained by considering the average length of the end to end vector $\mathbf{r}_0$ of a network chain in an unstressed rubber. As we are only interested in the magnitude of $\mathbf{r}_0$ and not its direction, we evaluate $\langle r_0^2 \rangle$ the mean square value of $\mathbf{r}_0$. Now

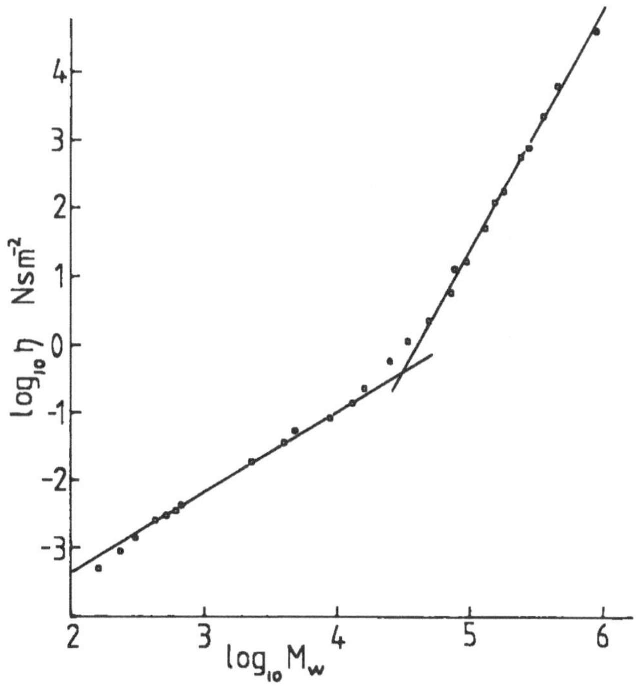


FIGURE 12. A log-log plot of the relationship between the low-shear-rate melt viscosity and molecular weight for PDMS at 20°C. The increase in the slope from 1 to 3.5 is due to the development of entanglements in the melt.

$$
\begin{aligned}
r_0^2 = \boldsymbol{r}_0 \cdot \boldsymbol{r}_0 &= (\ell_1 + \dots \ell_n) \cdot (\ell_1 + \ell_2 + \dots \ell_n) \\
&= \sum_{i=1,n} \ell_i \cdot \ell_i + \sum_{i \neq j} \ell_i \cdot \ell_j \\
&= n\ell^2 + \ell^2 \sum_{i \neq j} \cos\theta_{ij}
\end{aligned} \tag{11}
$$

The average value of the second summation is zero, because if the links are free to point in any direction, then the angle $\theta_{ij}$ between link i and link j takes all the values from 0 to 360° with equal frequency, so the average value of $\cos\theta_{ij} = o$. Hence

$$
\langle r_0^2 \rangle = n\ell^2 \tag{12}
$$

Thus, a chain of n links has a root-mean-square end to end distance of $\sqrt{n}\ell$, whereas it is fully extended length is $n\ell$. Hence it can be extended by a factor of $\sqrt{n}$ from its rest length. For $n = 25$, the network can be extended 500% before the network chains are fully extended.

## C. Melt State

The major difference between the melts of simple substances and those of polymers is the existence of entanglements between the polymer chains. If the viscosity of a particular polymer is plotted against molecular weight M on logarithmic scales (Figure 12), a break occurs in the relationship. For the lower molecular weight polymers (not used commercially), the viscosity is only proportional to M, whereas above a critical molecular weight, it becomes proportional to $M^{3.5}$, and some new factor must be introduced to explain this. The first thing to realize is the degree to which the random coils (Figure 13a) of neighboring molecules

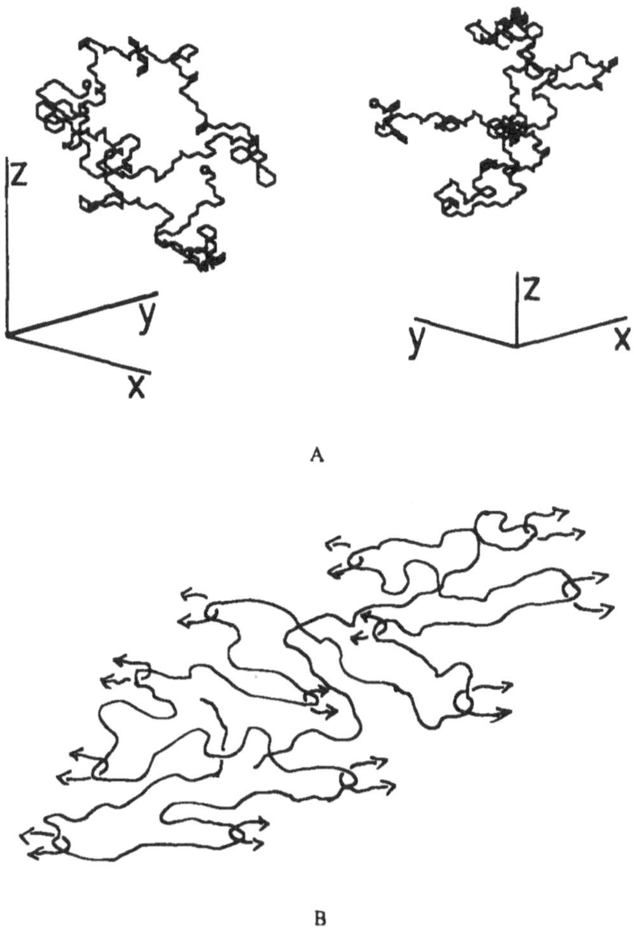


FIGURE 13. (A) Two projections of the instantaneous shape of a molecule in a polymer melt at rest modeled as a 400-step self-avoiding random walk in which each C–C bond is a *t*-, *g*-, or *g*-rotational isomer. The other molecules that fill the gaps in the coil are not shown for reasons of clarity; (B) the effect of shear flow in orienting the sections of the molecule between entanglements.

overlap. If the radius of gyration $r_g$ of a polymer molecule is measured in the melt (or in the glassy) state at rest, it is found to be proportional to the square root of the molecular weight, in agreement with Equation 12 for the freely joined chain model. The constant of proportionality is such that a PMMA molecule with $M = 10^6$ has $r_g = 25$ nm. Within a sphere of radius $r_g$, there is room to pack 40 such molecules, using the van der Waals' radii of the atoms for the calculation. Thus, each molecule of $M = 10^6$ is able to interact with at least 320 others whose centers occur within a radius of $2r_g$, and the number of entangled molecules is expected to increase with $M^3$.

Various elastic properties of polymer melts have been interpreted in terms of temporary networks, with entanglements of a limited lifetime replacing the permanent chemical crosslinks in rubbers. The entanglement networks are characterized by the average chain molecular weight between entanglement $M_e$, a typical value being 10,000 for PMMA. Thus, in any PMMA melt with $M > 30,000$ a roughly constant density of entanglements occurs. We now consider what happens to the melt during flow when molecules have to disentangle in order to flow past each other. Figure 13b shows that the sections of molecule between

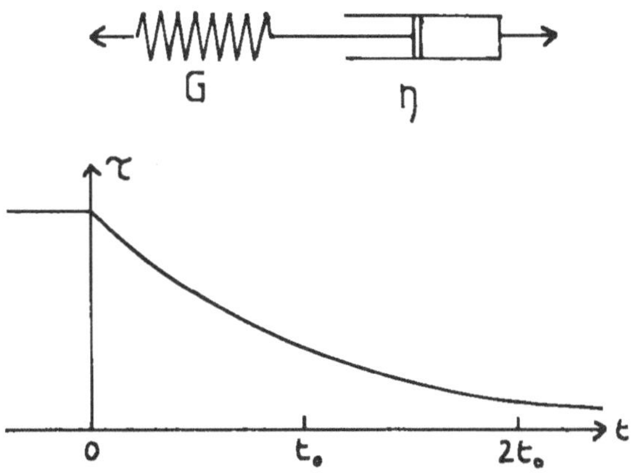


FIGURE 14. The Maxwell model for viscoelastic behavior and the predicted stress relaxation after the cessation of steady shear flow.

entanglements tend to become oriented in the direction of flow as a result of forces imposed by neighboring molecules. There are two consequences: molecular orientation occurs during flow and various elastic effects occur, most noticeably the die-swell when a stream of melt emerges from a die into air and increases in diameter. The molecular orientation is important because it may survive after the polymer has cooled into the solid state. This can be explained using the simple Maxwell viscoelastic model shown in Figure 14. The spring symbol represents the elastic nature of the liquid, with shear modulus G proportional to $M_e^{-1}$, and the viscous damper symbol representing the shear viscosity η of the liquid, which is proportional to $M^{3.5}$. The same stress acts on both elements, whereas the strains are added. When a constant shear flow rate is imposed, the stress is constant, and the spring extends a constant amount. When flow ceases, as in the injection molding process when the mold is full, the overall strain change is zero, but the molecular orientation relaxes, hindered by the viscous element. It can be shown that the shear stress τ relaxes exponentially with time t, s

$$\tau = \tau_0 \exp(-t/t_0) \tag{13}$$

where $t_0$ is the relaxation time = η/G.

Note that the relaxation time $t_0$ depends on $M^{3.5}$, so the consequences of using a very high molecular weight polymer are twofold; first, the high viscosity makes the output rate from processing machinery low; second, the high relaxation time (>>1 sec) leads to some of the flow-induced molecular orientation being frozen into the product. Therefore with polyethylene, for example, it is customary to use polymers with $M_w \simeq 10^5$ to give the best compromise between ease of processing and mechanical properties. Occasionally, when high wear resistance is paramount and difficult processing is acceptable, ultrahigh molecular weight polyethylenes with $M_w \simeq 2 \times 10^6$ are used.

### D. Glassy State

Although the main features of the structure of glassy polymers are established, there is still controversy about the details, in particular about the extent of short-range order. When a noncrystallizing polymer is cooled from the melt into the glassy state, there is no sudden volume contraction (Figure 15), only a change in the thermal expansion coefficient. The melt viscosity rises rapidly as the glass transition temperature $T_g$ is approached; consequently the relaxation time $t_0$ for molecular disentanglement also increases towards an effectively infinite value. For this and other reasons it is believed that the molecules still have the

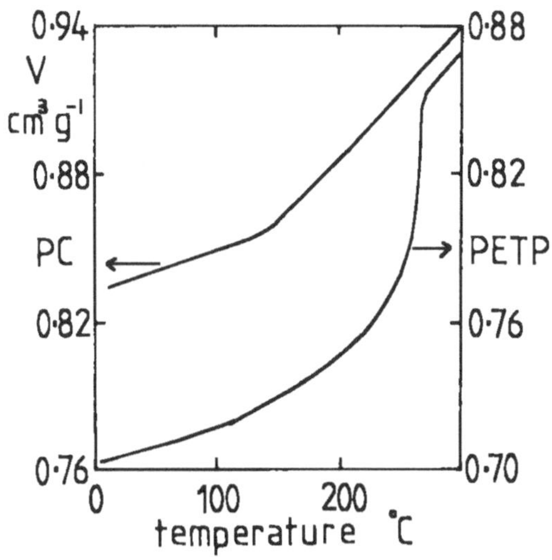


FIGURE 15. The specific volume of glassy polycarbonate and of semicrystalline PETP, as a function of temperature.

random coil configuration in the glassy state. Molecular shape changes can no longer occur and the molecules remain entangled. X-ray diffraction evidence leads to two main conclusions. First, glassy polymers contain no regular three-dimensional lattice on a scale >10 nm. This means that none of the deformation mechanisms associated with crystal lattices, such as dislocation movement during plastic deformation, are appropriate. Second, low angle X-ray diffraction experiments show that each polymer chain is surrounded by neighbors in close contact, but that these the intermolecular distances are subject to significant variation. At temperatures not too far below $T_g$ it is possible for small sections of four or five carbon atoms in the chain backbone to rotate in a "crank shaft" manner or for small side groups to rotate; these motions are frozen in at lower temperatures.

There is a great disparity between the weak van der Waals' force between molecules and the strong covalent bonds along the molecular backbones. On a scale greater than 10 nm the chain backbones lie in random directions; therefore the overall properties are isotropic. Elastic deformation is dominated by the weak van der Waals' forces; these allow neighboring chains to move axially relative to each other or separate slightly. The high value of Poisson's ratio for most polymers ($\nu$ = 0.35 to 0.4) shows that the former predominates.

Polymeric glasses are capable of yielding in some circumstances (Figure 16) so long as the molecular weight exceeds roughly four times the molecular weight between entanglements. Although the shape change appears to be permanent, it is recoverable on heating the polymer to just above $T_g$ while it is freely suspended. Thus when a glassy polymer yields, the sections of molecules between entanglements develop a preferential orientation; there is a limit to the strain that can be imposed before the section of chain becomes nearly fully extended (a tensile strain of 100 to 300%, depending on the polymer). However, the entanglement network cannot retract to its equilibrium form until the polymer is heated above $T_g$. This shape change phenomenon is also the reason why molecular orientation in injection moldings cannot be completely removed by heat treatment. The only way in which the molecules can relax to their equilibrium random-coil shapes is when they are heated into the melt; however, if this is done without constraining the shape of the product in a mold (economically not viable), the entropy elastic forces in the entanglement network will severely distort the shape of the product.

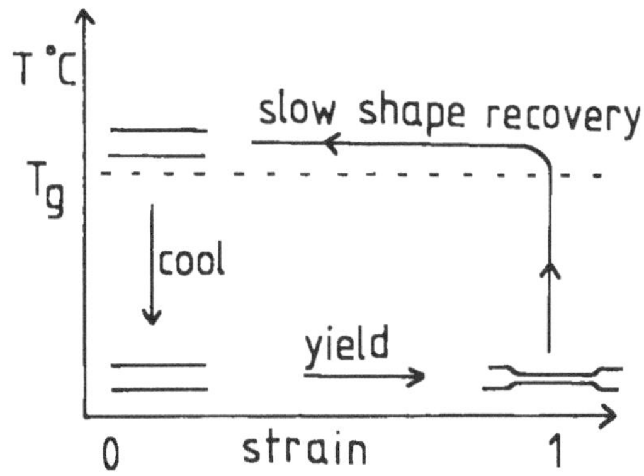


FIGURE 16. A deformation cycle in a glassy polymer. The specimen is plastically deformed at 20°C, then heated to just above $T_g$ without any shape constraints, whereupon it slowly recovers its original shape.

**E. Spherulitic Semicrystalline State**

Because the microstructure is complex, the molecular conformation will be described first, then the packing together of the molecules, and finally the shape and orientation of crystals in the spherulites. It is a reasonable approximation to consider the microstructure to consist of two phases, an amorphous one which is either glassy (Section III.D) or rubbery (Section III.B, there are no chemical cross-links, but the ends of molecules are trapped in crystalline regions), and a crystalline phase. In the crystals, the molecular conformation is a regular spiral or is nearly planar (Figure 17). In each molecule, the sequence of bond rotations repeats regularly, for example, in isotactic polypropylene it is $tg^+tg^+tg^+$ ... In the crystal, all the chain spirals lie parallel to the c (or chain axis) direction and the chains are packed together to fill space as effectively as possible. Figure 18 shows that in the ab projection of the polyethylene unit cell there are two settings of the planes in which the C-C-C zig-zags lie to achieve efficient packing. The crystal unit cell has covalent bonding in the c direction, but only van der Waals' bonding in the a and b directions. Consequently, the Young's modulus of the unit cell is highly anisotropic, the values being $E_a = 5$, $E_b = 8$, and $E_c = 255$ GNm$^{-2}$.

The shape of the crystals differs depending on the method of crystallization. In academic studies, when very dilute polymer solutions were crystallized very slowly, large lamellar (platelike) single crystals were prepared; in all commercial polymer products, however, crystallization occurs while the melt is cooling rapidly. In the latter case, small lamellar crystals, which may have curved surfaces, are formed in multilayer stacks interspersed with layers of amorphous material (Figure 19). The lamellae are still much broader (~1 μm) than they are thick (~20 nm). The angle between the c axis of the crystals and the normal to the lamellae surfaces can be up to 30°. The molecular lengths (for a polyethylene of $M_N = 10{,}000$ the mean molecular length ≃90 nm) are several times the thickness of the platelike crystals, so we need to know what happens to the molecules when they reach the crystal surface. For melt crystallized polymers, the relaxation time for the molecules in the melt to change their shape from entangled random coils to a regularly folded structure as in a chinese-cracker firework is much greater than the time available during crystal growth. Consequently individual molecules do not get incorporated in one crystal by folding regularly at the crystal's surface; instead, sections of the molecule get incorporated in different crystals with the intervening lengths forming part of the amorphous interlayer. Thus, the crystals act as physical cross-links for the amorphous material, and conversely, the amorphous interlayers transfer stress from one crystal to another.

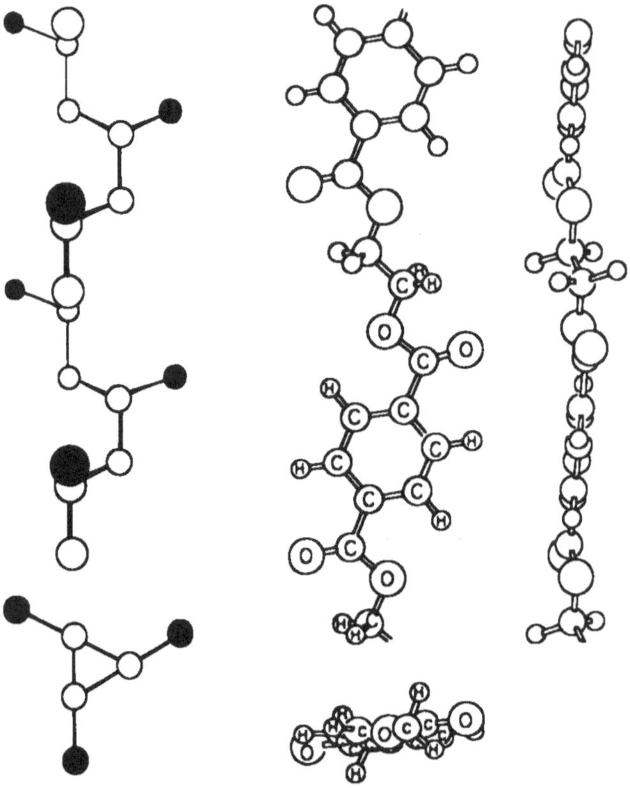

A B

FIGURE 17. Two projections of molecular conformations of (A) polypropylene (a spiral) and (B) PETP (nearly planar) in the crystalline state. (Reproduced with permission from John Wiley & Sons, New York.) ○ represents CH or $CH_2$ and ● represents $CH_3$ in the polypropylene molecule. (Figure B from Ward, I. M., *Mechanical Properties of Solid Polymers*, John Wiley & Sons, Chichester, 1971. With permission.)

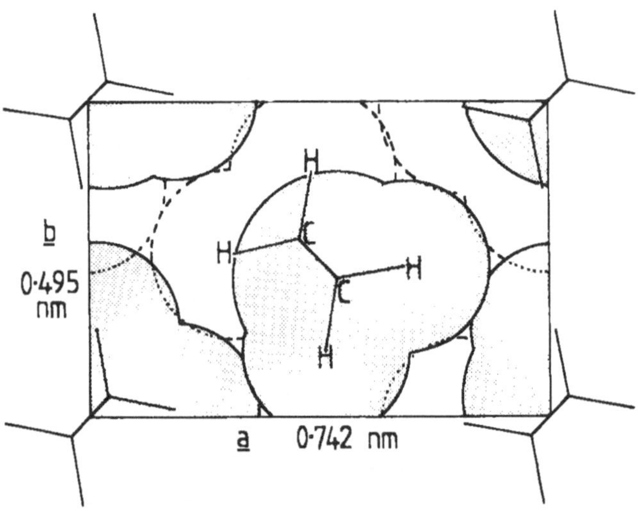


FIGURE 18. The polyethylene unit cell in the ab projection.

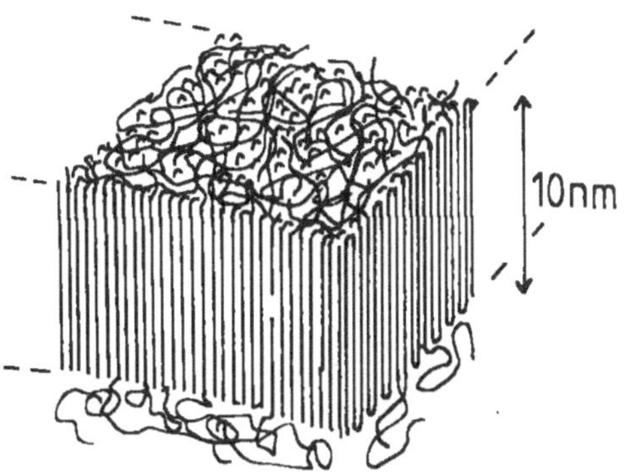


FIGURE 19. Part of a lamellar crystal showing some chain folding on the crystal surface and amorphous interlamellar material.

A polymer melt starts to crystallize when crystals nucleate at a number of randomly distributed sites. The nuclei cannot move as there are no convection currents in the highly viscous melt; consequently spherulites grow radially away from the nuclei. As the lamellae are not perfectly parallel in their stacks, further ones form at slight misorientations, so that the nucleus develops through a "wheat sheaf" stage to a growing sphere (hence the name spherulite). Finally the spherulites contact, leaving spherulite boundaries that are equiaxed. For polyethylene spherulites, the crystal b axes are radial, whereas the a and c axes are tangential, with often a regular variation in the a and c axis orientation with radial distance, that leads to a banded appearance when sections of the spherulite are observed with polarized light (Figure 20).

In considering the mechanical properties of spherulitic polyethylene, we need to bear in mind several main factors:

1. The overall percent crystallinity will largely be determined by the chain regularity (Section II.C) and will fix the relative thickness of the lamellar crystals and the amorphous interlayers. The morphology of interleaved plates of soft and hard material occurs in other composite materials such as the rubber/steel plate sandwich structures used for bridge bearings and in heavy vehicle suspensions.
2. As the mean spherulite diameters are usually found to be in the range 10 to 100 μm and each spherulite contains crystals of all possible orientations, the properties of the polyethylene will be isotropic on a scale of 1 mm and above, yet the deformation mechanisms will vary with angular position within the spherulite.
3. The amorphous material, being constrained between more rigid crystalline plates, will deform largely by shear (Figure 21). Excessive shear is, however, restricted by the multiply connected nature of the lamellar crystals. Consequently, strains exceeding 10% require the crystals to yield. It is much easier for the crystals to slip on planes that contain the c axes, as this avoids breaking any covalent bonds. It is very difficult to relate the macroscopic yield stress to the stresses required to deform the various parts of the highly complex microstructure.

**F. Fibrous Semicrystalline State**

The production of PET fibers consists of three main stages. The first is the melt-spinning of fibers that are amorphous and only slightly oriented. Between the crystal melting point of 260°C and the glass transition temperature of 67°C, the crystallization rate goes through a peak (Figure 22), but this maximum rate is much smaller than that for polyethylene, so it is possible to cool the emerging fiber so rapidly that it enters the glassy state without

FIGURE 20. Banded spherulites of PE seen by transmitted polarized light.

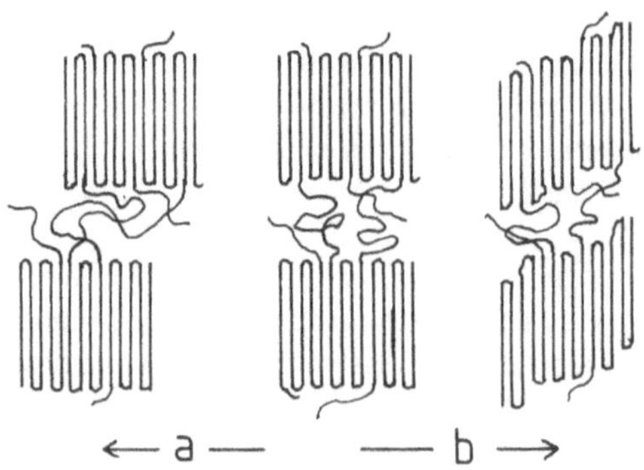


FIGURE 21. Possible deformation mechanisms in semicrystalline polymers. (A) Shear of the amorphous interlamellar material; (B) slip of the crystals on planes containing the chains, with slip direction c.

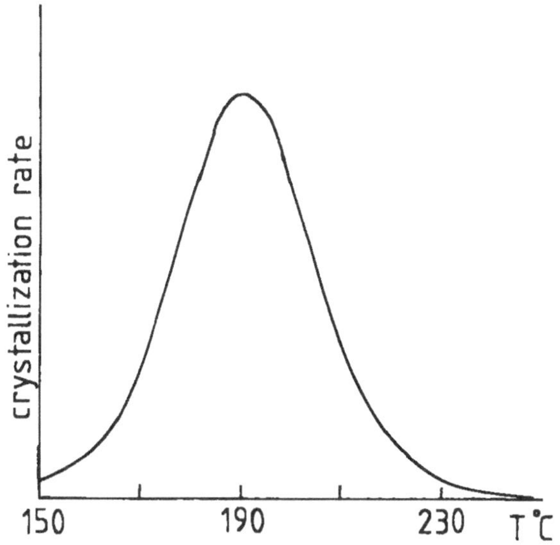


FIGURE 22. The crystallization rate of PETP as a function of temperature.

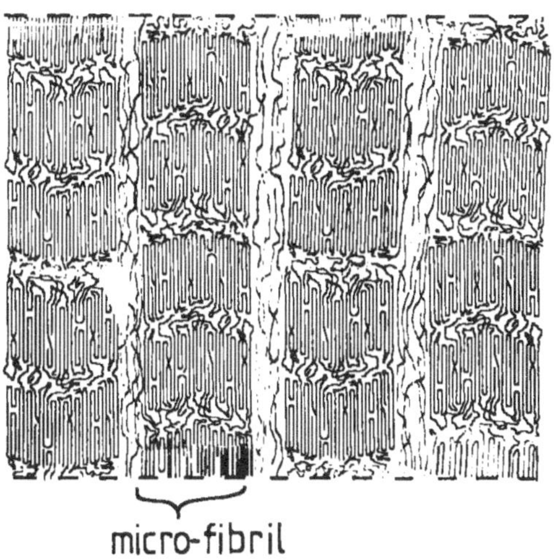


FIGURE 23. The microstructure of PETP fibers, showing the arrangement of crystals in the microfibrils.

crystallizing. The second stage is one of simultaneous stretching and crystallization, where the fiber is heated to around 100°C and drawn to 3 to 5 times the original length. The rubberlike orientation of the amorphous material encourages crystallization at a temperature that is near the lower limit of Figure 22; the consequence is that very small, highly oriented crystals are formed rapidly. The third stage is "heat-setting", where the fiber is heated to 100 to 220°C, while its length is usually kept constant. This allows the amorphous regions to reorient and possibly further crystallization can occur.

The microstructure of the fiber is shown in Figure 23. The fiber is made up of microfibrils that are ~10 μm long and 10 to 20 nm wide that lie nearly parallel to the fiber axis. Each microfibril consists of alternating folded chain crystals and amorphous glassy material repeating at intervals of about 15 nm. The fiber is about 40% crystalline and the orientation

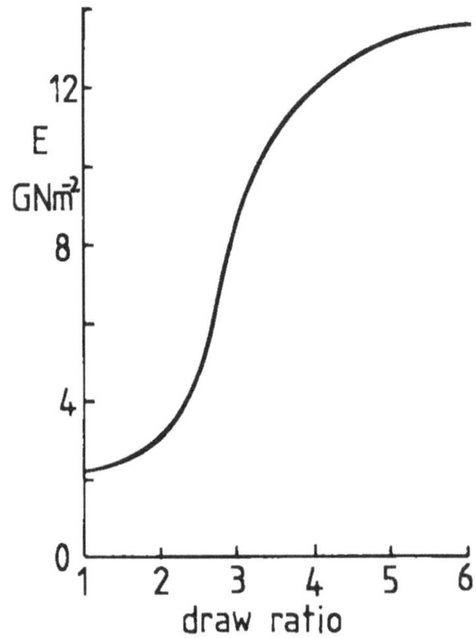


FIGURE 24. The variation of the elastic moduli of PETP fibers with the draw ratio.

of the crystals can be described by the average angle between the crystal c axis and the fiber axis, which is about 5°. The amorphous glassy regions are also oriented in the fiber direction. The effect of the hot-stretching process in increasing the Young's modulus of the fiber parallel to its length is shown in Figure 24. However, one side effect of lining up the PET chains towards the fiber axis is to make the fiber less stiff and less strong in the transverse direction, thereby increasing the anisotropy of the fiber.

The Young's modulus of drawn PET fibers is less than 10% of the Young's modulus in the c direction of the crystal. One reason for this is the large percentage ~60% of glassy phase which is only moderately oriented, of low modulus $\sim 4GNm^{-2}$, and loaded in series with the crystals. A second reason is the less than perfect orientation of the microfibrils, so that extension can occur by the microfibrils sliding past each other.

## IV. MECHANICAL PROPERTIES

The subdivisions in this section, and the order in which they are discussed, are a consequence of attempting to provide information for design purposes. Once an initial design of, for example, a polymeric prosthesis is made, it is necessary to check whether it will deform excessively under load, whether the stress levels in critical regions are below the yield stress of the material, and whether cracks can develop or grow during the designed life span. Because a wide range of polymers is being considered, Section IV.A explains the general level of their elastic properties before the further complication of time dependent behavior is introduced in Section IV.B. The molecular weight of a polymer has only a minor influence on its elastic and viscoelastic properties, however, it has a major influence on the fracture properties considered in Section IV.D and on whether failure is by crack growth or by the yielding processes described in Section IV.C.

### A. Elastic Moduli and Their Variation with Temperature

The range of elastic Young's moduli encountered with polymers is shown in Figure 25,

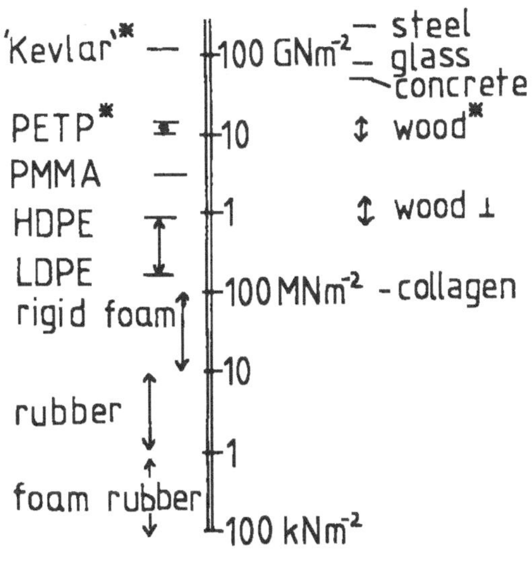


FIGURE 25. A chart showing the relative Young's moduli of polymers and other materials on a logarithmic scale; *, modulus parallel to the fiber direction.

together with the moduli of some other common materials. For some of the polymers it is possible to calculate the Young's modulus from a knowledge of the microstructure, but for others only upper and lower bounds can be set. Thus, for the rubbers a knowledge of the cross-link density and the theory outlined in Section III.B will suffice, although commercial rubbers often contain inorganic particle fillers that increase the modulus. For the glassy polymers such as PMMA at room temperature, it is possible to show that the variation of bulk modulus (relating volume changes to the imposed pressure) with the applied pressure is consistent with the bulk modulus being determined by the intermolecular van der Waals' forces. However, the strength of the van der Waals' bonds and the interatomic distances are not known with sufficient precision to predict the Young's modulus. Nevertheless, the Young's modulus of all uncross-linked glassy polymers at 20°C lies in the range 2 to 3.5 $GNm^{-2}$. A final benchmark is the Young's modulus, in the covalently bonded direction, of a fully extended polymer chain. This can be calculated, for example, for the c axis of the polyethylene crystal from a knowledge of the covalent bond stiffnesses. This level of 200 to 250 $GNm^{-2}$ has been nearly achieved by the recently developed aromatic polyamide fibers in which extended chain crystals are highly oriented in the fiber direction.

The moduli of the semicrystalline polymers, either like PE in a spherulitic form or like PET as an oriented fiber, are more difficult to predict. We can treat these as composite materials, with one phase being the anisotropic crystals and the other being the rubbery or glassy amorphous phase. With the minimum information of the volume fraction crystallinity $\phi_c$, it is only possible to say that the composite modulus E lies between the bounds

$$\phi_c E_c + (1 - \phi_c) E_a > E > (\phi_c/E_c + (1 - \phi_c)/E_a)^{-1} \quad (14)$$

where E is some average modulus of the crystals. For PE at 20°C, the upper and lower bounds are very far apart, reflecting the great difference between the moduli of the 2 phases, and the experimental moduli fall in between these bounds (Figure 26). We need to know the shape and connectedness of the two phases before any closer bounds can be drawn. Unfortunately, reliable information on the length to thickness ratio of the lamellar crystals is not available so it is difficult to find close bounds for the modulus. Besides, the next

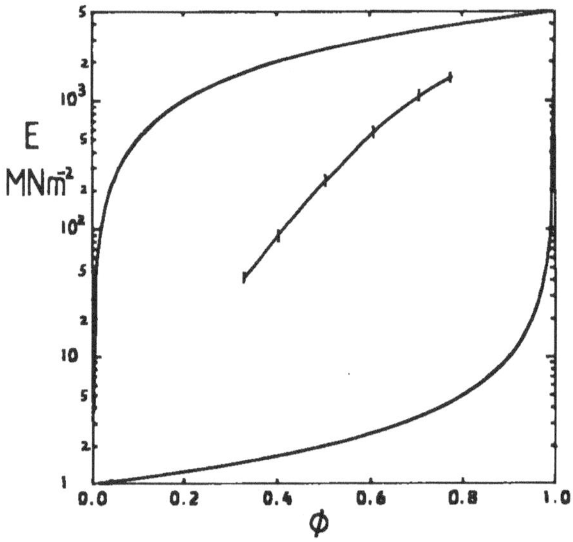


FIGURE 26. The variation in the Young's modulus, E, of polyethylene with the volume fraction crystallinity, ϕ, and the upper and lower bounds described by Equation (14).

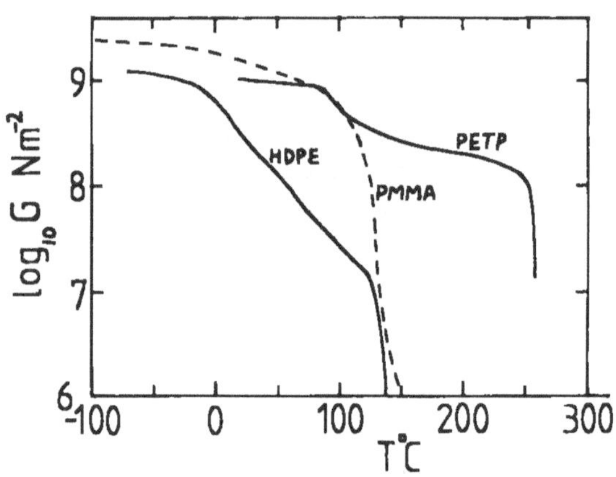


FIGURE 27. Variation with temperature of the shear modulus of PE, PETP, and PMMA.

section shows that the so-called "elastic" moduli are time and stress dependent. We can conclude from Figure 26 that there is neither a uniform strain nor a uniform stress situation in the spherulites, these assumptions corresponding, respectively, to the upper and lower bounds in Equation 14.

By combining the information in Figure 25 on elastic moduli with that in Figure 10 on the changes of state with temperature, we can predict that the elastic moduli will show strong temperature dependence in the vicinity of certain transition temperatures. Figure 27 shows the temperature variation of the shear modulus G (which is 33 to 40% of Young's modulus) for PE, unoriented PET, and PMMA. Although with PMMA the modulus falls rapidly to a very low level above $T_g$, this is not the case for semicrystalline PE and PET where there is a gradual decrease in modulus between $T_g$ and the crystalline melting point.

## B. Viscoelastic Behavior

A simple viscoelastic model (Figure 14) has already been used to explain the properties

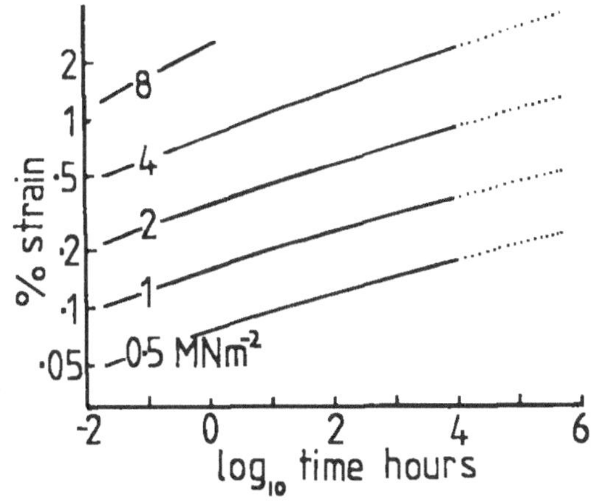


FIGURE 28. Data for the tensile creep of HDPE at 20°C.

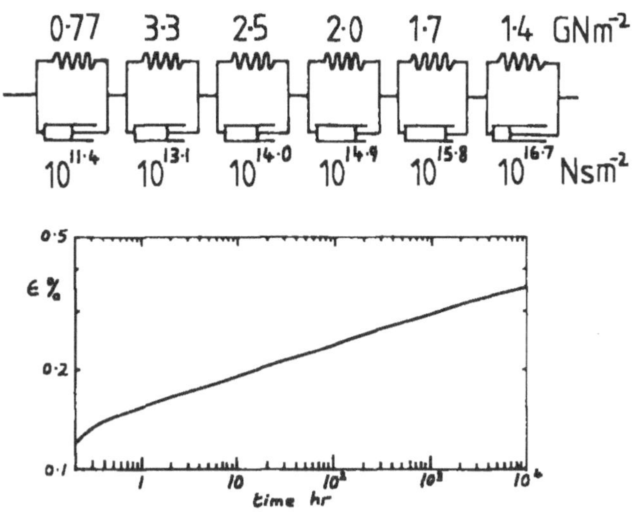


FIGURE 29. The type of generalized spring and dashpot model that can be used to model the creep of HDPE and the predicted creep curve for a creep stress of 1 $MNm^{-2}$.

of polymer melts. Here we consider whether or not more complex viscoelastic models are useful in predicting the phenomena of

1. Creep, where a steady stress is imposed, and the strain increases with time
2. Stress relaxation, where a constant strain imposed on a polymer, and the stresses decay with time
3. Other tests where both stress and strain are varying with time, possibly in a periodic way as in fatigue or "dynamic mechanical" tests

As an example of (1) Figure 28 shows the creep data for a range of tensile stresses applied to HDPE. Notice how the creep process which starts at short times (it takes ~1 sec to apply the load) continues to the limit of experimental patience. Figure 29 shows a linear viscoelastic model, together with the predicted creep curve. The difficulty with models of this type is that the creep process stops when the springs in each section are taking all the load; therefore a great range of retardation times ($= \eta_i/E_i$ for the ith section) is required to simulate the data in Figure 28. Unfortunately, two further, insurmountable difficulties occur. First, there is no way of relating the spring constants and viscosities in the model to particular features of the microstructure of PE, and second, such models with linear elements predict that the creep strain is proportional to the creep stress. Experimentally this is not the case — a doubling of the stress in Figure 28 from 4 to 8 $MNm^{-2}$ more than doubles the creep strain after 10 hr from 0.43 to 1.05%. Therefore viscoelastic models such as that in Figure 29 are of little practical use and separate measurements of viscoelastic behavior at different stress levels are essential.

The most common way in which creep data is used is called pseudoelastic design because it allows the use of the complete range of formulas for the deformation of beams, springs, etc. made of elastic materials. It is necessary to find a creep modulus E(t) to replace the Young's modulus E in such formulas. The maximum stress $\sigma$ in service (for example, on the outer layers of a bent beam) and the duration of this stress are calculated, and then the tensile creep data for this stress and time are inspected to find the creep strain e(t). Finally, the creep modulus is calculated as

$$E(t) = e(t)/\sigma \tag{15}$$

Figure 30 shows such creep modulus data for HDPE at 20°C.

Note that the modulus for 1 year under load and typical design stress of 4 $MNm^{-2}$ is only ~25% of the short-time (~1 min), low-stress moduli of the type shown in Figure 27. Furthermore, a safety factor must be used to allow for errors introduced by the pseudoelastic assumption.

Stress relaxation data are needed when plastics parts are to be assembled by interference fit (snap-fit) methods. For example, if a polyethylene tube of internal diameter 4.95 mm is forced over a copper pipe of 5.00-mm external diameter, a 1% hoop strain is imposed in the polyethylene. A similar strain would occur if the HDPE underwent a 60°C temperature change. Since stress relaxation data are not always available, it is usual to estimate the stress $\sigma(t)$ using the creep modulus for the appropriate imposed strain e, and

$$\sigma(t) = e\,E(t) \tag{16}$$

All the mechanical properties of plastics are time dependent; in particular the yield and fracture data discussed in the next sections will depend on the rate of testing, so it is essential to find test data appropriate to the time scale of the particular application.

### C. Yield and Crazing

The various yielding phenomena will be described first before considering the underlying molecular mechanisms or attempting to predict the yield stresses. The most common way of measuring yield stresses is via a tensile test, where only a single tensile component of stress acts in the material as the result of a constant imposed extension rate. The four typical polymers behave in quite different ways in this test at 20°C. The rubber PDMS extends by several hundred percent, but remains elastic, whereas PMMA remains elastic up to ~4% strain when brittle fracture occurs. With PE no significant permanent extension occurs until a neck forms; this is a region of length $\geq$ the specimen thickness of localized plastic deformation. The neck usually stabilizes and then propagates along the specimen at a drawing stress that is smaller than the initial yield stress. For oriented PET fiber, the onset of the yielding process is revealed by a change in the slope of the stress-strain curve and the fiber extends uniformly (Figure 31). It can be shown that a neck can only form if the engineering stress, defined as the applied tensile load divided by the original cross-sectional area, falls. The stabilization of the neck in PE, at a strain of 300 to 500%, is due to the orientation

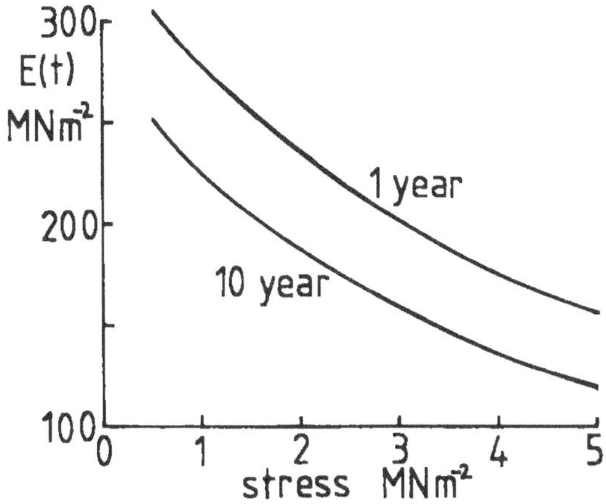


FIGURE 30. The creep modulus of Hostalen GM5010 HDPE at 20°C, as a function of creep stress.

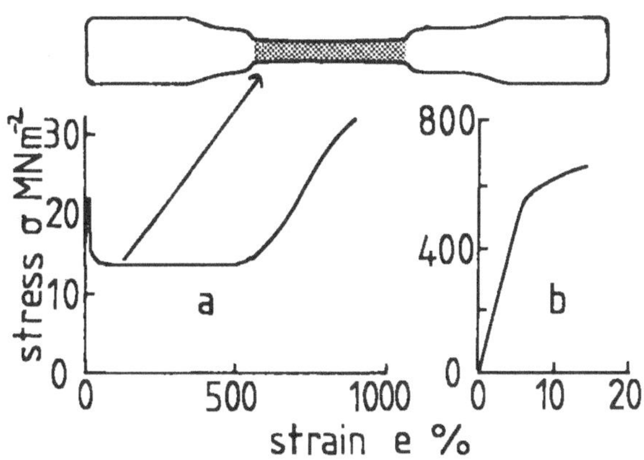


FIGURE 31. The tensile stress strain curve. (a) Unoriented PE; (b) two types an oriented PEPT fiber. MT, medium tenacity; HT, high tenacity.

hardening phenomenon in the highly strained material, which needs a stress higher than the initial yield stress before it will yield further. Orientation hardening should be distinguished from the work-hardening that occurs when metals are plastically deformed. The latter phenomenon, a result of the development of dislocation tangles in the metal grains, can be removed by heating the metal to 0.7 $T_m$ (where $T_m$ is the melting temperature in °K), whereupon the dislocation tangles diffuse away. Orientation hardening is a result of orientation developing in the amorphous phase of the polymer and possibly a breakdown of the spherulitic microstructure into a microfibrillar structure (Figure 23).

In bending or compression tests, it is rare to observe localization of plastic deformation. PMMA will yield in compression at 20°C; this is a result of the competing process of crazing not operating without a tensile component of stress. The phenomenon of crazing is easy to observe in transparent glassy polymers, but it can also occur in PE in some circumstances. Figure 32 shows that it is the result of highly localized yielding plus cavitation that occurs on planes that are perpendicular to the direction of the applied tensile stress. The crazes are only 1 to 10 μm thick, but they can easily extend 500 μm; they nearly always grow from free surfaces and appear to be nucleated by surface scratches and imperfections. Oriented fibrils of polymer span the gap between the two craze surfaces. The entanglement network

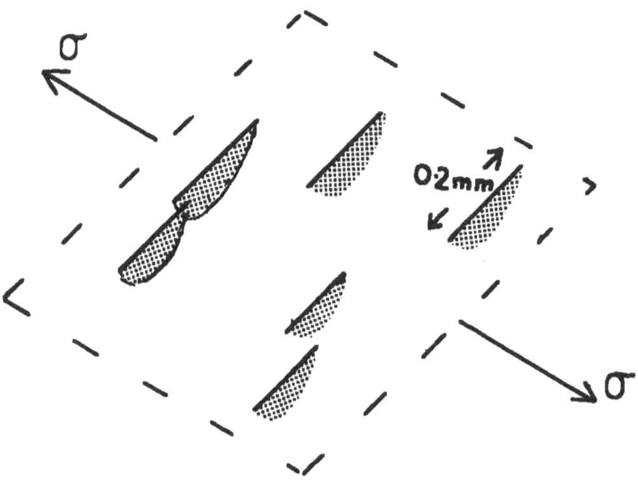


A

FIGURE 32. (A) A sketch of the geometrical shape of several surface crazes; (B) a transmission electron micrograph of a craze in polystyrene. (Reproduced with permission from *Proc. R. Soc. London, Ser. A*, Vol. 343.)

in glassy polymers is necessary for the formation of these fibrils and the process by which the craze thickens, by drawing in more polymer through the craze/bulk interface. The presence of roughly 50% air in the craze gives it a foamlike structure. Since craze growth involves the transformation of bulk polymer into this foamlike structure, tensile stresses are essential to open up the voids.

Yielding in polymers does not only happen at short times. A creep loading may eventually cause yielding and the delay time before this occurs is known as the creep-rupture time. Creep rupture data have been essential for the design of internally pressurized plastic pipes and Figure 33 shows such data for HDPE collected from tests on pipes. At 20°C the pipes fail by the wall, ballooning locally and then splitting, and it can be seen that as the creep stress is reduced, the creep rupture time becomes longer and longer. Tests at elevated temperatures have shown that there is also a competing brittle mode of failure, probably involving crazing, that causes a "knee" in the creep rupture curves. Changing to a higher molecular weight grade of PE can delay the onset of the brittle mode of failure and thereby allow higher design stresses in the pipe. A typical design stress for a HDPE used at 35°C for 5 years might be 5 $MNm^{-2}$, whereas the yield stress measured at 23°C in a tensile test lasting 1 min would be 22 $MNm^{-2}$.

The strain at yield for both glassy and semicrystalline polymers is high (5 to 10%) compared with those for metals (~0.1%), because there are no mechanisms, such as dislocation movement through a crystal lattice, that allow low strain yielding. The level of the yield stress is usually in the range E/80 to E/40, where E is Young's modulus. The consequences of the high yield strains are twofold. First, unless care is taken to design plastics products for the maximum rigidity (for example, by placing reinforcing ribs behind flat surfaces), it is likely that excessive deflections occur before any yielding takes place. Second, relatively large amounts of strain-energy can be stored in plastics and this energy is potentially available to provide the energy for crack propagation. Therefore, if crazes form then fail in a bent sheet of PMMA, it will shatter into many pieces as the strain energy is released. The means of avoiding such catastropic brittle fracture are discussed in the next section.

The link between yield stress and Young's modulus means that parameters such as the

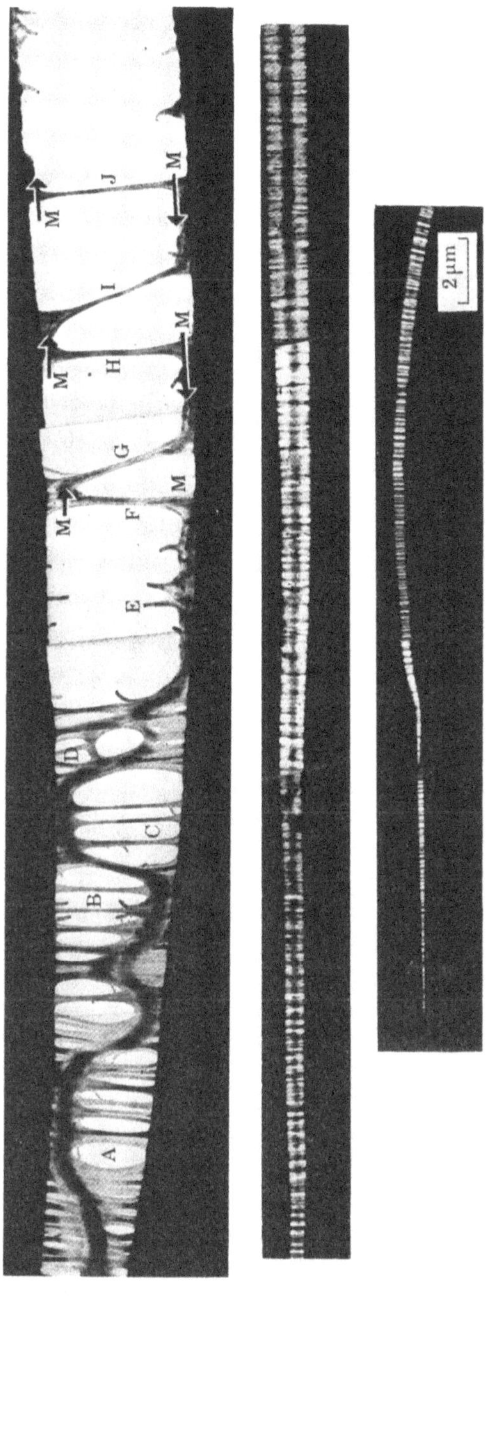


FIGURE 32B

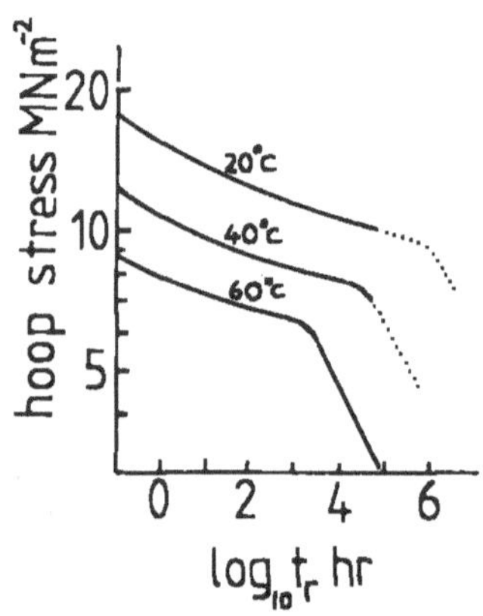


FIGURE 33. Creep rupture data for HDPE for a range of test temperatures.

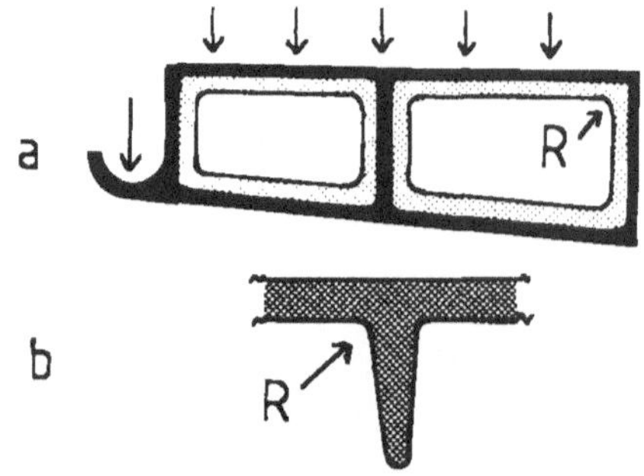


FIGURE 34. High stress regions in plastic products. (a) The corners of cut out holes in a shelf bracket; (b) where a reinforcing rib joins a flat sheet. The radius of curvature at these points should be at least half the sheet thickness to keep the stress concentration factor below 1.5.

percent crystallinity in PE and the degree of orientation in PET that influenced the modulus (Figures 26 and 24) have a similar effect on the yield stress. Also the fall-off in E between $T_g$ and $T_m$ for these semicrystalline polymers (Figure 27) means that the yield stress is strongly temperature dependent in this temperature range.

### D. Fracture

It has been assumed so far that polymeric products have been designed so that they do not deform excessively as a result of elastic strains, and yielding does not occur across the whole cross-section. We now need to consider the ways in which a crack can be initiated and propagate at loads that are insufficient to cause general yielding. There are many ways in which cracks can be initiated.

**Faulty product design** — An example is the radiusing of corners that occurs at changes of cross-section or around holes in sheet (Figure 34). Unless the radius of the corner is $\geq t/2$, where t is the sheet thickness, there will be a significant elastic stress concentration at the corner. With a ductile material like PE, the stress concentration will be reduced by localized yielding, but this has its dangers in that triaxial stresses will be set up at the tip of the yielded zone and these may cause crack nucleation.

**Faulty processing** — Faulty processing can lead to clumps of undispersed pigment or air bubbles in the product, both of which are potential sites for crack growth. Section III.C mentioned the injection molding process in which the melt flows rapidly into the mold, then is cooled rapidly. The molecular orientation, that may be frozen into the molding, provides weaker planes containing the direction of melt flow, and the rapid cooling may set up residual shrinkage stresses in which the surface is left in compression whereas the central layers of the component have tensile stresses parallel to the surface.

**Effects of usage apart from the imposed loads** — Contact with sharp objects can cause surface scratches in the relatively soft polymer. These scratches can then act as points of stress concentration from which crazes can grow. There are a number of environmental stress cracking (e.s.c) agents, e.g., alcohols for both PMMA and PE, and surface active agents, such as detergents for PE, that strongly reduce the stresses required for crazes to form. Given sufficient time (~1 month for 3 mm thick) ethanol will diffuse right into PMMA and convert it into a leathery solid having a $T_g$ = 30°C, but it rapidly reduces the tensile stress required crazing to below 10 $MNm^{-2}$. Finally there is surface degradation as a result of UV radiation in sunlight and/or oxidation which can produce brittle layers on the surface of PE.

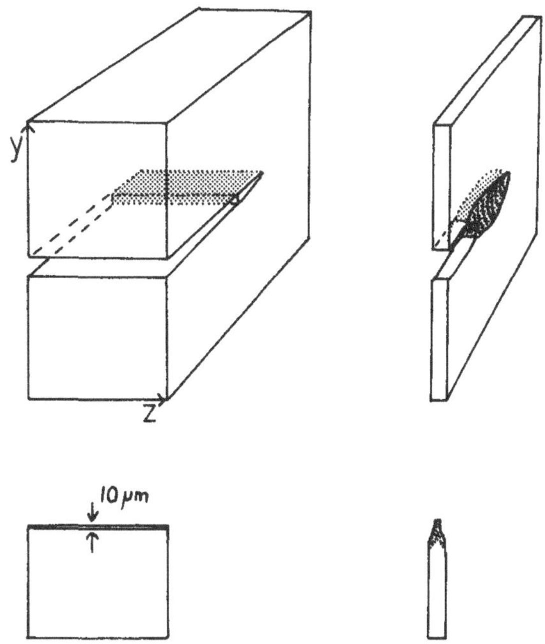


FIGURE 35. The geometry of yielding at a crack tip, represented by extreme cases. (A) There is no change in the length of the yielded zone (shown shaded) in the z direction. Craze formation at a crack tip in PMMA is an example of this plane strain yielding; (B) the sheet necks and becomes thinner in the yielded zone, thereby reducing to zero any stress in the z direction. The plane stress case is common for thin sheets of PE and PEPT; yz sections, cut from the specimens after the crack has propagated, are also shown.

**Unexpected load patterns** — Variation in load with time is known as fatigue loading and this has a particularly deleterious effect on the fibrillar structure of crazes which are more compliant than the bulk polymer and can break down, creating a crack. High speed or impact loading causes the yield stress to be much larger than in conventional tensile tests, and it may then exceed the less strain-rate-dependent stress for crazing.

Whether or not a crack once created will propagate easily depends not only on the polymer, but also on the thickness of the product. There are two extreme cases of crack growth (Figure 35). In the first, the crack is preceded by a craze, but otherwise the material ahead of the crack is unoriented bulk polymer. In the second, conventional yielded material is ahead of the crack; the yielded zone extends through the thickness of the sheet, and there is molecular orientation across the direction of crack growth. So long as neither the craze length nor yielded zone length approaches either the crack length or the ligament length (the distance the crack has to travel before failure is complete), the concepts of fracture mechanics can be applied. In particular the polymer can be said to have a critical stress intensity factor $K_c$ for crack growth. $K_c$ describes the magnitude of the elastic stress field around the crack tip excluding the yielded zone and the value $K_{Ic}$ for the "plane stress" situation in Figure 35b. In practice, the situation is a combination of those shown in Figure 35a and b; near the free surface plane stress conditions apply and the depth to which this is true is approximately given by

$$r_y = \frac{1}{2\pi}\left(\frac{K_{Ic}}{\sigma_y}\right)^2 \tag{17}$$

where $r_y$ is the radius of the plane stress yielded zone and $\sigma_y$ is the yield stress. Therefore if the sheet is thinner than 2 $r_y$, plane stress conditions should apply throughout the thickness, whereas if it is thicker than 10 $r_y$, the tougher surface layers are insignificant. Typical values for HDPE are $K_{1c}$ = 2 $MNM^{-1.5}$, $\sigma_y$ = 25 $MNm^{-2}$; so $r_y$ = 1.0 mm, and for PMMA, $K_{1c}$ = 1.5 $MNm^{-1.5}$, $\sigma_y$ = 80 $MNm^{-2}$; so $r_y$ = 0.06 mm. Most injection molded products have thicknesses in the range 1.5 to 5 mm; if they are made of HDPE, designing against crack propagation will not reduce the allowable loads significantly, however, for PMMA the thicknesses used are almost always sufficient to make plane strain fracture possible.

Finally the marked effects of polymer molecular weight on fracture properties should be reemphasized. For PE in particular increases in molecular weight lead to more marked orientation hardening after yield; they increase the resistance to crack initiation by fatigue or e.s.c, and they increase the value of $K_{1c}$.

# *Selected Applications*

Chapter 4

# POLYMERS OF NATURAL ORIGIN AS BIOMATERIALS. 1. FIBRIN

**Géza Kerényi**

## TABLE OF CONTENTS

I. Introduction ........ 92

II. Definitions ........ 92
A. Fibrinogen ........ 92
B. Fibrin Formation ........ 92
C. Insoluble Fibrin ........ 93

III. Preparation of Fibrin ........ 93

IV. Properties of Fibrin ........ 93
A. Physical ........ 93
B. Chemical ........ 94
C. Biological ........ 94
1. Biocompatibility ........ 94
2. Biodegradation ........ 94
3. Antigenicity ........ 94
D. Physiological ........ 95
1. Hemostasis ........ 95
2. Wound Healing ........ 95
3. Other Physiological Properties ........ 95

V. Structure Property Relationships ........ 95

VI. Fibrin in Medicine and Surgery ........ 96
A. Fibrinogen Clotted *In Situ* ........ 96
B. Fibrin Powder ........ 97
C. Fibrin Foam ........ 97
D. Fibrin Film ........ 97
E. Bioplast® ........ 97
1. Preparation ........ 97
2. Properties ........ 98
3. Clinical Applications ........ 99

References ........ 105

# I. INTRODUCTION

Involved in hemostasis and wound healing and then degraded by the organism, fibrin has been used as a hemostatic agent and as an absorbable implant material in a number of surgical situations. As the substance appearing in wounds, fibrin has been known for a long time. At the end of the 17th century, Malpighi realized that blood clotting is caused by the precipitation of a substance which was first called fibrin by Chaptal about 100 years later. Scanning a textbook at random one finds the following statement from 1890: "Escaping the living organism blood will soon clot ... and the protein constituent called fibrin is responsible for this. Fibrin is dissolved in the blood but it becomes insoluble outside the blood vessels ... Pure fibrin has neither colour nor flavour and it resembles albumin."[1] This is a remarkably correct description for a time when little was known about proteins in general. Even today, fibrin can be defined in simple terms as the material formed when fibrinogen clots under physiological conditions.

# II. DEFINITIONS

## A. Fibrinogen

Fibrinogen plays a physiological role in the mechanism of hemostasis by forming a dense fibrin structure around the blood platelets which aggregate at the site of a lesion in the vessel wall. Biologically, fibrinogen can be defined as the plasma protein which is clotted by thrombin and is normally present in plasma in concentrations of 200 to 500 mg/100 m$\ell$ of plasma. Both human and bovine fibronogen have a molecular weight between 330,000 and 340,000.[2,3] The amino acid composition of both bovine and human fibrinogen has been published and the differences between the human and bovine material are small.[4,5]

In addition to amino acids, the molecule also contains carbohydrate. Fibrinogen, produced in the liver, has an in vivo half life of 4 to 6 days. Electron micrographs show the molecule to be composed of three modules 50 to 70 Å in diameter connected by a thin strand about 15 Å thick.[6] The length of the molecule varies between 230 and 400 Å depending on pH.[7] When clotting fibrinogen, thrombin, a proteolytic enzyme, releases two molecules each of fibrinopeptide A and B having a molecular weight of 1500 and 2400, respectively.[8] Electric birefringence studies suggested that the two A peptides are situated at each end while the B peptides are symmetrically located about the center of the fibronogen molecule.[9] In 1968, a model of the molecule with the peptides at both ends was proposed,[10] but the idea of a modular structure with a central "knot" region including all four fibrinopeptides is more generally accepted today.[11] Sulfitolysis cleaving the disulfide bridges involving all the 56 to 58 half cystine residues of fibrinogen yields three components ($\alpha$, $\beta$, and $\gamma$) with molecular weights between 47,000 and 63,500. This suggests the presence of three pairs of chains or that fibrinogen normally exists in the dimeric form[12] although attempts to obtain the monomeric form have failed. The dimeric protein satisfies the formula $(A\alpha, B\beta, \gamma)_2$. Based on fragmenting fibrinogen with CNBr and proteolytic enzymes, an arrangement of the chains in the molecule has been proposed.[13] Fibrinogen structure and function have recently been summarized by Doolittle.[11,14] Another concise work on the subject is the publication of the First Martinsreed Workshop on Fibrinogen.[15]

## B. Fibrin Formation

One of the few in vivo polymerizations takes place when fibrinogen is converted into fibrin. Fibrin formation can be visualized as three separate reversible steps:

a) proteolysis: $\text{fibrinogen} \underset{}{\overset{\text{thrombin}}{\rightleftharpoons}} \text{fibrin monomer (f)} + \text{fibrinopeptides}$

b) polymerization: $nf \rightleftharpoons f_n$

c) clotting: $mf_n \rightleftharpoons \text{fibrin}$

In proteolysis, thrombin cleaves an arginyl-glycine bond when releasing the negatively charged fibrinopeptides. The loss of negative charge initiates polymerization. Incidentally, polymerization can also be induced by the methylation of fibrinogen by diazomethane providing a reduction of negative charge.[16] The release of peptide A is followed by that of peptide B. Also peptides AP and Y, analogs of the A peptide, are released.[17] Several different types of interchain interactions may be involved in the polymerization of the fibrin monomer (for a detailed discussion see References 18 and 19). The polymerization can be reversed by urea.

### C. Insoluble Fibrin

Insoluble fibrin is the term usually applied to the strands of insoluble protein formed under physiological conditions as a result of the action of thrombin and plasma factor XIIIa or fibrin stabilizing factor on fibrinogen. XIIIa is a plasma transglutaminase catalyzing a condensation reaction between the ε-amino group on a lysine residue on one chain of fibrin, with the γ-carbonyl of a glutamine residue, on an adjacent chain, during which ammonia is released and a new covalent bond is formed. The formation of fibrin, which is insoluble in 5 *M* urea, is accompanied by the appearance of ε-(γ-glutamyl) lysine dipeptides in enzymatic digests of fibrin.[20,21] Fibrin formed in the absence of XIIIa (e.g., purified fibrinogen + thrombin) is soluble in 5 *M* urea.[22]

Low angle X-ray diffraction[23] and electron microscopic studies[24] have shown that there is no fundamental difference in the arrangement of the amino acid residues in fibrin as compared with fibrinogen, fibrin having a cross-striation with a half period of approximately 230 Å due to a staggered, side-by-side arrangement of the fibrin monomer molecules.

## III. PREPARATION OF FIBRIN

It is beyond the scope of this book to describe the many methods reported for fibrin preparation. What they all have in common is that fibrinogen is clotted to yield the polymeric product in the presence or absence of plasma factor XIIIa. It is basically the purity of fibrinogen that differs from one method to another and it ranges from blood plasma to the cyrstalline protein. Insoluble fibrin can be produced by simply adding a $CaCl_2$ solution to blood plasma or Cohn fraction I containing most of the fibrinogen. Even the fibrinogen-rich plasma fraction contains sufficient prothrombin and factor XIII for the clotting process to take place. Also present are activating agents (factors V and Xa) transforming prothrombin into calcium dependent thrombin. This in turn clots fibrinogen and activates the inert precursor factor XIII into XIIIa (or fibrin stabilizing factor or fibrinoligase) which also requires calcium to cross-link the urea soluble intermediate into insoluble fibrin. The product is finally washed to remove contaminants. Using highly purified fibrinogen, thrombin, and factor XIIIa in the presence of calcium is at the other end of the scale of the many methods to prepare insoluble fibrin. Brief references will be made to preparation methods in the subsections describing fibrin products used in medicine and surgery.

## IV. PROPERTIES OF FIBRIN

### A. Physical

Fibrin is distinguished from fibrinogen chiefly by its insolubility under physiological conditions. Fibrin prepared in the absence of factor XIIIa and dissolved in urea has the same viscosity, sedimentation, light scattering, and optical rotation properties as fibrinogen in the same solvent. Its fingerprint is also identical except for the fibrinopeptide part. Electron micrographs show an end-to-end and a side-to-side alignment of fibrin units in the clot. The corresponding cross bands have a periodicity of about 230 Å. Due to cross-linking, the insoluble fibrin formed in the presence of factor XIIIa cannot be dissolved in urea and other similar solvents. Macroscopically, the appropriately purified and ground fibrin is an easily running yellowish white powder of no odor. It requires no special conditions of storage and is thermally stable below 170°C.

### B. Chemical

Fibrinogen contains the common unsubstituted amino acids. In addition, tyrosine-*O*-sulfate and phosphoserine are present. The proportion of hydrophilic amino acids is high. The sugar residues amount to 4 to 5%. The amino acid composition of fibrinogen has been reported elsewhere.[4,5] The amino acid analysis of fibrin, within the experimental error, is the same as for fibrinogen. Fibrin may contain 10 to 20% less carbohydrate components than fibrinogen. Fibrin has been chemically modified by different methods. These include radioiodination for diagnostic use, methylation with diazomethane[16] and formaldehyde,[25] grafting with synthetic polymers,[26] enzyme immobilization on fibrin,[27] etc.

### C. Biological

#### *1. Biocompatibility*

Implanted fibrin powder, fibrin foam, and compression molded implants as well as fibrin clotted *in situ* by introducing fibrinogen and thrombin have been shown biocompatible in a large number of studies.[28-38] There was no fever, excessive inflammation, or undue tissue reaction in animal experiments and subsequent clinical applications. Histological sections showed the normal cellular response found in tissue surrounding any biodegradable material, the situation being similar to the absorption or organization of an autogenous blood clot. The dominant cell type is the polymorphonuclear leukocyte. Lymphocytes are rare and giant cells do not appear. Foam cells are seen with progressive absorption. When solid implants as opposed to powder or foam are examined, the appearance of a fibrous capsule is also observed. With respect to toxicity, fibrin is a safe material, too. Fibrin breakdown products are well-established as being toxicologically innocuous.

#### *2. Biodegradation*

The biodegradation of fibrin products has also been studied by several authors.[30,32-34,37-43] The elimination of fibrin from the organism is partly an enzymatic (fibrinolysis), partly a cellular process (phagocytosis). Some authors suggest that fibrin formation and thus also fibronolysis are part of the normal fibrinogen catabolism. Several reports have shown fibrin to be fully absorbable. Various resorption rates have been found for different fibrin products ranging from a few days to several months. Resorption usually proceeds at a nearly even rate. Studies with $^{125}$I-labeled fibrin molds showed complete elimination of the breakdown products, mainly through the urine. Progressive absorption of the material by disappearance of radioactivity at the implant site was noted. Apart from the mass and the specific surface of the implant, the rate of absorption is determined by a number of factors, including the species and general condition of the recipient and the blood supply of the implant site. In the 1940s, methods of treating fibrin films with aldehydes and chromium salts were described which prolong the time of absorption without appreciably changing its histological pattern. The rate of absorption of implants for specific applications is reduced by cross-linking with formaldehyde today. However, a short treatment in a dilute aldehyde solution is sufficient for all practical purposes.

#### *3. Antigenicity*

A foreign protein may well be expected to induce an immune response. The antigenicity of fibrin has been extensively studied.[44-49] Fibrin suspension, foam, and solid implants of various sources were implanted into different animals in these studies. Fibrin was introduced intramuscularly, subcutaneously, intraperitoneally, intraosseally, as well as into the abdominal cavity and the marrow cavity. Reimplantations were performed. The appearance of antibodies was detected by the double diffusion technique and indirect hemagglutination and the delayed cellular reaction was examined by an intracutaneous test. No anaphylactic shock could be precipitated. Fibrin from a different species induces an immune response which can be minimized by heat treatment. The antibodies produced by heat-treated (150°C, 2 hr) fibrin can only be detected by serological methods if incomplete Freund's adjuvant is added to the antigen and a booster dose is applied several weeks after the first implantation. Otherwise the weak antigenicity of heat-treated fibrin can only be detected in the intracutaneous test. These observations have been interpreted as justifying the clinical application of heat-treated ox fibrin. Indeed, no symptoms of an immunological nature have been reported by clinicians performing ox fibrin implantations, including repeated uses.

### D. Physiological

#### *1. Hemostasis*

In larger wounds, the mechanism of hemostasis seems to be determined by platelet adhesion, vascular contraction, and the efficiency of fibrin formation. The appearance of the clot provided an apparently satisfactory explanation for the control of bleeding. The occurance of prolonged bleeding in cases of defective clotting seems to confirm this, but there remains a great deal to be learned about the role of fibrin formation in hemostasis.

#### *2. Wound Healing*

Apart from a hemostatic function, fibrin is believed to promote wound healing as suggested in many textbooks of pathology. Accelerated healing was reported after applying fibrin powder on wounds.[50-52] Fibrin has been thought to have a nutritive value in the healing process,[53,54] although the complete elimination of breakdown products from fibrin implanted in the s.c. tissue[43] is contradictory to this assumption. In other views, fibrin acts as a scaffold promoting the ingrowth of cells.[53-58] The role of fibrin in wound healing is far from being proven. However, it is a fact that even trivial injuries in hemophilic subjects are often very slow to heal. Repair is also decelerated in the absence of XIIIa, when there is no possibility of fibrin cross-linking.[59,60]

#### *3. Other Physiological Properties*

A bactericidal effect of fibrin has been reported.[53] The relative rates of fibrin deposition and removal have been suggested to be important in the maintenance of vascular integrity.[61,62] A general role has been attributed to fibrin in the recognition of foreign cells or particles by the organism.[63]

Not all of these views on the physiological properties of fibrin have been generally accepted. This is not really surprising against the following summary of knowledge on these questions. "What for instance, is the real function of coagulation in the haemostatic mechanism? Is clotting concerned in the repair of wounds and resistance to infection? Has it any part in the metabolic or immunological reactions of the blood? ... there are no satisfactory answers to any of them. In some instances, there is indirect evidence, from which at least plausible hypotheses can be constructed, in others only speculation is available."[64] While many details have been clarified in the past 20 years, the above statements remain basically valid.

## V. STRUCTURE PROPERTY RELATIONSHIPS

The structure of fibrin is characterized in Section II. Its physical, chemical, and biological properties and physiological effects are described under the subsequent subtitles. As with most natural polymers, the knowledge of structure and properties is mainly descriptive in character. While it is routine practice to influence the properties of many synthetic biomaterials by changing their composition or structure, not even a sound working knowledge is available on the correlation between the structure and properties of fibrin. For instance, the amino acid sequence of fibrinogen and that of the fibrinopeptides cleaved by thrombin are known, but it would be hard to pinpoint the part or parts of the fibrinogen molecule which are responsible for clotting. The blood clotting mechanism is farily well-known today, but a better understanding of its relation to fibrinogen structure is still lacking, although it could be helpful in drug research aimed at preventing thrombosis.

Metals and their alloys as well as synthetic polymers can often be modified to alter their properties. Such work is frequently based on an exact knowledge of structure-property relations. This cannot be so with most natural polymers where such knowledge is not available. In the case of fibrin, not even all properties are quite well-understood. In fact, hypotheses or speculation are only available on some physiological effects. Even speculation concerning structure-property relationships would therefore be too far-fetched today.

It is then understandable that there is limited experience on varying properties by changing fibrin structure. Fibrin solubility is influenced by the action of factor XIIIa, antigenicity is minimized by heat treatment, and the absorption rate of implants is reduced by treatment with formaldehyde, but these phenomena are only observed and recorded rather than understood in quantitative terms. With a view to possible future applications grafted poly-(vinylene carbonate)-fibrin has been prepared. This polymer modified fibrin has reduced swelling and increased mechanical strength.[26] Radioiodination, methylation, and enzyme immobilization on fibrin can be mentioned as further examples of modifying the chemical structure of fibrinogen or fibrin. Here again, it is not yet possible to predict the changes in properties caused by a specific modification of structure.

One probable reason why structural changes in fibrin as a biomaterial have not been studied in greater detail is the fact that any modification may change a whole set of properties. This would then call for extensive in vitro studies and animal experiments before clinical trials could be attempted. The amount of work and costs involved is certainly not justified when fibrin is a well-established hemostatic agent and implant material largely due to its natural properties.

## VI. FIBRIN IN MEDICINE AND SURGERY

The most significant asset of fibrin is its origin from a blood plasma protein and hence its obvious tissue and blood compatibility, lack of toxicity, and other detrimental effects. Reputed to be involved in hemostasis and wound healing, fibrin has been applied as a hemostatic agent. As a biodegradable material, fibrin has been used for temporary tissue replacement and it is the only absorbable implant material which has been commercialized to date. The following subchapters give a summary review of fibrin products and their uses in medicine and surgery.

### A. Fibrinogen Clotted *In Situ*

Fibrinogen is separated from other plasma proteins using routine purification methods (precipitation, chromatography, electrophoresis, etc). Skin transplants have been fixed by impregnating the transplant and the site with solutions of human or bovine thrombin and human fibrinogen.[65-73]

A simultaneous injection into the renal pelvis of dissolved thrombin and human fibrinogen is an interesting solution of coagulum pelviolithotomy or the removal of small stones entrapped in the clot.[74-82] As the production of human fibrinogen and thrombin was discontinued in 1949 in the U.S., bovine fibrinogen and thrombin were used for the same purpose.[83]

Fibrinogen alone or combined with thrombin has been used as a tissue adhesive in various opthalmic operations.[84-87] Fibrinogen has also been applied as an extrapleural filling preventing complications following pulmonary resection for tuberculosis.[88] More recently, a human fibrin adhesive kit* has been reported useful in ophthalmology,[87] vascular surgery,[89] maxillofacial surgery,[90,91] neurosurgery[92] and orthopedics,[93] as well as a hemostatic in various operations.[94-100] Some caution is warranted due to possible hepatitis infection.

### B. Fibrin Powder

Purified fibrinogen is clotted by thrombin and the product is ground after drying[101] or Cohn fraction I is coagulated with $CaCl_2$ to yield insoluble fibrin which is washed, dried, and powdered.[102] Freeze-dried fibrin powder can be prepared from fibrin foam.[101]

Freeze-dried fibrin powder and fibrin foam are powerful hemostatic agents due to their large specific surface, their ability to be soaked with blood and platelets, and their capacity to reversibly adsorb thrombin.[103] These physiological hemostatics have been used in themselves or combined with thrombin and sometimes antibiotics.[103-106] Human and heat deantigenized bovine fibrin displays the same hemostatic performance. Rapid healing and epithelization was noted after applying fibrin powder on wounds.[50-52] A coagulum of bovine fibrin, the patient's venous blood, thrombin, and antibiotics was used to fill up bone defects in chronic osteitis and osteomyelitis as long as hospital microflora permitted primary wound closure following the operation.[107,108] This was the first large-scale clinical implantation of heat-treated bovine fibrin (about 300 cases, including the repeated use of fibrin in many instances). One advantage of the coagulum was a gradual release of the antibiotics over a prolonged period of time. The coagulum has recently been used for filling up bone defects after jawbone cystectomy[109] and for postextraction hemostasis during coumarin anticoagulant therapy.[110,111]

### C. Fibrin Foam

A dilute fibronogen solution is whipped into foam, clotted with thrombin, and freeze-dried before heat treatment[112] or the dissolved fibrinogen is clotted simultaneously with the addition of hydrogen peroxide, whereby the evolving oxygen converts the fibrin into foam.[45] Similar to other hemostatic agents with a large specific surface (Surgicel®, Oxycel®, Gelfoam®, etc.), human and bovine fibrin foam, a physiological hemostatic with or without thrombin, has been reported useful in several fields of surgery.[29,35,36,104,113-118] Fibrin foam has also been used in the treatment of depressed scars[119] and as a filling material for defects in the surgery of the salivary gland.[120]

### D. Fibrin Film

A film of fibrin is produced by clotting Cohn fraction I with thrombin and removing the liquid by pressure.[121] The main field of application was in neurosurgery. Fibrin films were used to replace the dura[122-124] and to protect peripheral nerve sutures.[34,125-127] Limited use was reported in burn treatment.[124,128] Films were also used to eliminate perforations between the maxillary sinus and the oral cavity.[128] Human and bovine fibrin films performed equally.[129]

### E. Bioplast®

Bioplast® is the registered trademark for fibrin implants developed by Gerendás for temporary tissue replacement.[56] Availability** and current uses justify a more detailed treatment.

#### *1. Preparation*

Fibrinogen is precipitated from blood plasma using ethanol at a low temperature. Insoluble

* Immuno AG, Vienna, Austria.

** Under the trade name Biethium® by Ethicon Ltd., Edinburg, Scotland.

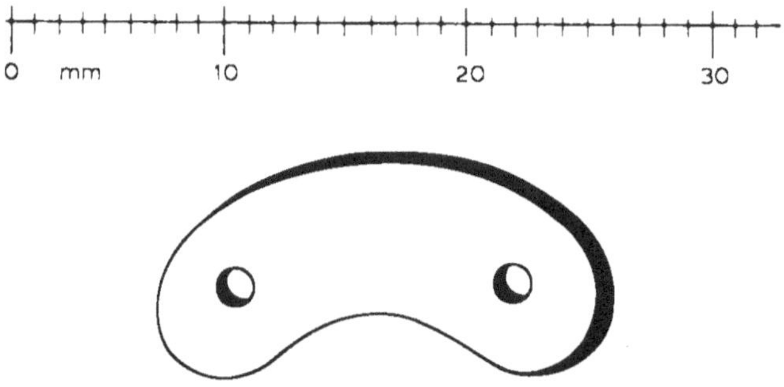


FIGURE 1. Bioplast® fibrin stress incontinence implant. (Reproduced from *Biomaterials*, 1, 1, 1980; by permission of the publishers, IPC Business Press Ltd., Guildford, Surrey, U.K.)

fibrin is produced by adding $CaCl_2$. It is washed, dried, and ground. Bovine fibrin powder is heated to 150°C to destroy or reduce antigenicity. Glycerol is used as an additive for molding at 120 to 140°C at a pressure of 10 to 50 MPa. The rate of absorption is controlled by pretreatment with formaldehyde.[102] Packaging is followed by sterilization with gamma irradiation.

*2. Properties*

For molding, 20 to 50% glycerol is added to fibrin as a softening agent or pseudoplasticizer. Glycerol content does not affect the in vivo behavior of Bioplast®, a yellowish-brown, translucent, and flexible plastic-like substance (Figure 1).

The glycerol content of Bioplast® influences its mechanical strength which may vary over a comparatively broad range as shown below for the implant material containing 20 to 50% glycerol.

| | |
|---|---|
| Pressure strength | 14—76 MPa |
| Tensile strength | 6—83 MPa |
| Bending strength | 1—35 MPa |
| Elongation at break | 8—66% |

The mechanical strength can be significantly increased by combining Bioplast® with various minerals. However, attempts to produce absorbable implants for osteosynthesis have failed as these composite materials swell up and soon lose much of their strength in contact with water or biological fluids. Preliminary results on grafted poly-(vinylene carbonate)-fibrin with reduced swelling and increased mechanical strength have been reported.[26]

Both fibrin and Bioplast® are thermally stable below 170°C as shown by thermal analysis (Figure 2).[130]

Pretreatment with formaldehyde used to control the rate of absorption decreases swelling in an aqueous medium, but does not otherwise influence the physical properties of Bioplast® (Figure 3).[131]

$^{60}Co$ gamma irradiation decreases swelling and tryptic digestibility above 10 Mrad, but the dose used for sterilization, 2.5 Mrad, does not cause cross-linking (Figures 4 and 5).[132]

Biologically, Bioplast® is capable of being absorbed by polymorphonuclear digestive action.[56]

Ox fibrin antigenicity is reduced by heat treatment at 150°C. No clinical symptoms of an immunological response are seen even after a repeated implantation of the heat-treated bovine material.[48]

The substance is biocompatible. Electron microscopic studies have shown the absorption mechanism to be identical with that of blood clot organization.[42]

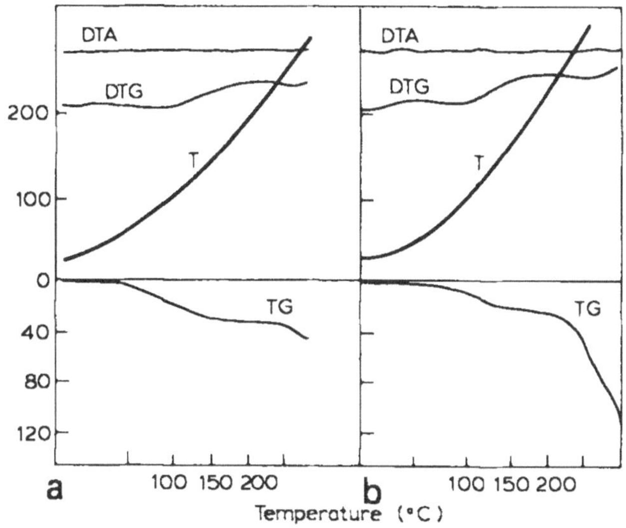


FIGURE 2. Thermal analysis (a) fibrin powder; (b) Bioplast®. (Reproduced from *Biomaterials*, 1, 1, 1980; by permission of the publishers, IPC Business Press Ltd., Guildford, Surrey, U.K.)

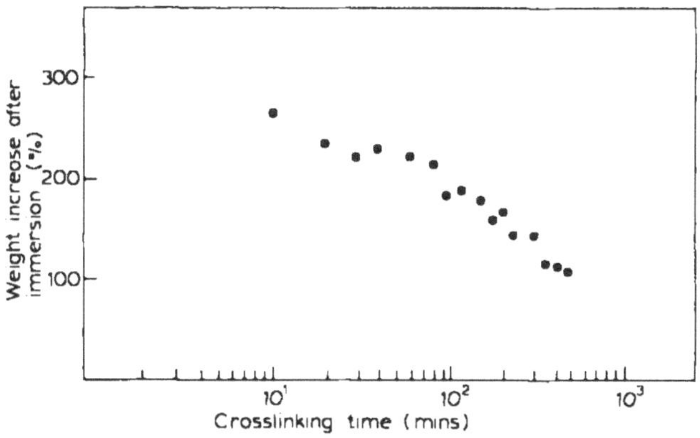


FIGURE 3. Swelling of formaldehyde-treated Bioplast®. Incubation: 150 min in aqueous solutions at room temperature. (Reproduced from *Biomaterials*, 1, 1, 1980; by permission of the publishers, IPC Business Press Ltd., Guildford, Surrey, U.K.)

The resorption of the $^{125}$I-labeled Bioplast® proceeds at an even rate (Figure 6). Elimination is complete and occurs mainly through the kidneys while the implant site is gradually invaded by host tissue.[43]

While the rate of absorption can be reduced by cross-linking with formaldehyde, a short pretreatment (20 to 60 min at room temperature in a 1% formaldehyde solution) is sufficient for all practical purposes.

### 3. *Clinical Applications*

**Arthroplasty** — Bioplast® was first used in hip arthroplasty. The ankylosed joint of coxotuberculosis patients was mobilized and a fibrin cup was placed on the femur head. Long-term mobility and load bearing was attributed to physicotherapy and the fibrous cartilage gradually replacing the implant.[133-135] The restoration of mobility in ankylosed hip

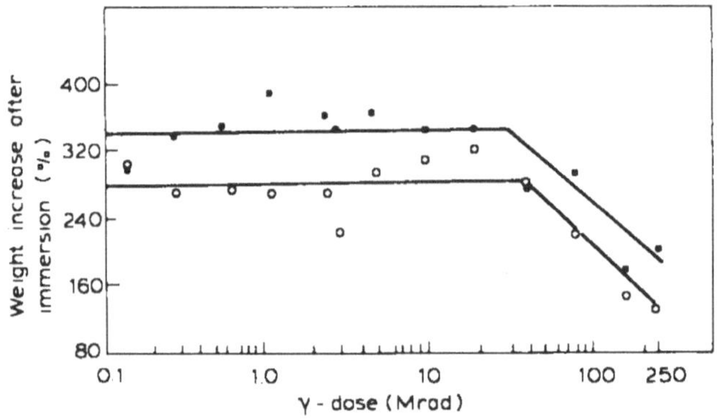


FIGURE 4. Swelling of $^{60}Co$ irradiated Bioplast® Incubation: 150 min in aqueous solutions at room temperature; ○, human fibrin; ●, bovine fibrin. (Reproduced from *Biomaterials*, 1, 1, 1980; by permission of the publishers, IPC Business Press, Ltd., Guildford, Surrey, U.K.)

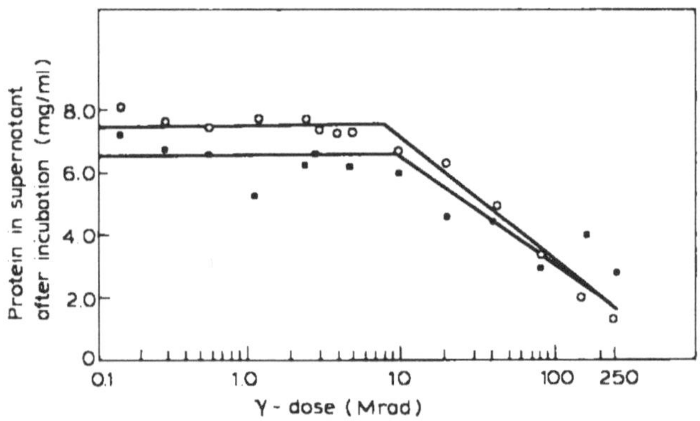


FIGURE 5. Tryptic digestibility of $^{60}Co$ irradiated Bioplast®. Incubation: 24 hr at 37°C, pH = 8, 05; ○, human fibrin; ●, bovine fibrin. (Reproduced from *Biomaterials*, 1, 1, 1980; by permission of the publishers, IPC Business Press Ltd., Guildford, Surrey, U.K.)

joints affected by tuberculosis was a remarkable achievement at a time when total hip prostheses were not yet generally available. Arthroplasty of the temporomandibular joint was also performed using Bioplast® interposition.[136]

**Retinal detachment** — Scleral buckling with a Bioplast® rod was performed in retinal detachment.[137,138] A few weeks later, when the implant had degraded, the eyeball regains its preoperative shape. The method was thus more successful in traumatic cases than in detachment caused by an elongated eyeball in myopia.

**Closure of perforations** — Perforations of the nasal septum were closed by implanting a Bioplast® plate, tailored to size, between the two mucosal layers. Postoperative mucosal growth is seen on both implant surfaces mainly with smaller defects when the mucosa is not markedly atrophic.[139] Antro-alveolar perforation may occur as a complication of tooth extraction. The oro-antral communication could be stopped by covering the alveolus with a Bioplast® film fixed by a mucoperiosteal lobe within the first 36 hr.[140]

**Current uses in ophthalmic surgery** — Current applications include the substitution of Bioplast® film for injured conjunctiva. On the 2nd to 4th day after implantation the site becomes veined. From the 5th to 6th day, it is increasingly difficult to distinguish the implant from its environment. After 30 days the implant cannot be recognized even with a slit lamp.

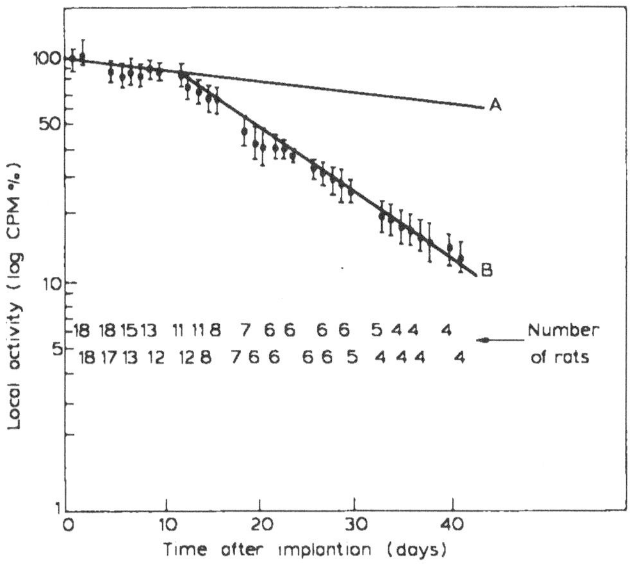


FIGURE 6. Resorption of $^{125}$I-labeled Bioplast®. (A) $^{125}$I-decay; (B) local activity. (Reproduced from *Biomaterials*, 1, 1, 1980; by permission of the publishers, IPC Business Press Ltd., Guildford, Surrey, U.K.)

The composition of tear proteins returns to normal in about 8 days or as soon as after free conjunctival grafting.[141-143] Bioplast® sheets are used to cover skin defects around the eyes and to replace part of the eyelid following the extirpation of tumors. In a few weeks, new epithelial tissue replaces the gradually absorbing implant. The new tissue has the elasticity, cosmetic appearance, and functional properties of adjacent skin.[141,142,144]

**Stress incontinence in the female** — Implants are commercially available for the treatment of female urinary stress incontinence, for liver hemostasis and suture protection, and for the repair of difficult hernias. A new and simple method of treating urinary stress incontinence of the female is to fix a bean-shaped Bioplast® implant (Figure 1) vaginally to the bladder neck and upper urethra. After exposing and mobilizing the urethra, the vesico-urethral junction, and the bladder, the latter is separated from the cervix. The anterior vaginal wall is separated from the vesico-vaginal fascia. The implant is placed and fixed transversally to the vesico-vaginal fascia, at the vesico-urethral junction (Figure 7). It is then imbedded by the fascial layers being plicated and brought together in the midline using catgut (Figures 8 and 9). A longitudinal line of fascial infolding catgut stitches are then inserted as far laterally as possible (Figures 10 and 11). The implant is pushed backward and upwards under the proximal part of the bladder elevating the bladder. Finally, the redundant vaginal wall is cut away and the edges are brought together by mattress sutures (Figure 12). Following this procedure, posterior repair is performed as a fundamental part of the operation. The presence of the implant restores normal urethral shape and position. Long-term continence is thought to be maintained by the fibrous tissue replacing the implant.[145,146]

**Liver hemostasis** — In hepatic surgery, effective control of hemorrhage and bile leakage is obtained using Bioplast® liver suture buffers which broaden the base of the suture and prevent suture material cutting through liver substance. The buffers are elongated buttons, 14 × 6 × 1.4 mm in size with holes and lateral notches which accommodate the sutures at both ends. In wedge excision biopsy, the buffers are placed on either side of the proposed biopsy site at the free edge of the liver or on the dome. They are fixed using chromic catgut passed through the holes on the buffer ends and then through the liver substance. For biopsies at the free edge, the buffers are placed to allow wedge excision, but on the dome they are

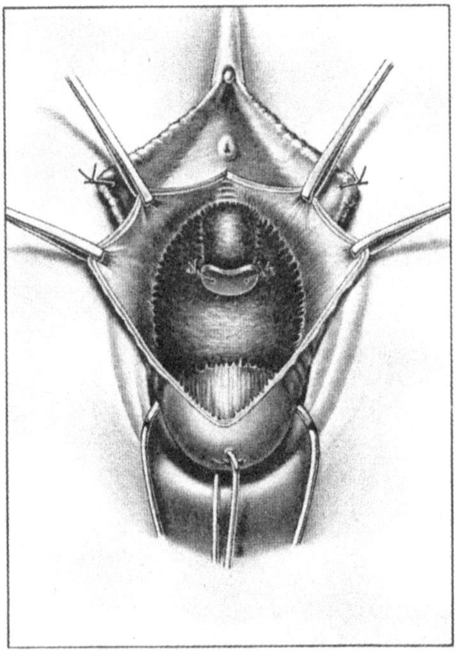

FIGURE 7. Bioplast® stress incontinence implant fixed at the vesico-urethral junction. (Courtesy of Ethicon, Ltd., Edinburgh.)

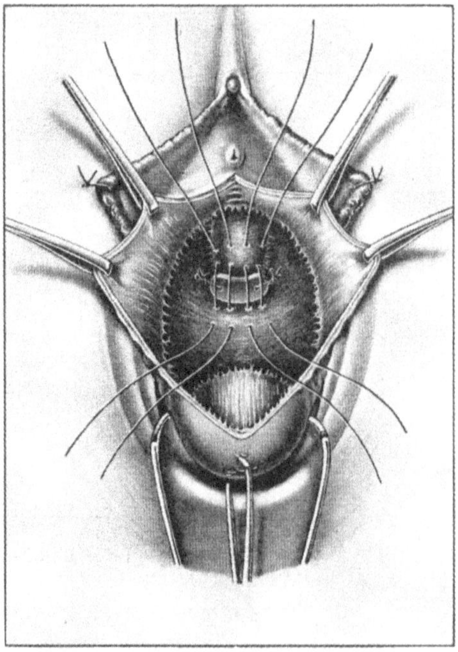

FIGURE 8. Sutures are inserted to imbed the implant by fascia. (Courtesy of Ethicon, Ltd., Edinburgh.)

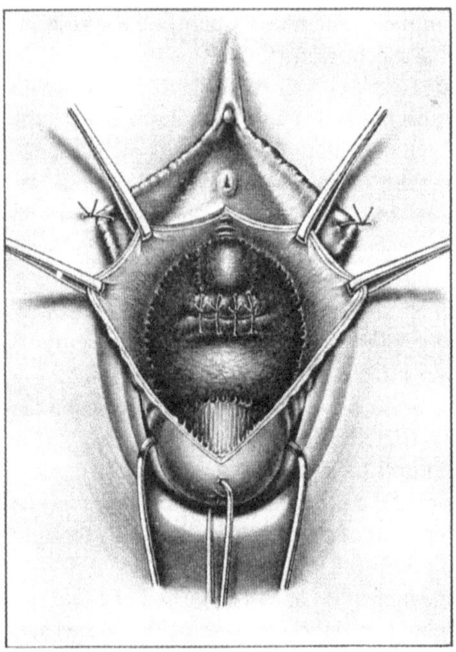

FIGURE 9. The implant is imbedded by the fascial layers. (Courtesy of Ethicon, Ltd., Edinburgh.)

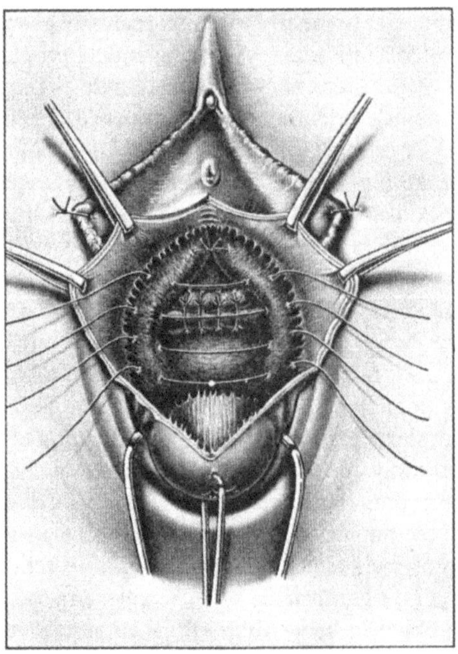

FIGURE 10. Lateral introduction of a longitudinal line of catgut sutures. (Courtesy of Ethicon, Ltd., Edinburgh.)

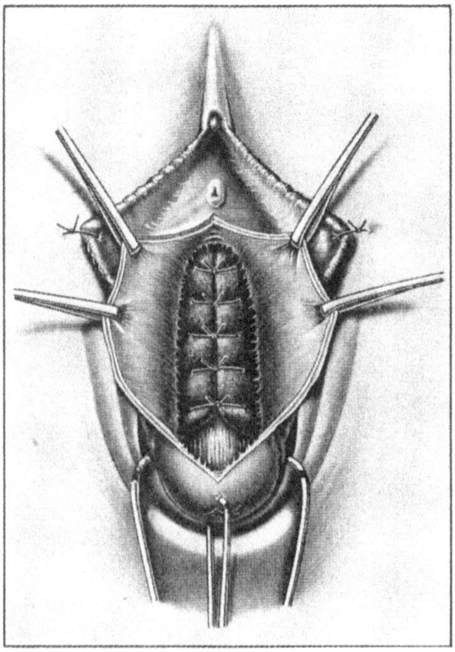

FIGURE 11. Fascial infolding stitches are inserted. (Courtesy of Ethicon, Ltd., Edinburgh.)

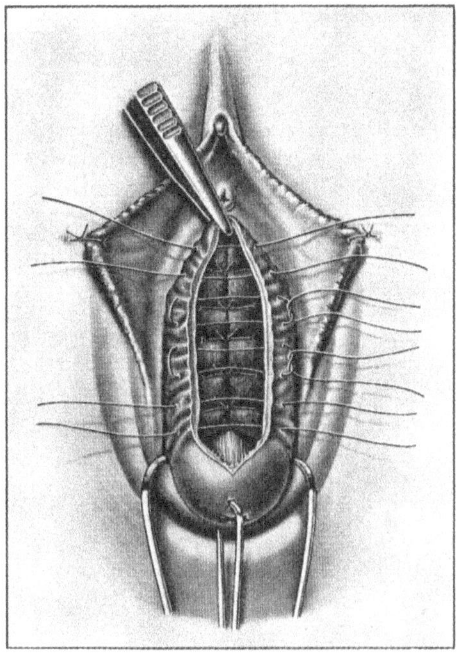

FIGURE 12. Edges are brought together by mattress sutures. (Courtesy of Ethicon, Ltd., Edinburgh.)

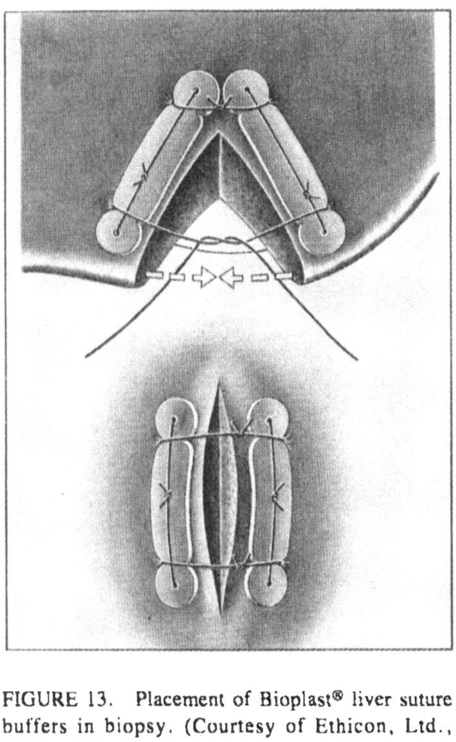

FIGURE 13. Placement of Bioplast® liver suture buffers in biopsy. (Courtesy of Ethicon, Ltd., Edinburgh.)

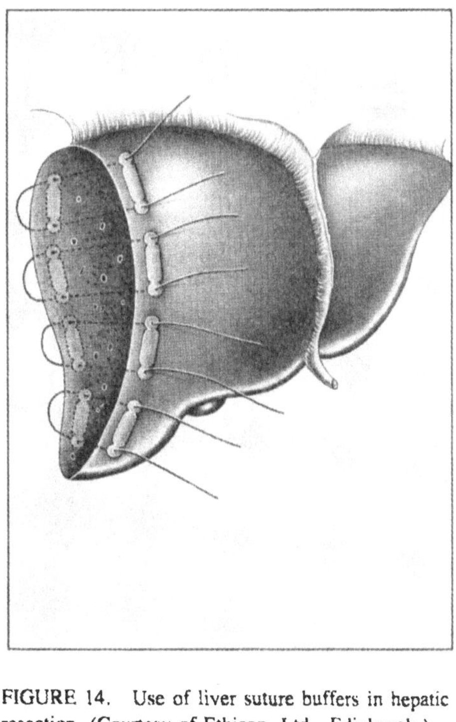

FIGURE 14. Use of liver suture buffers in hepatic resection. (Courtesy of Ethicon, Ltd., Edinburgh.)

placed in parallel (Figure 13). The buffers are also useful for the compression closure of exposed areas of liver following formal hepatic resection for tumor or trauma (Figure 14). The exposure of intrahepatic ducts can be associated with troublesome hemorrhage, which

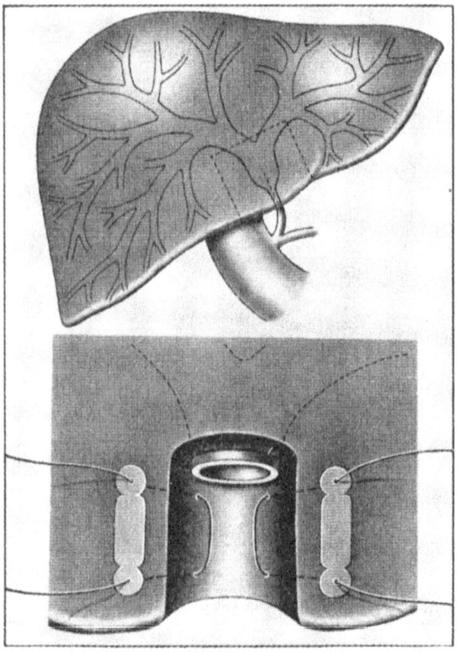

FIGURE 15. Control hemorrhage of liver suture buffers in the quadrate lobe. (Courtesy of Ethicon, Ltd., Edinburgh.)

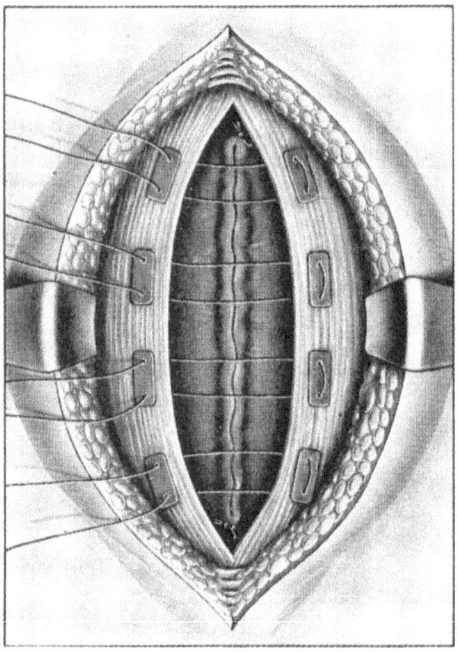

FIGURE 16. Bioplast® buttresses in the repair of large incisional hernias. (Courtesy of Ethicon Ltd., Edinburgh.)

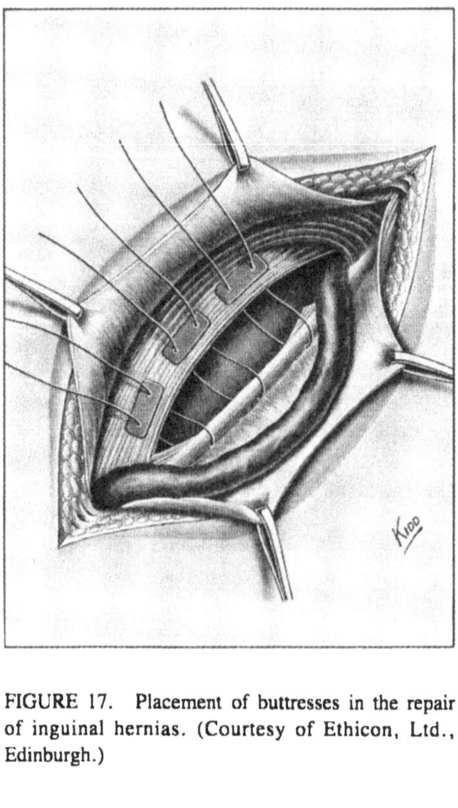

FIGURE 17. Placement of buttresses in the repair of inguinal hernias. (Courtesy of Ethicon, Ltd., Edinburgh.)

can be controlled by liver suture buffers (Figure 15). Successful applications of liver suture buffers have been reported.[147-150]

**Difficult hernia** — Buttresses measuring 18 × 7 × 2 mm, with a hole near each end, are used to facilitate placement of sutures in the repair of difficult hernias. They can be used by placing two rows along each side of the defect, as in large incisional hernias (Figure 16) or they may be placed along the transversalis fascia and sutured to the inguinal ligament in the repair of inguinal hernias (Figure 17). These implants are especially useful where it is important to have suture compression without tissue necrosis and without adding bulk to suture volume.[151,152]

## REFERENCES


1. **Balló, M.,** Elimi vegytan, 5th ed., Franklin, Budapest, 1890, 131.
2. **Caspary, E. A. and Kekwick, R. A.,** Some physicochemical properties of human fibrinogen, *Biochem. J.*, 67, 41, 1957.
3. **Shulmann, S.,** The size and shape of bovine fibrinogen: studies of sedimentation, diffusion and viscosity, *J. Am. Chem. Soc.*, 75, 5846, 1953.
4. **Henschen, A. and Blombäck, B.,** Amino acid composition of human and bovine fibrinogen and fibrin, *Ark. Kemi*, 22, 347, 1964
5. **Mihalyi, E., Small, P. A., Jr., and Cooke, J. P.,** The amino acid composition of bovine fibrinogen, *Arch. Biochem.*, 106, 229, 1964.
6. **Hall, C. E., and Slayter, H. S.,** in *Proc. 5th Int. Congr. for Electron Microscopy*, Breese, S. R., Jr., Ed., Academic Press, New York, 1962, 2:1.
7. **Hall, C. E.,** Electron microscopy of the fibrinogen molecule and fibrin clot, *Lab. Invest.*, 12, 998, 1963.
8. **Gladner, J. A., Folk, J. E., Laki, K, and Carrol, W. R.,** Thrombin-induced formation of co-fibrin, *J. Biol. Chem.*, 234, 62, 1959.
9. **Haschemeyer, A. E. V.,** A polar intermediate in the conversion of fibrinogen to fibrin monomer, *Biochemistry*, 2, 851, 1963.
10. **Blombäck, B., Blombäck, M., Henschen,A., Hessel, B., Iwanaga, S., and Woods, K. R.,** N-Terminal disulphide knot of human fibrinogen, *Nature (London)*, 218, 130, 1968.
11. **Doolittle, R. F.,** Fibrinogen and fibrin, in *The Plasma Proteins*, Vol. 2, 2nd ed., Putnam, F. W., Ed., Academic Press, New York, 1975, chap. 3.
12. **Johnson, P. and Mihalyi, E.,** Physicochemical studies of bovine fibrinogen, *Biochim. Biophys. Acta*, 102, 467, 1965.
13. **Blombäck, B.,** Fibrinogen-structure and function, in *Proc. 16th Int. Congr. Hematol.*, Kyoto, 1976, Excerpta Medica, Amsterdam, 1977, 24.
14. **Doolittle, R. F.,** Structure and function of fibrinogen, *Horiz. Biochem. Biophys.* 3, 164, 1977.
15. Shape and structure of fibrinogen. First Martinsried Workshop on Fibrinogen. Pehr Edman in memoriam, *Henschen A. Thromb. Haemost.*, 41, 633, 1979.
16. **Osbahr, A. J.,** The polymerization of fibrinogen under the influence of diazomethane modification, *Biomaterials*, 1, 100, 1980.
17. **Blombäck, B.,** The fibrinopeptides, *Thromb. Diath. Haemorrh. Suppl.*, 20, 201, 1966.
18. **Laki, K.,** *Fibrinogen*, Marcel Dekker, New York, 1968.
19. **Blombäck, B., Hogg, D. H., Gardlund, B., et al.** Fibrinogen and fibrin formation, *Thromb. Res.*, 8, Suppl. 2, 329, 1976.
20. **Matacic, S. and Loewy, A. G.,** The identification of isopeptide cross-links in insoluble fibrin, *Biochem. Biophys. Res. Commun.*, 30, 356, 1968.
21. **Pisano, J. J., Finlayson, J. S., and Peyton, J. S.,** Cross-link in fibrin polymerized by factor XIII (γ-glutamyl) lysine, *Science*, 160, 892, 1968.
22. **Robbins, K. C.,** Study on conversion of fibrinogen to fibrin, *Am. J. Physiol.*, 142, 581, 1944.
23. **Stryer, L., Cohen, C., and Langridge, R.,** Axial period of fibrinogen and fibrin, *Nature (London)*, 197, 793, 1963.
24. **Hall, C. E. and Slayter, H. S.,** The fibrinogen molecule: its size, shape and mode of polymerization, *J. Biophys. Biochem. Cytol.*, 5, 11, 1959.

25. **Tyihák, E., Gerendás, M., Kerényi, G., and Patthy, A.**, Examination of basic amino acids in formaldehyde treated fibrin preparations, in *Proc. 15th Hungarian Meet. Biochem., Miskolc*, Hungarian Chemical Society, Budapest, 1975, 21.
26. **Kovács, G., Fedorova, N., Kiss, L., and Killy, T.**, Properties of polymer modified fibrin, in *Proc. 1st World Biomaterials Congr.*, Baden near Vienna, European Society of Biomaterials, Vienna, 1980, 2.27.
27. **Dillon, J. G., Wade, C. W. R., and Daly, W. H.**, Enzyme immobilization on fibrin, *Biotechnol, Bioeng.*, 18, 133, 1976.
28. **Ingraham, F. D. and Bailey, O. T.**, Clinical use of products of human plasma fractionation; use of products of fibrinogen and thrombin in surgery, *JAMA*, 126, 680, 1944.
29. **Bailey, O. T. and Ingraham, F. D.**, Use of fibrin foam as hemostatic agent in neurosurgery; clinical and pathological studies, *J. Clin. Invest.*, 23, 591, 1944.
30. **Morrison, P. R. and Singer, M.**, Note on absorption rates of fibrin films in tissue, *J. Clin. Invest.*, 23, 573, 1944.
31. **Ingraham, F. D., Bailey, O. T., and Cobb, C. A., Jr.**, Use of fibrin film as dural substitute and in prevention of meningocerebral adhesions; further studies and clinical results, *JAMA*, 128, 1088, 1945.
32. **Blaine, G.**, Experimental observations on use of absorbable and non-absorbable plastics in bone surgery, *Br. J. Surg.*, 33, 245, 1946.
33. **Bagdy, D., Áfra, D., and Gerendás, M.**, Marhaplazmából készitett fibrinfilmek felszivódása, *Kiserl. Orvostud.*, 5, 176, 1953.
34. **Áfra, D., Bagdy, D., and Gerendás, M.**, Experimentelle Untersuchung der Resorption von Fibrinfilmen und ihre Anwendung in der neurochirurgischen Praxis, *Acta Med. Acad. Sci. Hung.*, 11, 1, 1957.
35. **Áfra, D.**, Vérzéscsillapitás fibrinhabbal idegsebészeti mütétekben, *Orv. Hetil.*, 99, 1421, 1958.
36. **Áfra, D.**, Blutstillung mit Fibrinschaum bei neurochirurgischen Operationen, *Zentralbl. Neurochir.*, 20, 78, 1960.
37. **Horváth, N.**, Vergleichende experimentelle Untersuchungen mit in die Markhöhle eingelegten Bioplast und Metallimplanten, *Z. Exp. Chir.*, 5, 336, 1972.
38. **Capperauld, I., Lawrie, P., and French, D.**, Properties of bovine fibrin absorbable implants, *Surg. Gynecol. Obstet.*, 144, 3, 1977.
39. **Barnhart, M. I.**, Importance of neutrophilic leukocytes in the resolution of fibrin, *Fed. Proc. Fed. Am. Soc. Exp. Biol.*, 24, 846, 1965.
40. **B Kovács, A., Somogyvári, K., and Gerendás, M.**, Untersuchungen der Resorption von Bioplastplatten, *Acta Vet. Acad. Sci. Hung.*, 15, 91, 1965.
41. **Still, W. J.**, An electron microscopic study of the organization of experimental thromboemboli in the rabbit, *Lab. Invest.*, 15, 1492, 1966.
42. **Szalontai, S.**, Investigation of the resorption of Bioplast by electron microscopy, in *Proc. 9th Hungarian Meet. Biochem.*, Veszprém, Hungarian Chemical Society, Budapest, 1968, 83.
43. **Török, B., Halmágyi, Gy., Lantos, J., and Kerényi, G.**, Resorption of $^{125}$I-labelled fibrin Bioplast from the dorsal muscle of rats, *Biomater. Med. Devices Artif. Organs*, 3, 205, 1975.
44. **Glynn, J. H. and Richardson, J. H.**, Antigenic properties of fibrin films and foams prepared from human and from bovine blood plasma, *J. Immunol.*, 53, 143, 1946.
45. **Ebbecke, U.**, Die Herstellung von Fasern und Membranen aus Fibrin, ihre Bedeutung für die normale und pathologische Anatomie und für die Chirurgie, *Dtsch. Med. Wochenschr.*, 77, 1623, 1952.
46. **Szilágyi, T. and Bagdy, D.**, Die Antigen-Eigenschaften von Fibrin, *Z. Immunitaetsforsch. Immunbiol.*, 112, 403, 1955.
47. **Horváth, N., Somogyvári, K., Somos, P., and Surján, J.**, Experimental studies on the antigenic action of Bioplast preparations, *Acta Vet. Acad. Sci. Hung.*, 19, 191, 1969.
48. **Horváth, N., Surján, J., and Somogyvári, K.**, Studies on antibody response after implantation of fibrin Bioplast preparations, *Ann. Clin. Res.*, 7, 54, 1975.
49. **de Man, K., Westbroek, D., and Brummelhuis, H. G. J.**, Antigenic properties of Bioplast., *Int. J. Oral Surg.*, 7, 564, 1978.
50. **Bergel, S.**, Ueber Wirkungen des Fibrins, *Dtsch. Med. Wochenschr.*, 35, 663, 1909.
51. **B Kovács, A., Somogyvári, K., Gerendás, M., and Gräf, S.**, Experimentelle Untersuchungen eines firbinhaltigen Wundstreupulvers, *Acta Vet. Acad. Sci. Hung.*, 20, 195, 1970.
52. **Várkonyi, V., Radó, Gy., and Soós, Gy.**, Az ulcus cruris konzervativ terápiájának ujabb lehetősége, *Borgyogy. Venerol. Sz.*, 54, 121, 1978.
53. **MacCallum, W. G.**, *A Textbook of Pathology*, 7th ed., W. B. Saunders, Philadelphia, 1940, 213.
54. **Astrup, T.**, Blutgerinnung, Fibrinelyse und Wundsheilung, *Med. Welt*, 42, 1972, 1959.
55. **Witte, S.**, Die Wundheilung bei hämorrhagischen Diathesen, in *Proc. 8th Congr. Eur. Soc. Haematol.*, Vienna, S. Karger, Basel, 1961, 350.
56. **Gerendás, M.**, Fibrin products as aids in hemostasis and wound healing, in *Fibronogen*, Laki, K., Ed., Marcel Dekker, New York, 1968, chap. 13.

57. **Park, J. B.,** *Biomaterials,* Plenum Press, New York, 1979, 133.
58. Improved wound healing through fibrin adhesion?, proceedings, MMW, 121, 743, 1979.
59. **Duckert, F., Jung, E., and Shmerling, D. H.,** A hitherto undescribed congenital haemorrhagic diathesis probably due to fibrin stabilizing factor deficiency, *Thromb. Diath. Haemorrh.,* 5, 179, 1961.
60. **Biel, H., Bohn, H., Ronneberger, H., and Zwisler, O.,** Beschleunigung der Wundheilung durch Faktor XIII der Blutgerinnung, *Arzneim. Forsch.,* 21, 1429, 1971.
61. **Jensen, H.,** Dynamic concept of fibrin formation and lysis in relation to haemorrhage (capillary permeability) and to thrombosis, *Exp. Med. Surg.,* 14, 189, 1956.
62. **Copley, A. L.,** Neue Auffassungen über Hämorrhagie, Hämostase und Thrombose, *Arzneim. Forsch.,* 11, 114, 1957.
63. **Wilkins, D. J.,** *The Reticuloendothelial System and Immune Phenomena,* Diluzio, N. R., Ed., Plenum Press, New York, 1971, 77.
64. **Biggs, R. and Macfarlane, R. G.,** *Human Blood Coagulation and Its Disorders,* 3rd ed., Blackwell Scientific, Oxford, 1962, 344.
65. **Michael, P. and Abbot, W. D.,** Use of human fibrinogen in reconstructive surgery, *JAMA,* 123, 279, 1943.
66. **Sheehan, J. E.,** Plasma fixation of skin grafts, *Am. J. Surg.,* 65, 74, 1944.
67. **Harris, H. I.,** Heterogenous skin grafts by coagulum contact method, *Am. J. Surg.,* 65, 315, 1944.
68. **Tidrick, R. T. and Warner, E. D.,** Fibrin fixation of skin transplants, *Surgery,* 15, 90, 1944.
69. **Cronkite, E. P., Lozner, E. L., and Deaver, J. M.,** Use of thrombin and fibrinogen in skin grafting; preliminary report, *JAMA,* 124, 976, 1944.
70. **Clark, A. M., Milne, G. R., and Todd, J. P.,** Fixation of skin grafts with human plasma and thrombin, *Lancet,* 1, 498, 1945.
71. **McEvitt, W. G.,** Experiences with fibrin fixation methods of skin grafting; clinical evaluation, *J. Mich. Med. Soc.,* 44, 1347, 1945.
72. **Braun, F., Holle, J., Kovac, W., Lindner, A., and Spängler, H. P.,** Untersuchunden über die Replantation autologer Vollhaut mit Hilfe von hochkonzentriertem Fibrinogen und Blutgerinnungsfaktor XIII, *Wien Med. Wochenschr.,* 14, 213, 1975.
73. **Staindl, O.,** Tissue adhesion with highly concentrated human fibrinogen in otolaryngology, *Ann. Otol. Rhinol. Laryngol.,* 88, 413, 1979.
74. **Dees, J. E. and Fox, H.,** Properties of human fibrinogen coagulum — preliminary report, *J. Urol.,* 49, 503, 1943.
75. **Dees, J. E.,** Use of intrapelvic coagulum in peylolithotomy; preliminary report, *South Med. J.,* 36, 167, 1943.
76. **Dees, J. E.,** Fibrinogen coagulum as aid in operative removal of renal calculi, *J. Clin. Invest.,* 23, 576, 1944.
77. **Dees, J. E.,** Use of fibrinogen coagulum in peylolithotomy, *J. Urol.,* 56, 271, 1946.
78. **Harrison, J. H. and Trichel, B. E.,** Experiences with fibrin coagulum in pyelolithotomy, *J. Urol.,* 62, 1, 1949.
79. **Hoffman, H. A.,** Coagulum pyelolithotomy, *Am. J. Surg.,* 79, 598, 1950.
80. **Stoll, H. G.,** Koagulum-Pyelolithotomie, *Z. Urol. Nephrol.,* 52, 610, 1959.
81. **Patel, V. J.,** Coagulum-Pyelolithotomie, *Urologe,* 13, 168, 1974.
82. **Smith, P. J., Roberts, J. B., and Dunn, M.,** Coagulum pyelolithotomy, *Br. J. Urol.,* 50, 570, 1978.
83. **Moore, T. D. and Sweetser, T. H., Jr.,** Coagulum pelviolithotomy; an improved technique, *J. Urol.,* 67, 579, 1952.
84. **Parry, T. G. W. and Laszlo, G. C.,** Thrombin technique in ophthalmic surgery, *Br. J. Ophthalmol.,* 30, 176, 1946.
85. **Edmund, J. and Gregersen, E.,** Experimental detachment of the choroid: a new possibility of treatment in retinal detachment, *Acta Ophthalmol.,* 42, 269, 1964.
86. **von Grósz, I., Orbán, T. and Réz, F.,** Biologische Wundsicherung der Starausziehung, *Klin. Montasbl. Augenheilkd.,* 147, 818, 1965.
87. **Freyler, H. and Klemen, U.,** Fibrinklebung in der Hornhautchirurgie, *Albrecht von Graefes Arch. Klin. Exp. Ophthalmol.,* 207, 27, 1978.
88. **Kekin, E. S.,** Ekstraplevral'naia plombirovka fibrinogenom kak metod profilaktiki plevralnykh oslozhnenii posle rezekstsii legkikh, *Grudn. Khir.,* 4, 47, 1979.
89. **Kletter, G., Matras, H., and Dinges, H. P.,** Zur partiellen Klebung von Mikrogefässanastomosen im intrakraniellen Bereich, *Wien Klin. Wochenschr.,* 90, 415, 1978.
90. **Matras, H., Jesch, W., Kletter, G., and Dinges, H. P.,** Spinale Duraklebung mit Fibrinkleber, *Wien Klin. Wochenschr.,* 90, 419, 1978.
91. **Matras, H.,** Fibrin clot sealants in maxillofacial surgery, in *Proc. 1st World Biomaterials Congr.,* Baden near Vienna, European Society for Biomaterials, Vienna, 1980, 4.4.1.

92. **Dinges, H. P., Matras, H., and Kletter, G.,** Watertight closure of spinal dura using a new fibrin clot sealing technique, in *Proc. 1st World Biomaterials Congr.*, Baden near Vienna, European Society for Biomaterials, Vienna, 1980, 4.4.4.
93. **Rupp, G. and Stemberger, A.,** Fibrinklebung in der Orthopädie, *Med. Welt*, 29, 766, 1978.
94. **Braun, F., Henning, K., Holle, J., Kovac, W., Rauchenwald, K., Spängler, H. P., and Urlesberger, H.,** Erfahrungen mit einem biologischen Klebesystem (Fibrin) bei Versorgung von Nierenparenchymwunden, *Zentralbl. Chir.*, 102, 1235, 1977.
95. **Gastpar, H.,** Tonsillektomie bei blutungsgefändeten Patienten, *Fortschr. Med.*, 19, 1277, 1977.
96. **Henning, K., Rauchenwald, K., and Urlesberger, H.,** Anwendung eines Fibrinklebers in der Nierentraumatologie, *Helv. Chir. Acta*, 44, 329, 1977.
97. **Nowotny, Ch. and Wutka, P.,** Anwendung von Fibrinkleber zur Blutstillung nach Zahnextraktionen bei Patienten mit angeborenen und erworbenen Gerinnungsstörungen (Vortrag 1.5.14), *Blut*, 38, 53, 1979.
98. **Spängler, H. P.,** Gewebklebung und lokale Blutstillung mit Fibrinogen, Thrombin und Blutgerinnungsfactor XIII, *Wien Klin. Wochenschr.*, 88, (Suppl. 49), 4, 1976.
99. **Spängler, H. P., Braun, F., Holle, J., Moritz, E., and Wolner, E.,** Die lokale Anwendung von Fibrinogen und Kollagen zur Blutstillung in der Herzchirurgie, *Wien Med. Wochenschr.*, 7, 86, 1976.
100. **Tilsner, V., Kalmar, P., and Pokar, H.,** Anwendung eines Fibrinklebers zur Verhinderung von Blutungskomplikationen nach Operation thoraler Aneurysmen, (Vortrag 1.5.9), *Blut*, 38, 53, 1979.
101. **Gerendás, M. Bagdy, D.,** Eljárás Fibrinszivacs és Nagy Tisztaságu Kristályos Firinogén Előállitásfa, Hungarian Patent 141, 438, 1952.
102. **Gerendás, M.,** Improvements in or Relating to Prostheses, U.S. Patent 3, 523, 807, 1970.
103. **Bagdy, D., Gerendás, M., Winter, L., and Benedek, T.,** Application of bovine fibrin foam and of a mixture of thrombin and fibrin powders as hemostatic agents, *Acta Physiol. Acad. Sci. Hung.*, 2, 493, 1951.
104. **Winter, L., Benedek, T., and Bagdy, D.,** Experimentelle und klinische Anwendung der aus Rinderplasma hergestellten Fibrinprodukte. III. Klinische Anwendung von hämostatischen Fibrinprodukten, *Zentralbl. Chir.*, 78, 469, 1953.
105. **István, L., Feszler, Gy., Sztudinka, Gy., and Gerendás, M.,** Gastrointestinalis vérzések csillapitása thrombin-fibrin készitménnyel, *Orv. Hetil.*, 105, 219, 1964.
106. **Bagdy, D., Gerendás, M., and Winter, L.,** *Thrombin-Fibrinprodukte und ihre therapeutische Anwendung*, G. Fisher, Jena, 1963, 152.
107. **Winter, L.,** Management of chronic osteomyelitis with a coagulum of autogenous blood and penicillin and thrombin, *J. Int. Chir. (Brussels)*, 11, 510, 1951.
108. **Winter, L. and Papp, S.,** *Die Osteomyelitis und ihre Behandlung*, G. Fischer, Jena, 1964.
109. **Kovács, B. and Kerényi, G.,** Bioplast fibrin coagulum in large cystic defects of the jaw, *Int. J. Oral Surg.*, 5, 111, 1976.
110. **Kovács, B., Tóth, K., and Kerényi, G.,** Post-extraction hemostasis during coumarin anticoagulant therapy with a locally applied coagulation-active substance, *Int. J. Oral Surg.*, 5, 3, 1976.
111. **Siegle, M., Blümel, G., Stemberger, A., Fritsche, H. M., and Haas, S.,** Problemlose Zehnextraktion bei Marcumarpatienten durch Fibrinklebung, *Zahnaerztl. Mitt.*, 68, 993, 1978.
112. **Kirk, R. E. and Othmer, D. F.,** *Encyclopedia of Chemical Technology*, Vol. 2, Wiley-Interscience, New York, 1948, 572.
113. **Ingraham, F. D. and Bailey, O. T.,** Use of products prepared from human fibrinogen and human thrombin in neurosurgery; fibrin foams as hemostatic agents; fibrin films in repair of dural defects and in prevention of meningercerebral adhesions, *J. Neurosurg.*, 1, 23, 1944.
114. **Ingraham, F. D., Bailey, O. T., and Nulsen, F. E.,** Studies on fibrin foam as hemostatic agent in neurosurgery, with special reference to its comparison with muscle, *J. Neurosurg.*, 1, 171, 1944.
115. **Woodhall, B.,** Fibrin foam as hemostatic agent in rehabilitation neurosurgery, *JAMA*, 126, 469, 1944.
116. **Weiner, L. and Wald, A. H.,** Fibrin foam and thrombin as used in surgical removal of large fibromyxoma of mandible, *J. Am. Dent. Assoc.*, 33, 731, 1946.
117. **Bagdy, D., Gerendás, M., Winter, L., and Benedek, T.,** Experimentelle und klinische Anwendung der aus Rinderplasma hergestellten Fibrinprodukte, *Zentralbl. Chir.*, 77, 848, 1952.
118. **Winter, L., Benedek, T., and Bagdy, D.,** Marhaplasmából előállitott fibrintermékek felhasználása, *Orv. Hetil.*, 93, 50, 1952.
119. **Spangler, A. S.,** Treatment of depressed scars with fibrin foam — seventeen years of experience, *J. Dermatol. Surg. Oncol.*, 1, 65, 1975.
120. **Jung, H. and Kley, W.,** Fibrinschaum als Defektfüller in der Speicheldrüsenchirurgie, *Arch. Klin. Exp. Ohren Nasen Kehlkopfheilkd.*, 204, 105, 1973.
121. **Ferry, J. D. and Morrison, P. R.,** Fibrin clots, fibrin films and fibrinogen plastics, *J. Clin. Invest.*, 23, 566, 1944.
122. **Bailey, O. T. and Ingraham, F. D.,** Fibrin films in neurosurgery, with special reference to their use in repair of dural defects and in prevention of meningocerebral adhesions, *J. Clin. Invest.*, 23, 597, 1944.

123. **Bailey, O. T., Ingraham, F. D., Neuhauser, E. B. D., and Cobb, C. A. Jr.**, Fibrin film in neurosurgery, further studies; insertion of fibrin film between sutured dura and intact leptomeninges; effect of roentgen therapy on tissue reactions to fibrin film, *J. Neurosurg.*, 4, 465, 1947.

124. **Filatov, A. N.**, New therapeutic preparations of Leningrad Institute of Blood Transfusion as applied in surgical practice, *Vestn. Khir.*, 70, 34, 1950.

125. **Singer, M.**, Combined use of fibrin film and clot in end-to-end union of nerves; experimental study, *J. Neurosurg.*, 2, 102, 1945.

126. **Seddon, H. J. and Medawar, P. B.**, Fibrin structure of human nerves, *Lancet*, 2, 87, 1942.

127. **Tarlov, I. M. and Benjamin, B.**, Plasma clot and silk suture of nerves; experimental study of comparative tissue reaction, *Surg. Gynecol. Obstet.*, 76, 366, 1943.

128. **Hawn, C. V., Bering, E. A., Jr., Bailey, O. T., and Armstrong, S. H., Jr.**, Note on use of fibrinogen and thrombin in surface treatment of burns, *J. Clin. Invest.*, 23, 580, 1944.

129. **Bagdy, D., Gerendás, M., and Winter, L.**, *Thrombin-Fibrinprodukte und Ihre Therapeutische Anwendung*, G. Fischer, Jena, 1963, 184.

130. **Kerényi, G.**, Properties and applications of Bioplast, an absorbable surgical implant material from fibrin, *Biomaterials*, 1, 30, 1980.

131. **Gerendás, M. and Hedvig, Z.**, Bioplast térhálósitó kezelésének ellenörzése a duzzadóképesség alaphján, *Proc. 10th Hungarian Meet. Biochem., Szeged*, Hungarian Chemical Society, Budapest, 1970, 81.

132. **Kerényi, G. and Gerendás, M.**, The effect of $^{60}$Co gamma irradiation on Bioplast, *Proc. 15th Hungarian Meet. Biochem., Miskolc*, Hungarian Chemical Society, Budapest, 1975, 7.

133. **Zinner, N., Gerendás, M., and Biró, T.**, A new method of arthroplasty, *Acta Med. Acad. Sci. Hung.*, 7, 217, 1955.

134. **Zinner, N. and Biró, T.**, A new method of arthroplasty, *Rheumatism*, 15, 69, 1959.

135. **Kovács, P. and Gerendás, M.**, Arthroplasty using of Bioplast in tuberculous coxitis, *Acta Orthop. Scand.*, 30, 294, 1961.

136. **Frankl, Z., Gerendás, M., Guoth, J., and Prohászka, I.**, Recherches expérimentales sur la régéneration de l'articulation temporo-mandibulaire par arthroplastie avec et sans interposition de fibrine. Applications chirurgicales chez l'homme, *Bull. Group Int. Rech. Sci. Stomatol. Odontol.*, 20, 283, 1977.

137. **Grósz, E., Orbán, T., and Gerendás, M.**, L'implant de bioplaste — Nouvelle méthode opératoire du décollement de la rétine, *Presse Med.*, 75, 2714, 1967.

138. **Grósz, I., Vereb, K., and Kerényi, G.**, Scleral buckling with Bioplast fibrin in retinal detachment, *Acta Ophthalmol.*, 54, 408, 1976.

139. **Gyeney, L. and Kerényi, G.**, Bioplast fibrin implants in nasoseptal perforation, *Arch. Otorhinolaryngol.*, 218, 143, 1977.

140. **Hampel, F.**, Antroalveoláris perforatio zárása Bioplast felhasználásával, *Fogorv. Sz.*, 71, 10, 1978.

141. **Tapasztó, I. and Kerényi, G.**, Möglichkeiten und Vorteile der Anwendung von Bioplast bei ophthalmologischen Operationen, in *Proc. 75th Zusammenkunft Dtsch. Ophthalmol. Gesellschaft, Heidelberg*, 1977, J. F. Bergmann Verlag, Munich, West Germany, 1979, 301.

142. **Czigány, A.**, An introductory report on the use of Bioplast in opthalmic surgery, *Br. J. Ophthalmol.*, 61, 660, 1977.

143. **Tapasztó, I. and Kerényi, G.**, Bioplast fibrin film for conjunctival replacement, *Acta Ophthalmol.*, 57, 725, 1979.

144. **Tapasztó, I. and Kerényi, G.**, Skin replacement with Bioplast fibrin in ophthalmology, *J. Biomed. Mater. Res.*, 11, 799, 1977.

145. **Horn, B., Kover, J., and Marton, I.**, Treatment of stress incontinence by a fibrin Bioplast, *Br. J. Obstet. Gynaecol.*, 82, 61, 1975.

146. **Hammond, G.**, Incontinentia urinae gyógyitása Bioplast segitségével, *Magy. Noorv. Lapja*, 40, 37, 1977.

147. **Drobny, S., Dóczy, Á., and Gerendás, M.**, Eine neue Methode zue Leberresektion, *Chirurg*, 33, 551, 1962.

148. **Drobny, S.**, Nouveau procédé pour la résection du foie, *Rev. Hepatol.*, 15, 259, 1965.

149. **Wood, C. B., Capperauld, I., and Blumgart, L. H.**, Bioplast fibrin buttons for liver biopsy and partial heptatic resection, *Ann. R. Coll. Surg. Engl.*, 58, 401, 1976.

150. **Balasegaram, M. and Capperauld, I.**, Biethium buffers in the control of liver haemorrhage, *Malays. J. Surg.*, 3, 139, 1977.

151. **B Kovács, A., Somogyvári, K., and Gerendás, M.**, Das Verschiessen von Bauchwanddefekten mit Hilfe von Bioplasten, *Wiss, Z. Karl-Marx-Univ. Leipzig*, 16, 397, 1967.

152. **Capperauld, I.**, Compression moulded ox fibrin as a potentially valuable biomaterial, in *Proc. 1976 Meet. Eur. Soc. Artif. Organs*, Vol. 3, European Society of Artificial Organs, London, 1977, 151.

Chapter 5

# POLYMERS OF NATURAL ORIGIN AS BIOMATERIALS. 2. COLLAGEN AND GELATIN

**Michael Parkany**

## TABLE OF CONTENTS

I. Introduction .......... 112

II. Collagen .......... 112
- A. The Structure of Collagen .......... 112
- B. Definitions .......... 113
- C. Purification of Collagen .......... 113
- D. Preparation of Collagen .......... 113
- E. Preparation of Collagen Articles .......... 114
- F. Animal Experiments and Clinical Applications .......... 114

III. Gelatin .......... 114
- A. The Structure of Gelatin .......... 114
- B. Manufacture of Gelatin .......... 115
  - 1. Acid Processing .......... 115
  - 2. Alkaline Processing .......... 115
  - 3. Steam Extraction .......... 115
- C. Preparation of Gelatin Articles .......... 115
- D. Animal Experiments and Clinical Applications .......... 116

References .......... 116

# I. INTRODUCTION

Collagen and gelatin are discussed here together because of the close relationship between them. Gelatin is prepared from collagen on a large commercial scale to cover the needs of the photographic, food, and pharmaceutical industries. The increasing demand on the market motivated and stimulated intensive research work on collagen and gelatin. Electron-microscopy, X-ray diffraction, the ultracentrifuge, and achievements in the field of polymer chemistry largely contributed to a better understanding of the structure of collagen, the chemistry of gelatin forming, and the tanning processes. Reference is made here to the excellent reviews of Yannas[1] and Chien.[2] The monographs of Gustavson[3] and Veis[4] correlate the vast amount of information concerning collagen and gelatin. These monographs are not only referred to, but also used as starting points for all research work in this field.

In the practice of surgery, collagen sutures are widely used to sew together tissues, nerves, or blood vessels, because collagen is the only completely and readily absorbable suture material. Collagen sutures are absorbed or digested during the healing process by the tissues in which they are embedded. Their function is to serve either as a suture or a ligature or an implant, until such time as body repair has progressed to the point when they are no longer needed. Collagen and gelatin sponges are also used in surgery and may be left in the body cavity, as these sponges are slowly absorbed by surrounding tissues.

# II. COLLAGEN

Collagen, a proteinaceous substance, is present in the mammalian connective tissue. Tendons, hide, and bones of bovines and porcines are its most important industrial sources. The mammalian skin collagens display close homology from species to species; moreover, X-ray diagrams obtained from collagens isolated from a wide variety of animal species show a remarkable similarity.

Collagen is unique among the proteins because of its amino acid composition. It is the only mammalian protein containing large amounts of hydroxyproline and it is surprisingly rich in glycine and proline. Its hydroxylysine content is specifically characteristic of collagen. The high content of the amino (pyrrolidine ring) acids which have no hydrogen atom on the peptide bonds involving these residues explains the fact that intramolecularly hydrogen-bond stabilized structures, such as the α-helix, cannot be formed.

The sulfur content of collagen is very low. Crysteine is absent and methionine is the only sulfur-containing amino acid, so that disulfide bridges are not available to create the tertiary structure of the collagen system.

## A. The Structure of Collagen

The elementary basic molecular unit of collagen is tropocollagen. The tropocollagen macromolecules consist of three polypeptide chains coiled together in a long helix, about 10 to 15 Å in diameter and about 3000 Å long ($1 \text{ Å} = 10^{-10} \text{ m} = 0.1 \text{ nm}$).

Like its degradation product, gelatin, tropocollagen is completely soluble in acidified water. Since it is the natural building block in the production of collagen fibrils, it has been extensively used for producing reconstituted fibrous collagen products. The next higher organized state in which collagen has been known is the collagen fibril, which consists of long, thin strands comprising thousands of individual tropocollagen units. In this form, collagen is water insoluble. These fibrils associate to form the microscopic fibers present in natural substances, which comprise many thousands of fibrils bonded together. Collagen occupies a key position in the molecular architecture of higher animals, which enables it to be the main agent controlling the distribution of both external and internal forces.

### B. Definitions

1. Native collagen fibril — a thread-like collagen structure such as it exists in connective tissue
2. Tendon collagen fibril — a thread-like collagen structure derived from tendon (mainly beef or calf) that is round in cross-section. These fibrils, in the completely dehydrated state, measure about 500 to 2000 Å in diameter
3. Swollen tendon collagen fibril — a collagen fibril derived from tendon (mainly beef or calf) that has been swollen in an acidic solution; the diameters of swollen collagen fibrils range from less than 5,000 to about 90,000 Å
4. Monofilament — a single thread of oriented collagen fibrils as extruded through a single orifice in a spinnerette
5. Multifilament — a group of individual separate filaments extruded through a spinnerette
6. Strand — a group of filaments that have been united to form a unitary structure; the sutures for surgery are prepared from the above strands

### C. Purification of Collagen

Collagen in the native state is associated with impurities and must be separated from them. The success of a process employing collagen as a basic material often depends upon maintaining the collagen fibril structure throughout the process. Denaturation or degradation of the collagen structure should be avoided as much as possible, in order to keep its mechanical properties unchanged. In the patent literature, there are numerous methods for purifying collagen.

Zachariades[5] was the first to observe that some collagen dissolves in cold dilute acetic acid solutions from acid-swollen collagen fibers. The removal of some soluble components should be the first step to reduce antigenicity. The remaining antigenicity is believed to be due to the presence of structural glycoproteins linked to insoluble fibrous stromas of connective tissue. As such, structural glycoproteins are insoluble in 0.15 mol/ℓ trichloracetic acid at 90°C; the trichloracetic acid insoluble residue is a sort of measure of the antigenicity associated with collagen. For further purification, treatment with elastase and with borohydride is claimed to be successful.[6]

### D. Preparation of Collagen

In all industrial processes for the preparation of collagen dispersions, attempts are made to minimize the degradation of collagen molecules. The dispersion is used for the preparation of collagen articles. In the case of collagen sutures, prepared from collagen dispersions by extruding them into a precipitating agent, it is especially important to preserve the tensile strength of the monofilaments.

Sodium, potassium, and tetramethyl ammonium are claimed[7] to be suitable for the treatment of the finely divided connective tissue before the collagen is swollen in an aqueous acid medium. Other processes[8,9] use proteolytic enzymes active in the acidic range, combining the enzyme treatment and the acid extraction in one technological step. It is claimed that a collagen solution with a yield of 100% can be obtained and that the collagen can be reconstituted in fiber form.

Another invention[10] claims that a new microcrystalline or particulate colloidal form of collagen may be prepared in a physical state intermediate between that of swollen collagen fibrils and tropocollagen units. It is water insoluble, particulate and colloidal, and substantially free of molecular tropocollagen and water-soluble degradation products. The maximum dimension of these colloidal collagen particles in any one direction is less than 10,000 Å. Such particles consists of bundles of aggregated tropocollagen units and vary in length from 50 Å to under 10,000 Å. This form of collagen constitutes an aqueous nonelastic type gel in sharp contrast to the aqueous elastic or emulsoid type gels formed by tropocollagen and degraded products of collagen, such as gelatin, which thicken or exhibit substantial increases in viscosity on standing to yield rubbery masses. An improvement of this method[11] makes use of water-miscible organic solvents (dioxane, dimethylsulfoxide, isopropanol) and the concentration of collagen can be raised to about 25 to 35%. This enables the obtainment of stable dispersions of microcrystalline collagen in minimum 10% concentrations for the preparation of collagen filaments and films. When mixing a 0.5% aqueous dispersion of microcrystalline collagen with calcium phosphate and calcium fluoride, cartilage-like and bone-like structures were reportedly[12] obtained.

### E. Preparation of Collagen Articles

In surgery, sponges are used to control bleeding through the application of pressure and coagulating material such as thrombin. The preparation and use of fibrin foam has been reported earlier. For the same purpose, sponges are also made of collagen[13] and gelatin. Through adjustment of the technologies, sponges can be prepared for the same use, with similar characteristics, out of fibrinogen, collagen, and gelatin. Among these, gelatin is the cheapest, and human fibrinogen the most expensive.

These sponges have a very important advantage, as they can remain in the body cavity since they are nonantigenic and absorbable. In the sponges, the cavities are partly communicating, partly closed. It is claimed[14] that for wound healing, felt or fleece-like products give optimum results. These structures have open, communicating voids between the fibers. The fleece-like collagen products have been reported to have high hemostatic activity and high absorption capacity for body fluids, thus promoting the regeneration of tissues, especially bone tissues. According to this method, the purified collagen is precipitated and dissolved repeatedly; the last solution, which has 0.5 to 2% collagen is poured into circular glass dishes, then freeze-dryed, packed into polyethylene bags, and sterilized by irradiation with a dose of 2.5 Mrad of $\gamma$-rays. Absorbable sutures, surgical aids in the form of films, filaments, strands, and tubing can also be obtained from the dispersion of collagen fibrils.

### F. Animal Experiments and Clinical Applications

Collagen fleece has been reportedly[15] used in an experiment with rabbits for healing surgically created mandibular defects. Bone regeneration was quicker, improved, or complete, and collagen fleece was used later with success, clinically, in maxillo-facial surgery, to fill excised cystic bone cavities and small osseous fissures. Collagen fleece was found to provoke bone regeneration in experiments with rats and rabbits[16] and also in the clinical therapy[17] for filling bone cysts and osteomyelitic bone cavities, etc.

## III. GELATIN

### A. The Structure of Gelatin

Gelatin (gelatine) is prepared from collagen by boiling animal tissues with water, straining the liquid, and evaporating to dryness. Commercially, it is manufactured from bone and from hide.

Gelatin is the principal constituent of glue. Gelatin swells in cold water, but is insoluble in it. It dissolves in hot water to give a very viscous solution, and a solution containing 1% or more solidifies to a jelly on cooling. For the preparation of surgical prostheses, as described later, solutions containing 10 to 35% gelatin are used. Purified, medical-grade gelatin is much cheaper than collagen and can replace it in cases where the required mechanical properties of the articles made therefrom are moderate. Contrary to the processes described earlier in connection with the purification and preparation of native collagen, the preparation of gelatin destroys the secondary structure of collagen, hence the radical change in the important solubility and mechanical properties.

## B. Manufacture of Gelatin

The commercial manufacture of gelatin aims at obtaining maximum yields of soluble, relatively pure gelatins, with a predictable range of commercially desirable properties such as gel strength, viscosity, color, etc. All the processes consist of three steps: removal of noncollagenous components from the stock material, conversion of the purified collagen to gelatin, and recovery of the gelatin in dried form. The process must be arranged so as to avoid loss of yield due to hydrolysis and solubilization during purification.

### *1. Acid Processing*

Acid processing is usually applied to pig skins and ossein. The most important commercial acid process is the preparation of edible gelatin and of some of the medical grade gelatins from frozen pig skins. The skins are soaked in 3 to 5% inorganic acids (hydrochloric, phosphoric, sulfuric, etc.) for 10 to 30 hr. The excess acid is washed out. The pH of the swollen stock is about 4. The noncollagenous proteins of the skin (which are impurities from the point of view of gelatin manufacturing) have isoelectric points in this range (pH 4 to 5) and are most readily coagulated and separated during extraction, thus removing all the antigenic components.

### *2. Alkaline Processing*

Alkaline processing of collagen is the most widely used commercial process for the production of gelatin and glues. Any collagenous stock can be converted to gelatin by this means. The raw collagen stock is washed and thoroughly hydrated in cold water in large tanks. The excess water is drained from the stock and lime is added to the stock in sufficient amount so that when fresh water is added, a saturated solution of calcium oxide is formed. A sufficient excess of calcium oxide must be maintained to make up for the base consumed in the conditioning reactions. The hides are left in the tank for about 3 to 10 months, depending on the nature of the stock, the ambient temperature, the alkalinity of the lime liquors, etc. The preparation, e.g., photographic gelatins from calf skins, usually requires 8 to 10 months. When the liming is complete, the lime is washed out and the residual base is neutralized by washing with dilute hydrochloric acid and finally with a dilute solution of aluminium or zinc sulfate. The collagen stock is then extracted in a series of cooks at successively higher temperatures. The highest quality gelatins are obtained in the first few extractions.

### *3. Steam Extraction*

High pressure steam extraction is a commercial procedure for the production of gelatin and glue from bones. This can be carried out rapidly without prior pretreatment of the bones. The hydrolysis of the insolubilizing linkages is carried out at high temperature. Conversion of the collagen into a soluble form apparently takes place while the collagen is surrounded by its sheath of hydroxyapatite crystals. The gelatin is extracted at a somewhat lower temperature in order to avoid excessive degradation.

## C. Preparation of Gelatin Articles

Gelatin is easily soluble in hot water and therefore certain special technologies can be used which are not — or not easily — applicable in the case of fibrin or native collagen. The gelatin solution is prepared from gelatin powder (14 to 20 mesh) or from fine gelatin sheets. Usually they are soaked in cold distilled sterile water, left to swell for a few hours, then stirred in a water bath at 60 to ~ 80°C until a homogenous solution is obtained. The concentration of gelatin is usually of 5 to 20%; in special cases, in order to obtain more concentrated solutions (25 to 35%), a water bath of 90 to 100°C is required.

Depending on the desired mechanical properties of the article, glycerol or sorbitol can be added as plasticizers. Cross-linking agents are added to prolong the absorption time of the article. The solution solidifies, and after drying at ambient temperature (1 week or more depending on the thickness of the piece), the block can be cut and machined on a lathe. Bone-fixing screws have been obtained by this method.[18]

A much simpler technology is to dip the desired forms in the gelatin solution. A layer of gelatin solidifies on the surface of the form. The thickness of the gelatine layer can be adjusted (by the gelatin concentration, by the temperature of the solution, by repeating the dipping, etc.). After drying at ambient temperature, and hardening of the surface if desired, the forms are pulled out and the articles are cut to size.[19,20] Gelatine cups and tubes have been prepared by this method.

### D. Animal Experiments and Clinical Applications

Gelatin sponges are well-known in clinical practice under different trade names. Sutures prepared from gelatin have not, so far, the desired wet strength. Gelatin bone screws cannot bear the weight of the body and cannot replace metallic screws. The requirements of materials used for healing wounds or burns and for filling cavities, etc. are less problematic to fulfil.

In veterinary surgery, cross-linked gelatin cups are used in the surgical treatment of dysplasia in dogs.[19] Experiments on dogs using absorbable cross-linked gelatin tubes for the replacement of a resected urether were successful.[20] Gelatin has not shown antigenicity and, after fulfilling the task of temporary replacement, it absorbs completely.

## REFERENCES


1. **Yannas, I. V.,** Collagen and gelatin in the solid state, *J. Macromol. Sci. Rev. Macromol. Chem.*, C7, 49, 1972.
2. **Chien, J. C. W.,** Solid-state characterization of the structure and property of collagen, *J. Macromol. Sci. Rev. Macromol. Chem.*, C12(1), 1. 1975.
3. **Gustavson, K. H.,** *The Chemistry and Reactivity of Collagen,* Academic Press, New York, 1956.
4. **Veis, A.,** *The Macromolecular Chemistry of Gelatin,* Academic Press, New York, 1964.
5. **Zachariades, P. A.,** *C. R. Soc. Biol.*, 52, 182, 1900.
6. **Ethicon Inc.,** Method for Purifying Na[illegible]agen, British Patent 1,204,048, 1970.
7. **Ethicon Inc.,** Method for Swelling and [illegible]g Collagen, British Patent 1,198,310, 1970.
8. **Nikon Hikaku Kabushiki Kaisha,** Meth[illegible]olubilization of So-called Insoluble Collagen Fibers, British Patent 1,062,083, 1967.
9. **Hikon Hikaku Kabushiki Kaisha,** Method for Solubilization of So-called Insoluble Collagen by Proteolytic Enzyme, British Patent 1,119,342, 1968.
10. FMC Corporation, Colloidal Collagen Compositions and Methods of Preparation Thereof, British Patent 1,143,533, 1969.
11. FMC Corporation, Colloidal Collagen Compositions and a Method of Manufacturing Them, British Patent 1,156,361, 1969.
12. FMC Corporation, Improvements in or Relating to Strucutres, for Example Prosthetic Devices, British Patent 1,224,925, 1971.
13. Ethicon Inc., Article and Preparation Thereof, British Patent 942,226, 1963.
14. **Pentapharm, A. G.,** Sterile Surgical Collagen Product, U.S. Patent 4,066,083, 1978.
15. **Joos, U. and Ries, R. E.,** Collagen Fleece as a Biomaterial for Mandibular Defects, BES 71st Scientific Meeting, 3rd Conf. on Materials for Use in Medicine and Biology, Mechanical Properties of Biomaterials, Keele, 1978.
16. **Springorum, H. W. and Puhl, W.,** Bone Regeneration after Implantation of Collagen Fleece. (A) In Animal Experiments, BES 71st Scientific Meeting, 3rd Conf. on Materials for Use in Medicine and Biology, Mechanical Properties of Biomaterials, Keele, 1978.

17. **Puhl, W. and Spingorum, H. W.,** Bone Regeneration after Implantation of Collagen Fleece. (B) In Clinical Therapy, BES 71st Scientific Meeting, 3rd Conf. on Materials for Use in Medicine and Biology, Mechanical Properties of Biomaterials, Keele, 1978.
18. **Parkany, M. and Horváth, N.,** Absorbable surgical implants made of gelatinous-Bioplasts (Russian), *Ortop. Travmatol. Protez.*, 10, 73, 1972.
19. **Sági, L., Somogyvári, K., and Parkany, M.,** Rehabilitation of work dogs diseased in hip dysplasia by surgical application of a gelatin interpositum, *Acta Vet. Acad. Sci. Hung.*, 26(1), 33, 1976.
20. **Breier, M., Parkany, M., and Somogyvári, K.,** Experimental Investigations by Cross-linked Gelatin Tubes in Urological Surgery, BES 71st Scientific Meeting, 3rd Conf. on Materials for Use in Medicine and Biology, Mechanical Properties of Biomaterials, Keele, 1978.

Chapter 6

# BIORESORBABLE PLASTIC MATERIALS FOR BONE SURGERY

**Michel Vert, Pascal Christel, François Chabot, and Jean Leray***

## TABLE OF CONTENTS

I. Introduction .......... 120

II. Early History of Organic Bioresorbable Internal Fixation Devices .......... 120

III. Introducing Poly (α-Hydroxy-Acids) .......... 124

IV. Biocompatibility and Degradation .......... 127
- A. Resorption in Rats (Figure 1) .......... 127
- B. Resorption in Sheep (Figure 2) .......... 129
- C. Quantitative Evaluation of Degradation in Sheep .......... 133

V. Biofunctional Investigations .......... 134
- A. In Sheep .......... 134
- B. In Patients .......... 137

VI. Conclusions .......... 140

Acknowledgments .......... 141

References .......... 141

* Jean Leray is deceased.

## I. INTRODUCTION

The everyday reality of the bone surgeon is the repair of skeletons damaged either by disease or trauma. Methods, devices, and instruments for the internal fixation of bone fractures and for the replacement of injured or diseased joints have been developed as new concepts based on biomechanical principles have been elaborated. This favors the search for new high-quality materials to bring better solutions to specific problems.

In bone surgery, some problems such as prosthetic joint materials and prostheses' anchorage require biostable compounds. Other problems, however, whose solutions are based on the ability of the bone for self-repair, require only temporary aid, and better solutions would be found with the use of bioresorbable compounds, i.e., compounds capable of degradation in the biological milieu followed by resorption and elimination through natural pathways.

This is true for bone fractures whose repairing often needs internal fixation to align and maintain stability of bone fragments during the healing period. To date, osteosynthesis has been achieved using metallic parts. However, rigid metallic fixation prevents the rapid proliferation of periosteal callus and deprives the bone of the normal stress pattern because of the large difference in stiffness between cortical bone (E = 10 to 20 GPa) and metals (E = 100 to 200 GPa). The stiffer the osteosynthesis, the more frequent are delayed unions or nonunions. The long-term results of stress-protection show a mechanically inferior bone structure beneath the plate because of bone atrophy.[1] Refracture is therefore a resultant possibility after the removal of the plate which is often required in the posthealing period because of the pain elicited by local corrosive reactions, soft tissue friction, etc. The use of fixation with intermediate stiffness and properties closer to those of bone has been proposed as a solution to prevent stress-protection.[2,4] Such devices seem to be able to limit bone atrophy and porosis,[5] but do not eliminate the necessity of posthealing removal.

The most attractive solution consists of the use of internal fixation parts mechanically suitable to provide safe aid for the healing period, sufficiently flexible to permit healing by periosteal callus formation, and bioresorbable to ensure progressive transfer of stresses to the bone and obviate a second surgical procedure that is always uncomfortable for the patient and expensive for society.

A rather large number of compounds have been reported as being bioresorbable in the body: ceramics,[6] corals,[7] and organic macromolecular compounds.[8,10] However, ceramics and corals are too stiff and too brittle to fit the characteristics required for primary union.[11]

In contrast, polyesters of the poly-α-hydroxy-acid type were recognized early as serious candidates.[12,14] A great deal of potential work has been done. Nevertheless, it is likely that data reported in the open literature only constitute the visible part of the iceberg.

Since the subject of this book is heavily concerned with synthetic polymers as biomaterials, we should first take a moment to analyze and comment on the literature and on the macromolecular aspects of the compounds people have been working with before presenting our own work in the field. Although bioresorbability and mechanical properties are the basic properties requested of candidates, many others have to be added before being able to label a successful system. This is the main reason why we shall consider in a stepwise manner the various aspects (synthesis of the basic polymers, determination and control of their structural and physical characteristics, determination and control of their biological behavior, and compromise between advantages and shortcomings) in the making of internal fixation parts using poly-α-hydroxy-acids in view of their effective applications in humans.

## II. EARLY HISTORY OF ORGANIC BIORESORBABLE INTERNAL FIXATION DEVICES

The interest for the use of bioresorbable materials for suturing and fracture fixation has been increasing since 1965.

Investigations of the potential of poly (L-lactic acid) and poly(DL-lactic acid) were first reported by Kulkarni et al.[12,13] For powdered forms, degradation rates in vitro under homogeneous and heterogeneous conditions were evaluated by acid-base titrations and by radiolabeling for heterogenous conditions only. The kinetics of deesterification under homogeneous conditions were second order and an activation energy of 11 kcal/mol was calculated. The technique used for determining the biological degradation in vitro allowed the authors to conclude that poly(DL-lactic acid) degrades at a faster rate than poly(L-lactic acid). No specific histological responses were reported for both poly(L-lactic acid) and poly(DL-lactic acid).

In preliminary experiments, Kulkarni et al.[13] prepared suture material and 2-mm-thick films for repair of blow out fractures of the orbital floor of monkeys and created extruded rods of 3.2-mm diameter which were used as pins for reduction of mandibular fractures in dogs. L- and DL-polymer properties were analyzed and compared by differential thermal analysis, X-ray diffraction, specific viscosity, and tensile strength measurement. Degradation was studied in vitro in aqueous medium, in tissue culture with $^{14}C$ tagged PLA, and also in vivo after implantation.

Poly(DL-lactic acid) showed a melting point at 60°C and no detectable glass transition temperature. Poly(L-lactic acid) showed a melting point at 170°C and a second-order transition temperature at 110 to 115°C.

The rods did not exhibit reproducible strength. The authors assigned the discrepancies to the high crystallinity and high sharp melting point of the poly(L-lactic acid). Poly(L-lactic acid) appeared to be less soluble in tissue culture and less degradable than the DL-polymer due to a higher order of crystallinity. When implanted in vivo, DL-polymer was also found to degrade more rapidly than L-polymer. The tissue reaction around implants showed a minimal inflammatory response. After 8 months, the pins used for mandibular fracture reductions had virtually disappeared from the fracture site which healed normally. With $^{14}C$ labeling, PLA implants exhibited a 14% loss after 3 months with conversion into $CO_2$. The resorption rate appeared slow compared to polyglycolic acid.

In 1971, Cutright et al.[15] published results on mandibular fracture reduction in monkeys using transosseous ligatures of poly(lactic acid) suture materials. Animals were sacrificed from 2 to 12 weeks and fracture healing and degradation of poly(lactic acid) were evaluated through histology. The cross-sectional area of the sutures decreased from 8.9 $mm^2$ at 2 weeks down to 0.3 $mm^2$ at 12 weeks. After 12 weeks, early features of bony union appeared and the sutures became infiltrated by cellular connective tissue with fibroblasts, endothelial cells, mononuclear phagocytes, and giant cells. Sutures were progressively replaced by irregular bands of young collagen and vascular connective tissue. The tissue reaction was limited to the immediate perisutural area. The complete degradation of the sutures was not seen at 12 weeks.

In 1972, Getter and co-workers[16] reported on plates and screws used in the reduction of mandibular fractures in dogs. Four-hole plates and nonself-tapping thread screws were implanted in six dogs. The animals were killed between 4 and 32 weeks. During the surgical procedure, plates and screws were fused into a one-piece system with a warm soldering iron. This ensured stability of the fixation, thus avoiding micromovements at the interface between screws and plate. Fracture sites healed with secondary callus at 4 to 6 weeks. Scattered inflammatory cells infiltrated the fibrous tissue surrounding the plate. Degradation of screws was already observed with slight movement of the device. Inflammatory reaction remained low all along the observation period and at 32 to 40 weeks screws and plates were completely degraded. Microscopically, only slight debris from one screw was observed.

The same authors reported in 1972[17] the repair of blow out fractures of the orbital floor in monkey using poly(lactic acid) sheets, 1.5 mm thick, inserted into the floor. The animals were killed between 4 and 38 weeks and the repair was evaluated along with histology. The material appeared to induce a mild inflammatory response with early deposition of collagen at 4 weeks. At 8 weeks, a bony plate was in apposition to the fibrous connective tissue capsule surrounding the PLA sheet. No direct contact was observed throughout the healing period process. At 38 weeks, the sheet exhibited little degradation surrounded by a very thin fibrous capsule.

Soon after, poly(glycolic acid) was patented as a useful material to make solid prostheses such as pins, screws, plates, or thin sheets.[18] As an example, osteosynthesis of rabbit femora was carried out either by medullary rods or bone plates held in place with absorbable pins. At 3 weeks, the bone was partially knit and the PGA implants were essentially intact. By 6 weeks, the break in the bones was healed with the PGA plates showing signs of erosion. By 24 weeks, the plates were almost completely absorbed. According to others, results obtained in humans appear similar to those in animals.

Brady et al.[19] reported on the resorption rate and the route of elimination of poly(lactic acid) pellets implanted into the abdominal walls of rats. The samples weighed between 0.05 and 0.09 g and were labeled with $^{14}C$. Animals were sacrificed from 28 to 168 days and radioactivity of urine, feces, implantation site, kidney, liver, and lungs was measured. A total of 63.2% of the radioactivity remained at the implantation site at 168 days. According to this work, the DL-polymer showed the same resorption characteristics as those reported earlier for the L-polymer. This conclusion was rather contradictory when compared with the slower degradation of the L-polymer reported earlier[13] and confirmed later from short-term organ cultures of chick embryo liver.[20] In any case, the degradation rate of the poly(lactic acid) considered resulted in total resorption at 1 year and 5 months. The highest radioactivity was found in the kidney and most of the decrease in radioactivity was via respiratory $CO_2$ elimination.

Attempts to determine the degradation rate of polymers and copolymers of lactic acid (LA) and glycolic acid (GA) quantitatively were reported by Cutright et al. in 1974 for LA/GA copolymers in various ratios.[21] In order to test the effect of crystal rate formation on biodegradation, slow and fast cooling poly(L-lactic acid) and poly(glycolic acid) were also implanted. Cylinders of 1.6 × 1.75 mm were inserted in the cortex of the upper metaphyseal area of the femur in rats. They were sacrificed every 20 days up to 220 days. The rate of degradation was evaluated by histologically measuring height and width of the pellets. The poly(L-lactic acid) implants were 40% -degraded at 220 days with no difference between slow- and fast-cooled specimens. The LA/GA copolymer degradation rate was related to the GA content: 100% degradation at 180 days for 25% GA and 100% degradation at 120 days for 75% GA. Poly(glycolic acid) was only 18% degraded at 220 days. This result disagreed with the degradation characteristics of poly(glycolic acid) sutures. The authors noted that pellets of similar composition did not disappear simultaneously, except for poly(glycolic acid), and observed direct bone-implant contacts. They concluded that the degree of crystallinity did not affect biodegradation and that the range of resorption which varies according to the LA/GA ratio, encompassed the time duration required for fracture fixation. In this series of papers, Cutright et al. did not mention anything about variation with respect to mechanical properties and molecular weight. Furthermore, the initial physicochemical characteristics of these biodegradable materials have never been discussed. Reports on the use of poly(lactic acid) and poly(glycolic acid) polymers in bone surgery were subsequently discontinued by Cutright's group. One can conclude that the mechanical properties of their polymers were not suitable enough for long bone surgery.

Nevertheless, the group continued in its attempts to determine the differences in degradation rates between poly(L-lactic acid), poly(glycolic acid), and copolymers containing both acids in varying ratios.[22] Half-lives of the polymers and copolymers, as calculated from $^{14}C$ and $^{3}H$ radioactivity in the implant area at various time intervals, decreased from 5 months for poly(glycolic acid) to 1 week for the 50-50 LA/GA copolymer and rapidly increased to

6.1 months for poly(L-lactic acid). In tissue, fast-cured PGA had a half-life of 0.85 months. No difference in the rate of degradation was seen in soft tissue or bone, and no significant radioactivity was detected in urine, feces, or tissue samples. From this study, it was concluded that the control of implant degradation rates could best be attained by varying the LA/GA composition between 75 to 100% LA along with a corresponding 25 to 0% GA, thus providing a half-life range of the implant from 2 weeks to 6 months.

In a series of recent papers, Alexander and co-workers[23,25] have proposed absorbable polymers of filamentous carbon tissue scaffold for tendon and ligament replacements using carbon fibers coated with poly(L-lactic acid). Rows of 6000 fibers (6μm diameter) were coated by spraying with a dilute solution of poly(L-lactic acid) in chloroform. Various implantation sites were used: patellar tendon and medial collateral ligaments in dogs and Achiles tendon replacements in rabbits. Histological and mechanical evaluations were performed up to 12 months. Carbon fibers as well as a neighboring neotissue formation similar in nature to aligned tendon were seen histologically. Mechanical testing demonstrated properties similar to those of the control. The coating, which degraded slowly, protected the carbon fibers from migrating and reduced the intense inflammatory reaction. No data were available on the characteristics and degradation rate of the poly(L-lactic acid) used. The results were convincing enough to allow human implantation of this filamentous carbon composite for tendon, collateral, and cruciate ligament replacements.[26] On the other hand, Parsons with the same previous authors[27,28] has developed an absorbable composite plate of variable stiffness based on the same composite material combination; i.e., poly(L-lactic acid) matrix reinforced by carbon fibers. Two types of laminate plates were both processed by pressure molding above the polymer melting temperature. One was an angle ply laminate with a 15° ply angle relative to the long axis of the plate. The second one was a laminate with short fibers randomly orientated within each lamina. Fatigue tests in bending and in saline serum demonstrated a poor resistance of the angle ply plate. Only the second type of plate with short fibers was implanted in vivo in canine long bones with or without osteotomies. A total of 6 dogs were sacrificed at 8 weeks. One failure was observed. The tissue response was excellent and mechanical testing showed a 70% decrease in bending rigidity of the plate after 8 weeks. No data were available on PLA degradation. After those preliminary results, the adjustment of mechanical properties and degradation rate were investigated relative to different molecular weights, coating, and molding procedure.

Properties of biodegradable plates were also investigated by Lewis[29] using poly(L-lactic acid) combined with carbon or ceramic fibers. In vitro data indicated that satisfactory flexural properties could be maintained for about 6 weeks. However, no in vivo implantation results were reported.

This historic survey has been limited to papers dealing with bone surgery. However, much further work has been done on poly(α-hydroxy-acids) with respect to their potential as bioresorbable polymers for other biomedical uses such as sutures,[30,33] vascular prostheses,[34] sustained-release delivery systems,[35,38] and microencapsulation.[39]

The first conclusion which can be drawn from this historical survey may be that the data reported have been generally oriented towards the determination of biological characteristics (tissue response, degradation rate, resorption route, etc.). Only little attention has been paid to structural analysis and no effort has been given to describe carefully the materials used, though work has been reported in that direction recently.[40,43] This statement is well-supported by Table 1 which shows the discrepancies between data concerning the degradation rate of polymers derived from lactic acids of different origins and generally referred to as PLA. As one can see, the longer half-life reported for the reputedly most biostable poly(L-lactic acid) was approximately 6 months until 1981. In the last few years we have apparently obtained a better control of the physical and biological properties of poly(α-hydroxy-acids). In particular, poly(L-lactic acid) based bone plates exhibiting in vivo biostability for periods as long as 1 year have been reproducibly processed under conditions selected especially to retain the integrity of the macromolecules during the various steps of the processing of ready-for-implantation end products.[43]

We shall next present our results in this field, paying special attention to the discussion of our data with respect to data available in the open literature. However, comments on the nature of poly(α-hydroxy-acids) in focusing on their structural particularities will be presented beforehand.

## III. INTRODUCING POLY(α-HYDROXY-ACIDS)

Poly(α-hydroxy-acids) constitute a particular class of polyesters whose repeating units, $\{O–CO–CHR\}_n$, are derived from α-hydroxy-acids, HO–CHR–COOH. Though higher homologues are known, the more common poly(α-hydroxy-acids) are macromolecular compounds derived from glycolic (R=H) and lactic (R=$CH_3$) acids, the latter being chiral, i.e., having two enantiomeric forms, L and D, with similar intrinsic chemical properties, but opposite configurational structures. The macromolecules one can obtain by combination of glycolic (GA), L- and D-lactic acids (L-LA and D-LA) are presented in Table 2 together with the simplified nomenclature used in the remainder of the text.

Poly(α-hydroxy-acids) were initially synthesized from the corresponding α-hydroxy-acids by the simplest step-growth polymerization method.[44,45] The resulting polymers had low molecular weight and poor mechanical properties. Shortly thereafter, the most efficient method to produce high molecular weight polymers was shown to be the ring-opening polymerization of glycolide and lactide cyclic diesters.[46,48] It is of value to point out that the ring-opening polymerization of cyclic diesters has been shown to proceed through a particular mechanism in the sense that repeating units are basically added by pairs to the growing chains according to the following reaction.[49]

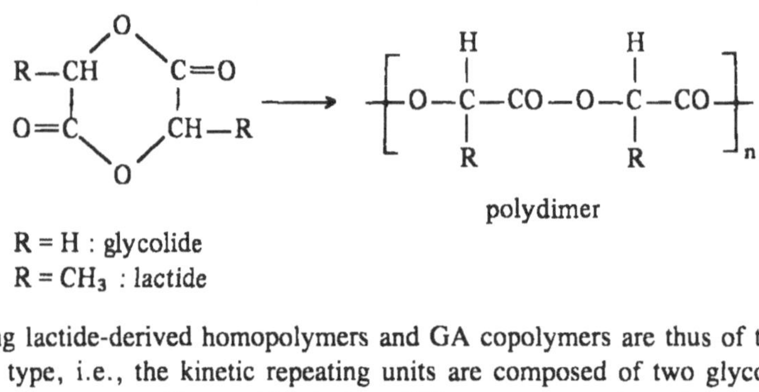


The resulting lactide-derived homopolymers and GA copolymers are thus of the polydiastereoisomer type, i.e., the kinetic repeating units are composed of two glycolic or lactic units, in the latter case with two identical or different asymmetric carbon atoms. It is of value to note that lactide exists as four different compounds depending on the L and D configuration of the lactic moieties forming the molecules (Table 3):[50] the LL and DD antipodes (L- and D-lactides, respectively), the DL isomer (meso-lactide), and the LL + DD solid combination (racemic- or DL-lactide). Since usual initiators (e.g., $SbF_3$, powdered Zn, stannous octoate) are configuration-respecting, lactic acid-containing poly(α-hydroxy-acids) having particular configurational structures can be obtained by polymerizing suitable mixtures of lactides. In the case of lactic acid sterocopolymers, the pair-addition mechanism can basically lead to different configurational structures for a given enantiomeric composition which depends on the composition of the feed in lactides. For example, poly(rac-lactide) shows a partially isotactic structure as evidenced by $^1H$ and $^{13}C$ NMR,[51,52] while poly(meso-lactide) shows a partially syndiotactic structure.[53] These examples show how greatly the structure of lactic-containing poly(α-hydroxy-acid) can be complicated insofar as configurations are concerned. However, reality is even worse. On one hand, it is known that glycolide

**Table 1**
**VARIOUS VALUES OF THE DEGRADATION RATE OF POLY(LACTIC ACID) AS DEDUCED FROM THE OPEN LITERATURE**

| Authors | Year | Compound | Method | Degradation (%) | Time (months) | Ref. |
|---|---|---|---|---|---|---|
| Kulkarni et al. | 1966 | PLA (L) | $^{14}C$ labeling | 14 | 3 | 12 |
| | 1971 | PLA (DL) | $^{14}C$ labeling | 100 | 8 | 13 |
| Cutright et al. | 1971 | PLA ? | Size evaluation | 100 | 4 | 15 |
| Brady et al. | 1973 | PLA (DL) | $^{14}C$ labeling | 37 | 5.5 | 19 |
| Cutright et al. | 1974 | PLA ? | Size evaluation | 45 | 7 | 21 |
| Miller et al. | 1977 | PLA (L) | $^{14}C$ labeling | 50 | 6.5 | 22 |
| Reed | 1978 | PLA (L) | GPC | 60 | 4 | 40 |
| Vert et al. | 1981 | PLA (L) | GPC | <10 | 13 | 43 |

**Table 2**
**POLY-α-HYDROXY-ACIDS**

| | |
|---|---|
| 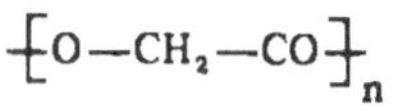 | PGA |
| 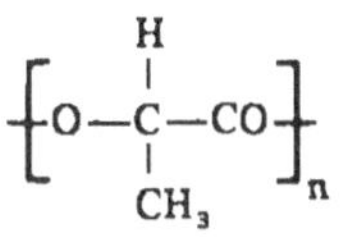 | PLA 100 |
| 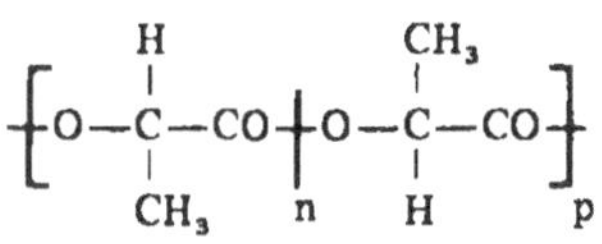 | PLA X ($X = 100\ n/(n + p)$) |
| 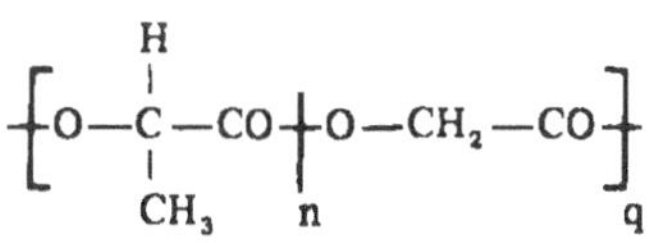 | PLA (100 − Y) GA Y ($Y = 100\ q/(n + q)$) |
| 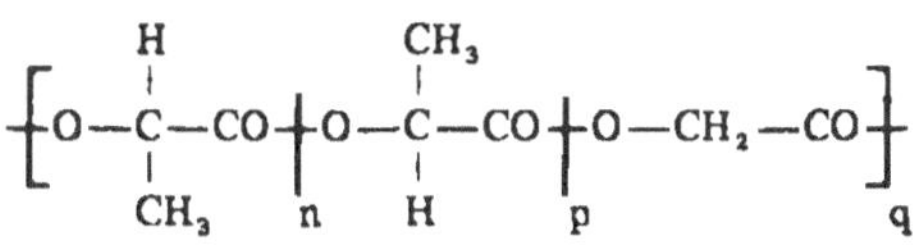 | PLA X GA Y ($Y = 100\ q/(n + p + q)$) ($X = 100\ n/(n + p + q)$) |

polymerizes much faster than lactides, a difference which leads to broad composition ranges in LA-GA copolymers.[41] DL-Lactide apparently polymerizes much faster than L-lactide,[53] though L and DL-lactides have the same reactivity since the enantiomeric composition of stereocopolymers are governed by the enantiomeric composition of the feed.[54] On the other hand, the occurrence of stereosequence redistribution due to transesterification reactions involving either active species[41] or the various ester bonds as suggested by Dittrich and Schulz[49] has been questioned by several authors.[51,53] The stereosequence redistribution has been recently confirmed for stereocopolymers initiated with powdered Zn using $^{13}C$ NMR and an enhanced resolution technique.[53]

In the historical survey, we have pointed to the lack of detailed identification of the polymers used and the scattering of the biological data as reported so far in the open literature. The chemical and configurational particularities mentioned above might by themselves account for the discrepancies. Actually, many other factors, i.e., processing and sterilization can also affect the physical and biological properties of the polymers and, a fortiori, those

**Table 3**
**LACTIDE DIASTEREOISOMERS**

| | |
|---|---|
| 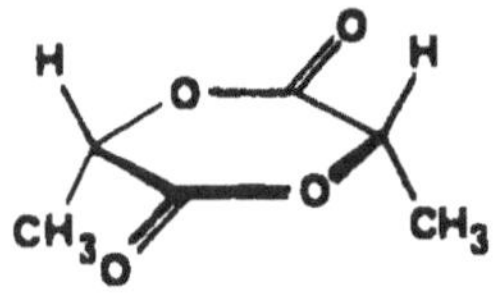 | L- lactide → isotactic polymer |
| 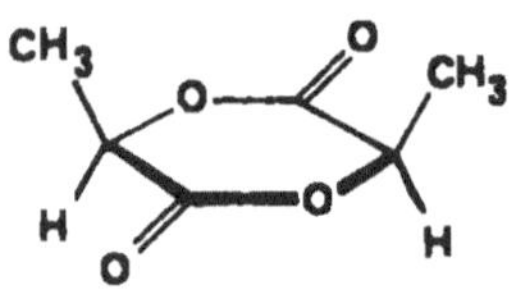 | D- lactide → isotactic polymer |
| 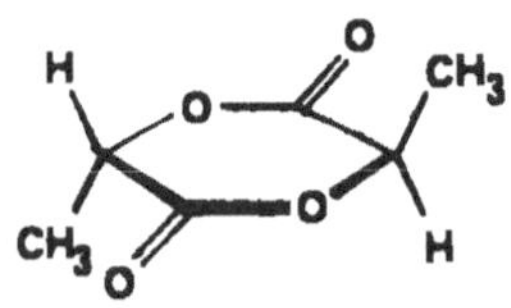 | Meso- lactide → partially syndiopolymer |
| 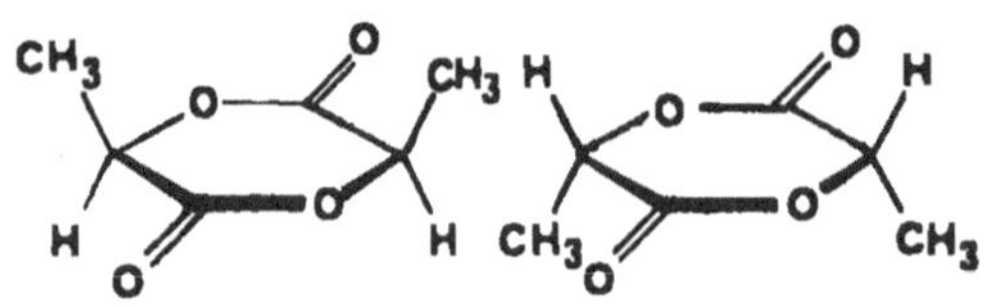 | Rac- lactide → partially isopolymer |

**Table 4**
**FACTORS WHICH CAN AFFECT THE MECHANICAL AND BIOLOGICAL PROPERTIES OF READY-FOR-IMPLANTATION PLA 100 IMPLANTS**

- Molecular weight
- Configurational structure
- Residual monomer and low molecular weight compounds
- Processing
- Annealing
- Sterilization
- Crystallinity

of ready-for-implantation end products. The principal factors responsible are listed in Table 4. It is of supreme importance to understand that none of these factors can be considered separately in defining a given poly(α-hydroxy-acid). For instance, the presence of low molecular weight compounds can be dramatic not only when these are initially present, but also if they are formed during processing or sterilization as a result of thermal, radiation, or chemical degradation of macromolecules. Accordingly, the scatter of biological data in the literature can be assigned to the neglect of one or several of these factors. This statement is supported by GPC analysis carried out in our group on PLA 50 compounds of various origins.[55] These points being emphasized, thermal and mechanical properties of some poly(α-hydroxy-acids) can be presented (Table 5) as they have been determined using nonstandard tests.

**Table 5**
**THERMAL AND MECHANICAL PROPERTIES OF POLY-α-HYDROXY-ACIDS OF DIFFERENT COMPOSITIONS**

| Polymer | Monomer mixture (mol %) Glycolide | L-lactide | Rac-lactide | $M_p$ (°C) | $T_g$ (°C) | E GPa | $\sigma_B$ (MPa) | ΔL/L (%) |
|---|---|---|---|---|---|---|---|---|
| PLA 100 | 0 | 100 | 0 | 174 | — | 3.6 | 58 | 2.1 |
| PLA 87.5 | 0 | 75 | 25 | —[a] | 60 | 3.4 | 46 | 1.6 |
| PLA 75 | 0 | 50 | 50 | —[a] | 60 | 3.3 | 46 | 3.2 |
| PLA 62.5 | 0 | 25 | 75 | —[a] | 59 | 2.8 | 41 | 2.9 |
| PLA 50 | 0 | 0 | 100 | —[a] | 59 | 3.2 | 48 | 8.7 |
| PLA 25 GA 75 | 75 | 25 | 0 | —[a] | 44 | — | — | — |
| PLA 50 GA 50 | 50 | 50 | 0 | —[a] | 44 | — | — | — |
| PLA 75 GA 25 | 25 | 75 | 0 | —[a] | 52 | — | — | — |
| PLA 12.5 GA 75 | 75 | 0 | 25 | —[a] | 43 | — | — | — |
| PLA 37.5 GA 25 | 25 | 0 | 75 | —[a] | 54 | — | — | — |
| PGA | 100 | 0 | 0 | 228 | | — | — | — |

[a] Amorphous compounds according to X-ray diffraction.

## IV. BIOCOMPATIBILITY AND DEGRADATION

We initially evaluated the tissue reaction around poly(α-hydroxy-acids) and the degradation of the polymers through histological means.

### A. Resorption in Rats (Figure 1)

The following compounds were tested,

- PLA X stereocopolymers, with X = 100, 87.5, 62.5, and 50
- PLA X GA Y copolymers, with X and Y values being, respectively, 0 to 100 (PGA); 25 to 75; 50 to 50; 75 to 25; 12.5 to 75 and 37.5 to 25. The two latter compounds being based on DL-lactide (for explanation see Table 2).

The resorption was monitored according to the protocol developed by Cutright and co-workers:[21] bilateral implantation in the cortex of the proximal tibial metaphysis of 2 × 2 mm cylinders in 94 Sprague Dawley rats. Histological features were evaluated on 5-μm-thick transverse sections of methacrylate embedded specimens at 1, 3, 7, 9, 15, and 20 weeks.[56] Sequential labeling of bone growth was achieved through injection of calcein blue, xylenol orange, and tetracycline, respectively, at 8, 11, and 15 weeks.

For all the PLA X stereocopolymers, no leukocyte reaction was observed and only a mild foreign body reaction with histiocytes and mast cells was detected. No lymphocytes or plasma cells were present. PLA 100 implants exhibited a fibrous encapsulation at 1 week which became totally ossified, but with no direct contact between cortical bone and implant after 7 weeks. At 20 weeks, implants were not degraded and the medullary part of the pellet was encapsulated by a bony layer which was not in direct contact with the polymer (Figure 1B). At 20 weeks, the degradation of PLA 50 did not differ from that of PLA 100 insofar as histology was concerned. The stereocopolymers also appeared to be intact. As in the case of PLA 100, a bony capsule had built up around every implant with a thin fibrous interface separating the osteoid line from the polymers. Cracks within the implants already observed at 9 weeks were certainly related to manufacturing processes. PLA X GA Y copolymers were evaluated at 15 and 20 weeks (Table 6).

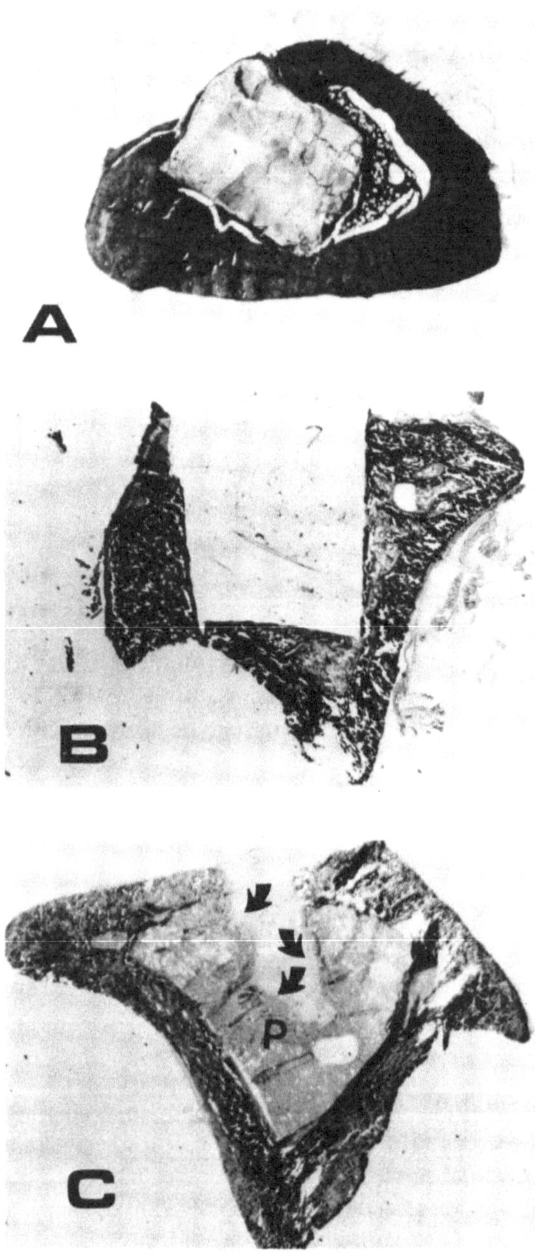


FIGURE 1. Transverse section of proximal tibial metaphysis of rats implanted with cylinders of polymers. (Magnification × 1.) (A) Pellet of PGA at 9 weeks with bony encapsulation and cracks; (B) PLA 100% L at 15 weeks — no degradation; (C) PLA 25/GA 75 at 15 weeks — note the vascular and fibrous invasion (arrows); P, pellet.

The degradation of copolymers was in sharp contrast to that of the previous stereocopolymers. At 15 weeks, implants were almost totally replaced either by fibrous or bony tissue (Figure 1C). At 20 weeks, the last traces of the PLA 50 GA 50 copolymer could only be located by means of the interrupted labeling line. This total resorption was regularly observed for every lactide-glycolide copolymer. The copolymers based on DL-lactide, instead of L-lactide, showed a higher degradation rate.

**Table 6**
**QUALITATIVE FATE OF IMPLANTED PLA X GA Y COPOLYMERS**

| Copolymers | | Composition of the monomer feed: LA %L | LA %DL | GA % | Degradation of the implants: 15 weeks | 20 weeks |
|---|---|---|---|---|---|---|
| PLA 25 | GA 75 | 25 | — | 75 | Implants replaced by fibrous tissues, few debris | Implants totally replaced by Cortical bone |
| PLA 50 | GA 50 | 50 | — | 50 | Few debris, implants replaced by bone and fibrous tissue | Implants totally replaced by bone |
| PLA 75 | GA 25 | 75 | — | 25 | Few debris, cortical bone | Cortical bone |
| PLA 12.5 | GA 75 | — | 25 | 75 | Cortical bone | Cortical bone |
| PLA 37.5 | GA 25 | — | 75 | 25 | Cortical bone | Cortical bone |

PGA implants were cracked at week 5. Fibrous tissue invaded the cracks. An important histiocytic reaction was observed containing giant cells and intracellular debris of the polymer. At 7 weeks, a bony capsule surrounded the implants, but no direct contact with the implant was found either in the marrow or in the cortex. A fibrous capsule lay between implant and bone and the cracks of the implants were invaded by bony trabeculae. At 9 weeks, the implants were not degraded and exhibited a few cracks which were also invaded by bone (Figure 1A). Direct bone contact with the pellets was observed at some places. The degradation of the implants which differed from one specimen to another was possibly related to cracks which may have been induced during the manufacturing process.

### B. Resorption in Sheep[57] (Figure 2)

There were 5 × 5 mm cylinders of polymers bilateraly implanted into the lateral cortex of 1-year-old male sheep femurs; 5 implants were inserted subperiosteally 20 mm apart in each bone. Animals were sacrificed at 6, 9, and 12 months and tissue response was evaluated on 5 and 100-μm-thick nondecalcified sections. The same polymers as in the previous rat experiments were used. PLA 100 showed a bony encapsulation around the medullary part of the implants with no direct contact between bone and polymer. After 1 year, histologically, the implant was hardly degraded, showed no change in form, and presented a fibrous invasion containing capillaries, giant cells, and mononucleated cells into the cracks. This low degradation rate is in contrast with the results of Cutright and co-workers.[21] The discrepancy may be due to the crystallinity of the PLA used in our experiments.[43]

Stereocopolymers (PLA 87.5, PLA 75, PLA 62.5) exhibited an increase in degradation rate as the D-content increased. The fastest degradation rate was apparently observed for PLA 50. After 6 months, the shape of the implant was quite modified with fibrous and bony tissue invasion. After 1 year, implants of PLA 50 disappeared almost totally. There was cortical bridging and some polymer debris, giant cells, and fibrocytes within the marrow. These results disagree with our previous observations on rats where no difference has been demonstrated between the different stereocopolymers. This discrepancy has to be attributed to an improvement in the manufacturing processes, thus showing the significance of the processing conditions on the properties of implants. The degradation of PLA X GA Y copolymers was already quite advanced at 6 months. The implants were replaced by woven cortical bone and polymer fragments surrounded by giant and mononucleated cells which

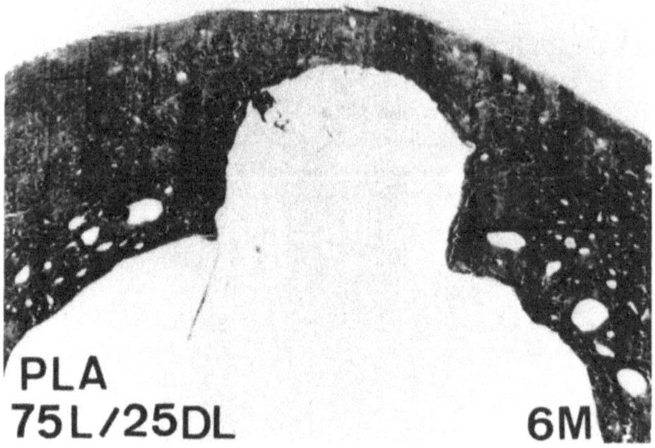


A

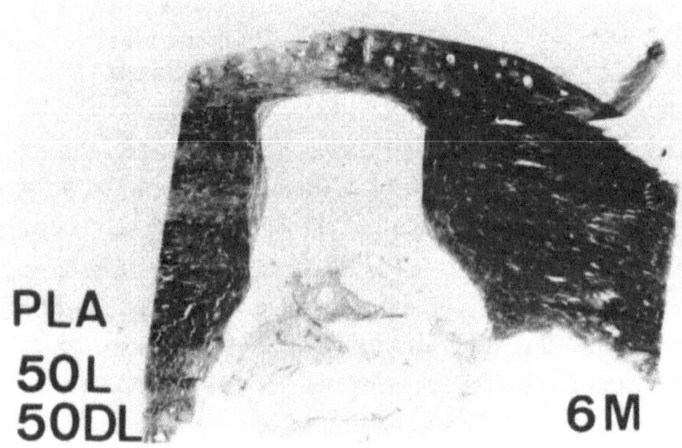


B

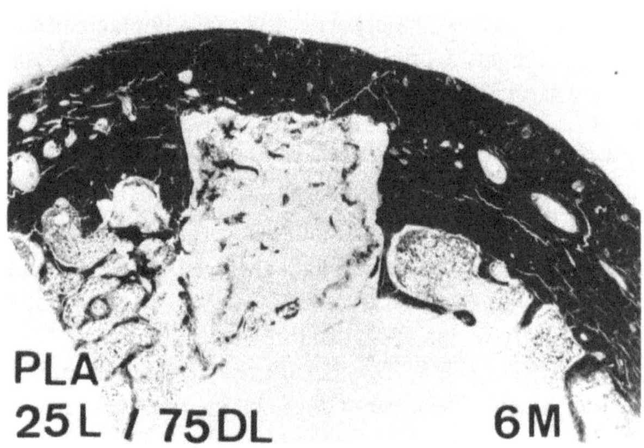


C

FIGURE 2. Histological pictures of 5 × 5 mm cylinders of stereocopolymers implanted into the femoral cortex in sheep; transverse sections. (Magnification × 1.) After 6 months (pictures 2A to 2D), degradation increases with percent of DL-lactide in the monomer feed (Table 5). After 1 year (pictures 2E and 2F), PLA 100% L is not histologically degraded, whereas stereocopolymers 25 L/75 DL is almost totally resorbed except for some medullary debris.

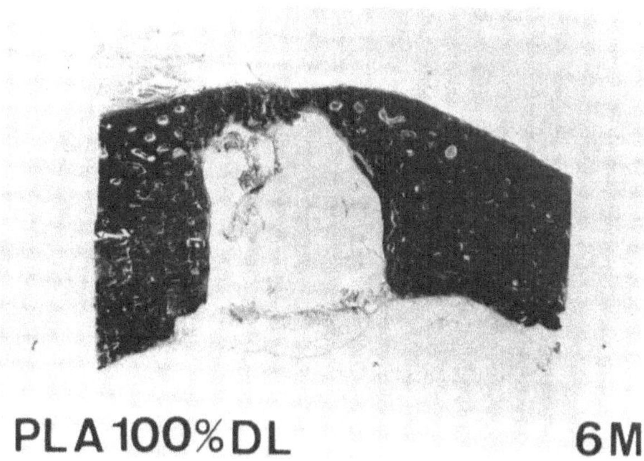


FIGURE 2D

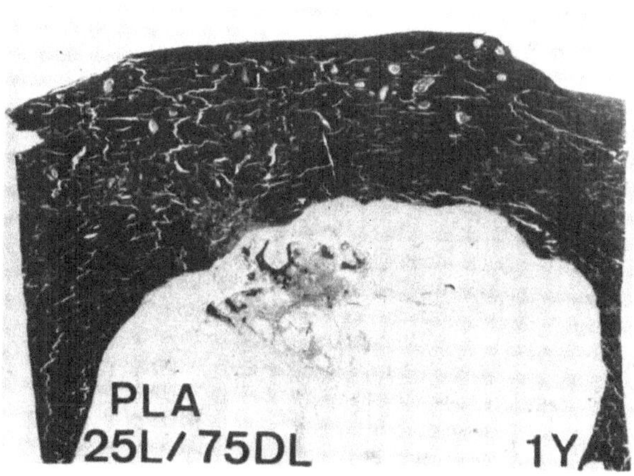


FIGURE 2E

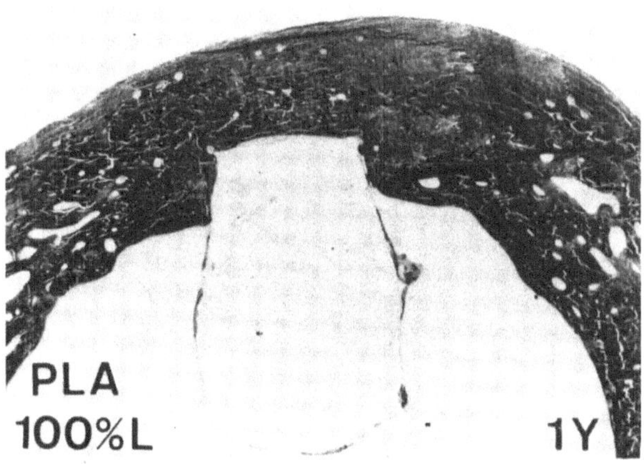


FIGURE 2F

could be found in the femoral medullary cavity (Figure 3). At 9 months, cortical repair was achieved and only some small debris of polymers situated in the marrow allowed one to locate the former implant. No difference could be demonstrated between the degradation of

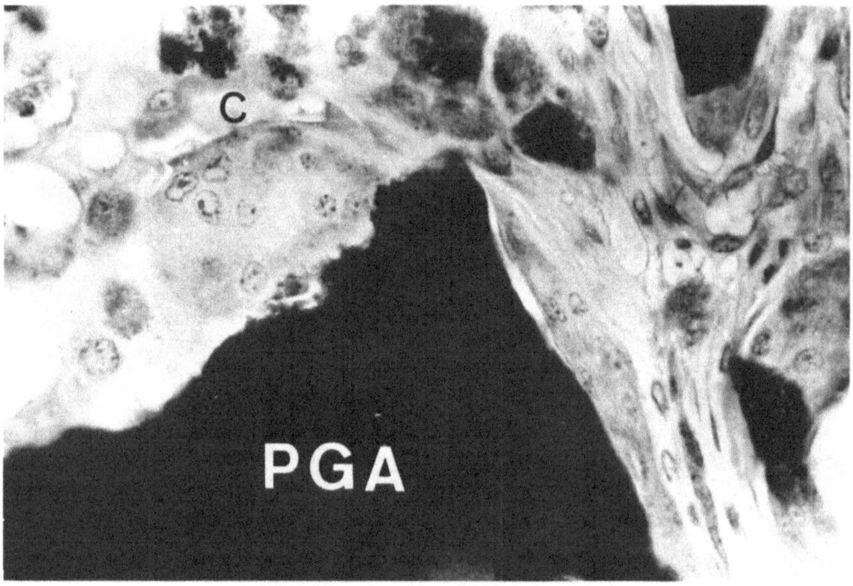


A

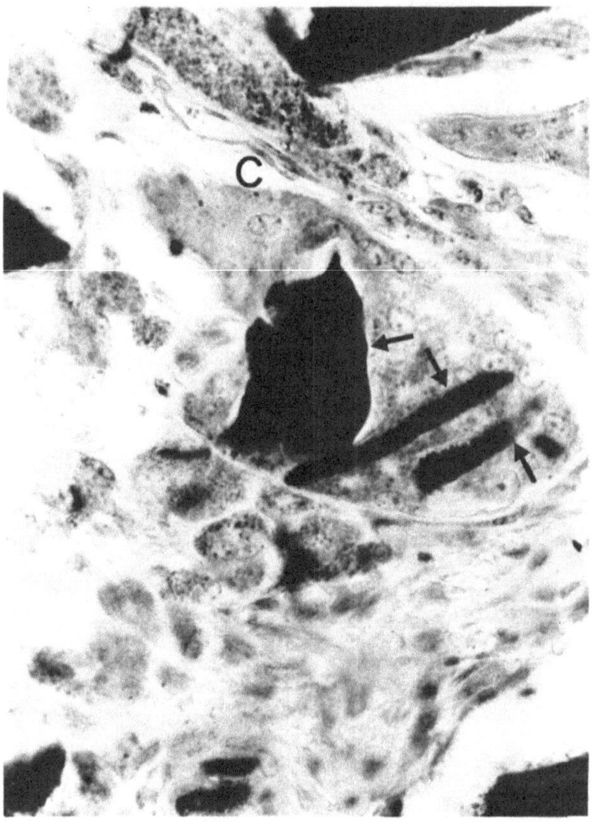


B

FIGURE 3. (A) PGA implant being resorbed by a giant cell (c). (Magnification × 130.) (B) Giant cell (c) with intracellular debris of PGA (arrows). Smaller debris may be observed within the surrounding cells. (Magnification × 130.)

copolymers as the histological evaluation was performed only at 6 and 9 months. However, these preliminary studies on biocompatibility and biodegradation confirm the results of Cutright's group except for PLA 100. The results can be summarized as follows.

**At 6 months** — All the polymers induced a mild foreign body reaction. The degradation rate of the stereocopolymers increases as the percentage in D-units increases. All the PLA X GA Y copolymers disappeared and were replaced by woven bone.

**At 9 months** — The PGA implants were cracked and invaded with giant cells and fibrocytes. All stereocopolymer implants are partially degraded; this increases with the content in D-units. A very thin fibrous capsule is always situated between the bony tissue and the implants.

**At 12 months** — Only PLA 100 implants are not degraded histologically. All the other copolymers or stereocopolymers are resorbed and replaced by woven bone.

Although the shape of PLA 100 implants is unchanged beyond a 1 year implantation time, a constant discoloration of the material which becomes whitish is noticed, thus showing a qualitative sign of the beginning of degradation and physicochemical property changes.

## C. Quantitative Evaluation of Degradation in Sheep

A total of 12,6-hole osteosynthesis plates (90 × 15 × 4 mm) were compression molded from PLA 100 and fixed on sheep tibias without osteotomies using stainless steel screws. They were retrieved at various intervals up to 2 years of implantation time. The molecular weight variation was measured on the whole sample, on the outer part lying underneath the surface, and on inner core material.

Even after 1 month, all retrived plates were broken with cracks initiated around the screw holes. The initial molecular weight was approximately $21 \times 10^4$ as monitored by GPC with respect to polystyrene standards in dioxane and approximately $12 \times 10^4$ based on viscosity.[58] No significant change in GPC molecular weight was detected at the outer plate surface after 4 to 5 months of implantation ($18 \times 10^4$). The molecular weight began to decrease after 1 year of implantation[53] to reach $5 \times 10^4$ with the beginning of a bimodal distribution after 2 years (Figure 4).

Measurements of impact strength on 1-hole plates, i.e., on nonstandard specimens, showed an initial value of 3.6 $kJ/m^2$ which remained relatively unchanged after 2 months in vivo (3.2 $kJ/m^2$). GPC and impact strength data converge to show that PLA 100 bone plates have an outstanding stability in the biological milieu as was qualitatively observed before for small-sized implants.

According to these results, PLA 100 now appears to be too biostable for bone surgery. We mentioned earlier that PGA- and GA-contained PLA copolymers degrade quickly in the biological milieu and thus cannot ensure safe internal fixation for 2 to 3 months, the period usually considered as necessary for fracture consolidation.[59] According to our experiments, only stereocopolymers seem to present sufficient resistance to biodegradation to preserve their initial properties for at least 2 to 3 months.[57] In order to optimize the mechanical and the biological performances of lactic acid stereocopolymers in vivo, tensile strength $\sigma_B$ and molecular weight distributions were studied for PLA X, with X varying from 100 to 50%. Small-sized specimens were implanted on sheep bone for up to 5 months. Eight different stereocopolymers, PLA 100, 100a, 98, 96, 96a, 92, 75, and 50 (where a corresponds to test bars annealed at 100°C for 5 hr), were prepared from highly purified L- and DL-lactides and cleared from their low molecular weight compounds by solid/liquid extraction with acetone or by disssolution/precipitation using dioxane/methanol when possible. As the crude polymers were very sensitive to thermal degradation, the specimens shaped for tensile testing (test bars 25 mm long, thinnest cross section 2 × 2 mm) were compression molded and rapidly cooled to room temperature. A total of 80 specimens were implanted on tibias and femurs of 5 sheep in order to minimize the biological spread data.[43,60]

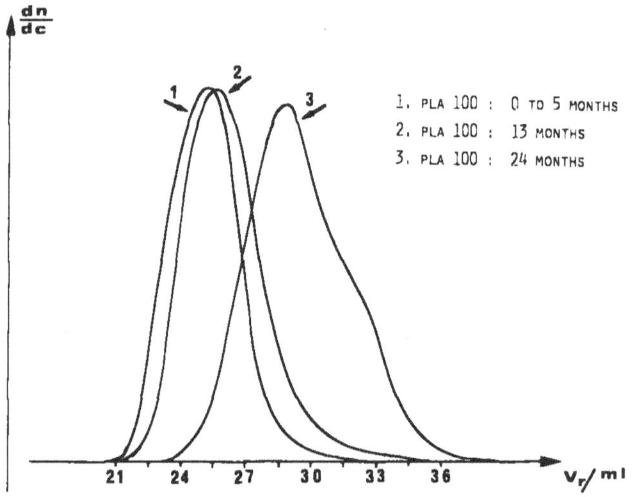


FIGURE 4. Degradation of annealed 6-hole PLA 100 bone plates as monitored by GPC in dioxane. $M_{GPC}$ = (1) 21 × $10^4$; (2) 18 × $10^4$; (3) 5 × $10^4$ taken at the peak.

Initial tensile strength increases regularly with X from 42 (PLA 50) to 64 (quenched PLA 100) and 70 MPa (annealed PLA 100). Figure 5 shows the aspects of the specimens after 1 and 2 months implantation before breaking. Figure 6 shows the variation of tensile strength to implantation time. After 1 month, no decrease was detected for PLA 100, 98, and 96. PLA 75 dropped from 48 to 15 MPa, while PLA 50 broke on removal from bones. At 2 months, only PLA 100 and PLA 100a did not present any decrease in tensile strength, size, molecular weight, and color changes. The other quenched specimens became whitish. In contrast, the color of the annealed PLA 96a specimens did not change. The PLA 92 specimens had lost approximately 20% of their initial $\sigma_B$, while the PLA 75 ones had lost all stiffness. At 5 months, the implants were fully encapsulated in new bone and most of them broke on removal from the adhering tissues. Figure 7 shows the corresponding variations of molecular weight distribution as monitored by GPC. After 5 months, PLA 100 remained almost unchanged. PLA 92 shows a slight decrease of molecular weight. In contrast, PLA 75 shows a significant decrease of molecular weight with the appearance of a bimodal molecular weight distribution, a trend more pronounced for PLA 50 whose mechanical properties were totally exhausted after 1 month. Bimodal GPC chromatograms have already been observed for degrading PGA- and GA-containing polymers.[40] On the other hand, these results agree with data reported by Fisher et al.[61] which show that alkaline degradation of lactide copolymers occurs primarily in amorphous regions, thus giving low molecular weight degradation products.

Though no crystallinity determination has been carried out so far, the comparison between quenched and annealed samples shows that both mechanical strength and degradation rates are strongly dependent on enantiomeric purity and on molecular weight distribution. However, crystallinity is definitely the main factor. Only materials with X higher than 92 exhibit degradation rates low enough for orthopedic surgery.

## V. BIOFUNCTIONAL INVESTIGATIONS

### A. In Sheep

Although the degradation rates of lactic acid stereocopolymers can be reasonably controlled

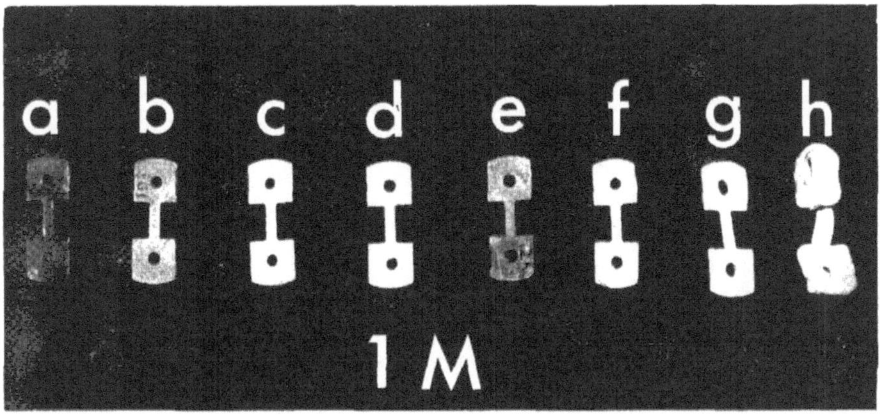


A

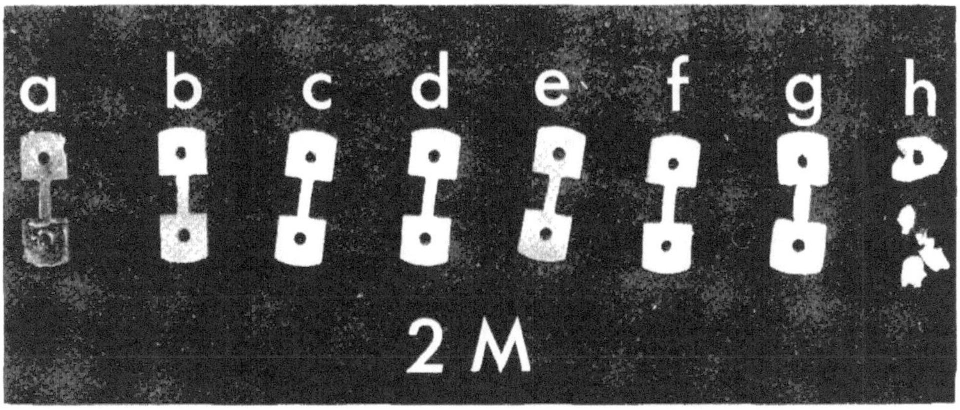


FIGURE 5. Gross appearance of mini tensile specimens after removal at 1 and 2 months. (a) PLA 100; (b) PLA 100 annealed; (c) PLA 98; (d) PLA 96; (e) PLA 96 annealed; (f) PLA 92; (g) PLA 75; (h) PLA 50. After 2 months, quenched specimens became whitish; in contrast annealed specimens (b and e) did not exhibit discoloration.

through enantiomeric purity, their mechanical properties still appear to be insufficient for bone surgery. From the literature, it is likely that only composite materials can provide enough strength for osteosynthesis in human fractures.[62] Attempts to use carbon fiber reinforced PLA bone plates have been recently reported.[28] This solution, however, is only partially suitable because of the nonresorbability of carbon fibers. In order to overcome this shortcoming, we have investigated the feasability of a completely resorbable composite material which associates high tensile strength PGA fibers (Mp = 228°C) to a PLA matrix (Mp = 175°C).[58,63] Composite plates composed of a PLA 100 matrix with several plies of two-dimensionally woven fabrics of PGA threads (0.1 mm in diameter) were processed by compression molding.[63] The fibers located around the plate holes, in conjunction with good adhesion between fibers and matrix, improved the impact strength of the matrix significantly (Table 7). Furthermore, compared with thermoset composite materials, these thermoplastic plates present a greater advantage in being heat deformable above approximately 80°C (Tg ≃ 60°C) during the surgical procedure. This deformability allows one to adapt the plate to the bone surface shape (Figure 8). Composite plates of similar size to the pure PLA 100

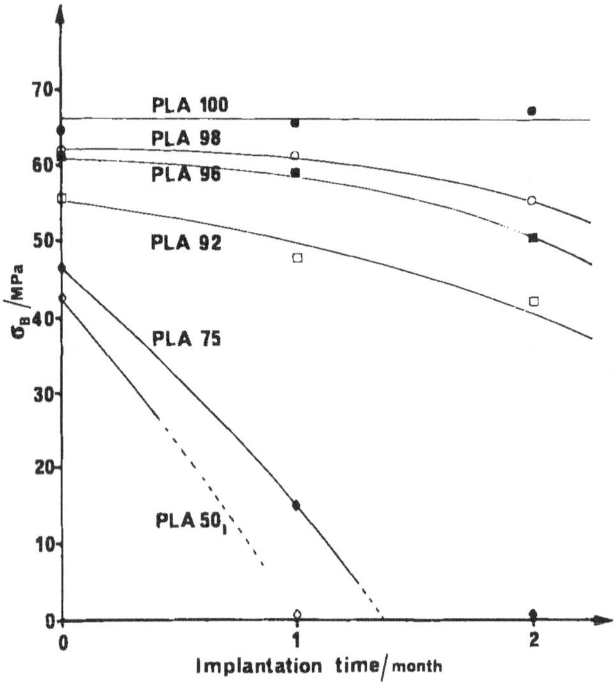


FIGURE 6. Variation of the tensile strength with time for the various stereocopolymers.

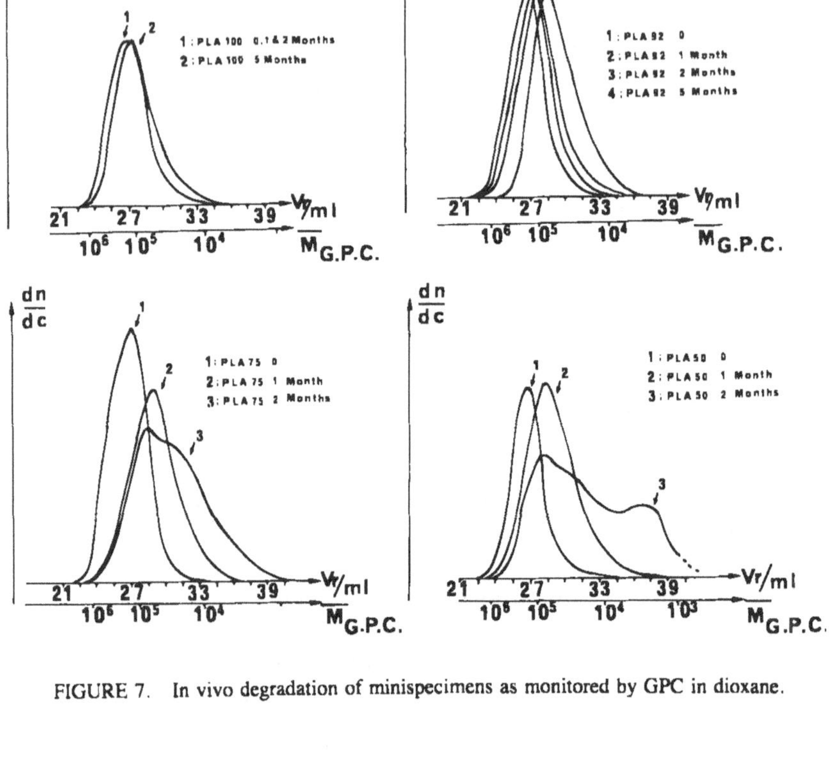


FIGURE 7. In vivo degradation of minispecimens as monitored by GPC in dioxane.

**Table 7**
**IMPACT STRENGTH (NOTCHED IZOD kJ/m²)**

| | | |
|---|---|---|
| PLA 100 (matrix) | 3.6 | (3.2 after 2 months) |
| PLA 100 Annealed (matrix) | 5.2 | |
| PLA 100 + 30% PGA mesh (composite) | 35.4 | |
| PLA 100 + 60% PGA mesh (composite) | 92 | |

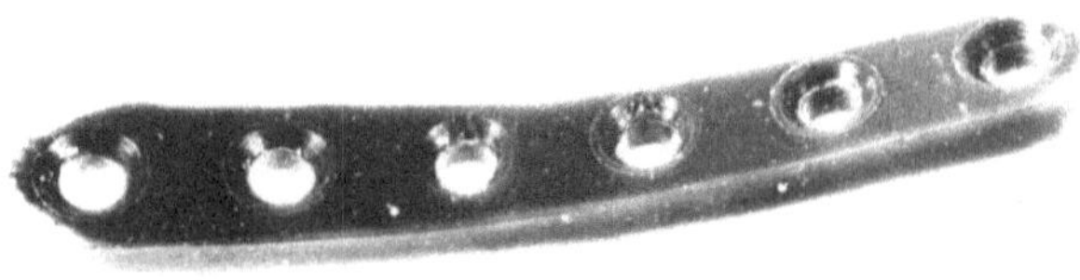

FIGURE 8. Composite osteosynthesis plate made of PLA matrix reinforced with PGA fibers. The plate has been heat-deformed to be adapted on the upper tibial metaphysis of a sheep.

plates studied before were fixed with stainless steel screws on seven sheep tibias, five without osteotomy and two with a midshaft osteotomy. The 5 plates without osteotomy, implanted between 1 to 7 months, did not break, but exhibited cracks initiated at screw holes where stress concentration occurs. Amid the two plates implanted on tibias with an osteotomy, the first plate broke 1 month later in spite of the plaster cast. The second osteotomy healed with a large amount of callus and, at sacrifice, the plate was found to be broken.

Although the mechanical improvement by fiber reinforcement of the PLA matrix is obvious, the composite material is still not strong enough, at this stage, to be considered suitable for the manufacture of long bone plates.

PLA 100 screws were also considered with implantation into tibia shafts for 1 to 4 months in sheep without osteotomy. Screws, shaped as usual nylon screws, were obtained by an injection molding technique. The purpose of the experiment was to evaluate histologically the interface between PLA thread and cortical bone. The holes, perforating both cortices, were drilled with a burr on an electric-driven hand piece. A metal tap of the same size as the screws was used to cut thread in the bone. Despite the lack of important mechanical stresses, the screws appeared to be already cracked after 1 month of implantation (Figure 9). A fibrous tissue layer was interposed between the thread and the bone. The thickness of this fibrous layer reached about 250 $\mu m$ at 1 month and decreased progressively to become an almost unicellular layer after 4 months. Giant cells were in direct contact with the polymer and some polymer debris was found intracellularly. The cracks were invaded by bony tissue at 2 months. After 4 months, only minor changes appeared in the shape of the screws. These results indicated the need for reexamination of the design of these biodegradable screws to allow for their clinical use. Furthermore, reinforcement with short PGA fibers should permit improving their mechanical resistance.

### B. In Patients

Implantation in animals being the source of problem because of their lack of cooperation, clinical applications of composite bone plates have been carried out in the case of reconstructive facial surgery.[64] Small size PLA/PGA composite plates (2 mm thick) reinforced by 2 plies of a PGA fabric were shaped in order to fit the dimensional requirements for the repair of mandible and skull fractures. These plates were cut from large disks (80 mm diameter; 2 mm thick) using a laser beam in order to have a regular cut-line without

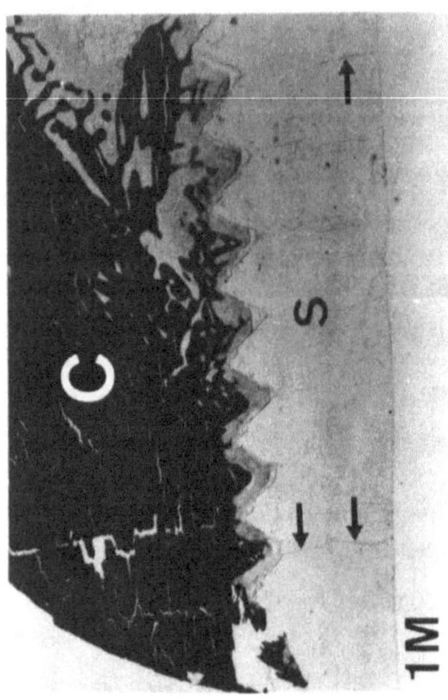


FIGURE 9A

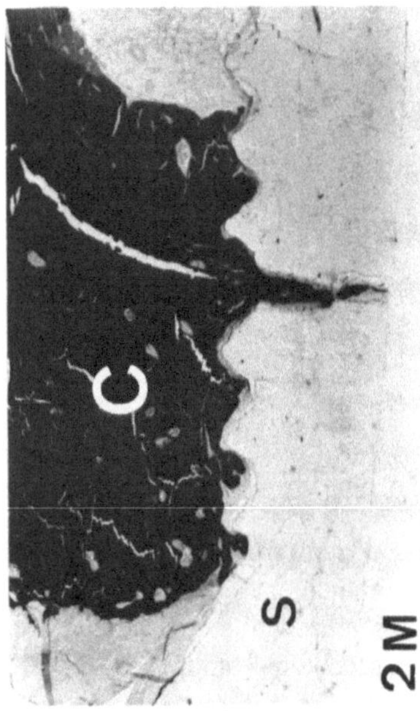


FIGURE 9B

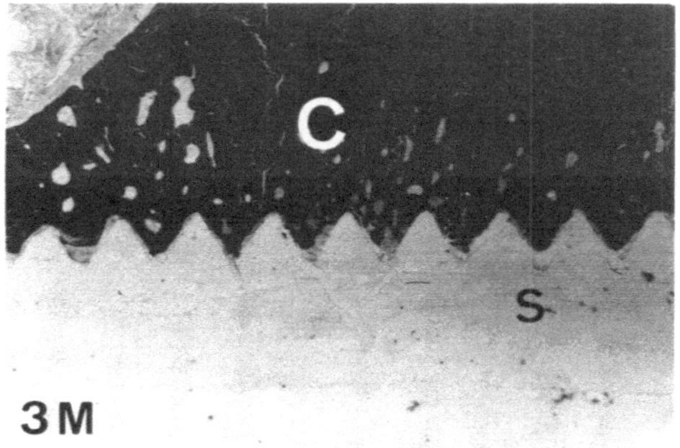


C

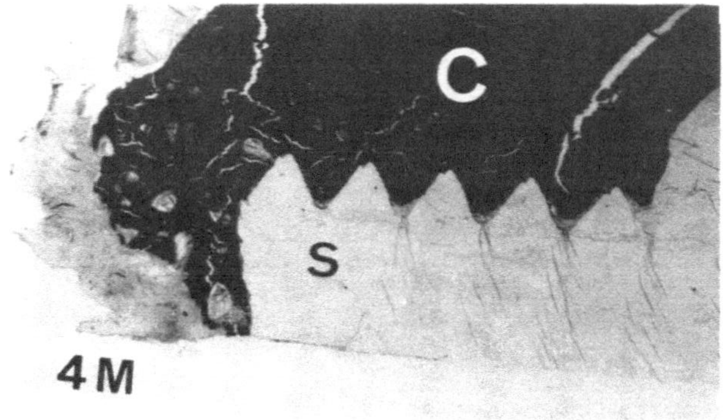


D

FIGURE 9. Histological appearance of PLA 100 screws. Transverse sections. (Magnification × 2.5.) (A) After 1 month, cracks may be already observed (arrow). Note the important fibrous tissue thickness (250 μm) interposed between cortical bone (c) and screw (s); (B) after 2 months, cracks have been invaded by new bone (arrow); the fibrous tissue layer decreases; (C) after 3 months the fibrous tissue becames very thin; (D) after 4 months the fibrous tissue is reduced to a unicellular layer; new bone lies at the bottom of the thread.

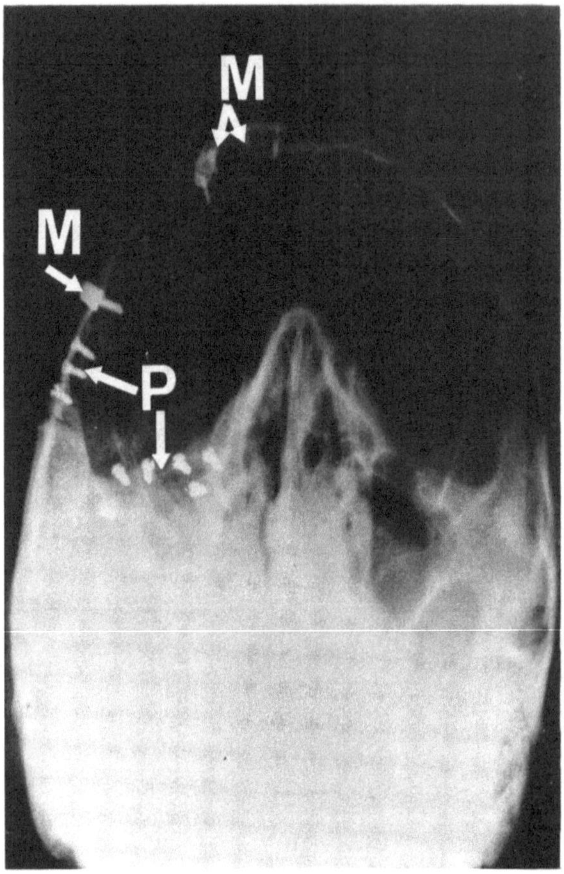


FIGURE 10. Patient with periorbital and skull fractures treated with metal (M) and composite PLA/PGA plates (P). Polymer plates which are radiotranslucent are fixed to bone with stainless steel screws.

microcracks. After warming with an electric hairdryer, the plates were shaped either pre- or peroperatively. With the plates in their sterilized bags, the shaping was done without any inconvenience.[64]

A total 25 patients have been treated within the last 2 years. The first clinical results are very favorable with an expected absence of acute or chronic tissue response. The matrix of these composite plates being of the PLA 100 type, the resorption of the material is still not completed, but evidence of partial degradation has been obtained, in particular through GPC measurements of the molecular weight of the PLA 100 matrix of retrieved plates.[64] Figure 10 shows plates which have been placed on periorbital and skull fractures. So far, all the fractures thus fixed healed normally without infection. Orthopedic applications are currently under investigation.

## VI. CONCLUSIONS

In this paper, we have tried to review critically the work (including our own) which has been done in the search for the use of poly(α-hydroxy-acids) in bone surgery. The present state of the knowledge in the field has led us to propose a new composite material composed of totally resorbable elements. When carefully processed, PLA 100 appears as a relatively biostable polymer. Fortunately, the degradation rate can be readily increased by the introduction of a small percentage of D-units within the poly L-chains. The introduction of D-units also results in a decrease of the melting temperature. This decrease appears to be of special interest for the manufacture of the composite with PGA fibers as it increases the difference in melting points between the matrix and the reinforcing fibers. The possibilities for optimizing its performance are still numerous and we believe that better mechanical properties can still be obtained. This should provide completely resorbable elements for osteosynthesis of slightly loaded fractures.

## ACKNOWLEDGMENTS

The authors thank MM. de Charentenay and Dewas (University of Technology of Compiègne) for mechanical and physical testings, Mrs. Hott for histology, MM. Masingue and Rouse (French Textile Institute-North) for weaving PGA, and Mrs. Henry-Amar for typing the manuscript. The authors are indebted to Dr. M. Audion and Prof. M. Michelet, who carried out the clinical experiments using their bioresorbable composite materials. This work was supported by D.G.R.S.T. Grants numbers 757 1183, 757 1184, 757 1185, 757 1186, 787 1021, and 797 0311.

## REFERENCES

1. **Moyen, G. J.-L., LaHey, P. J., Weinberg, E. H., and Harris, W. H.,** *J. Bone Jt. Surg.*, 60, 940, 1978.
2. **Akeson, W. H., Woo, S. L.-Y., Coutis, R. D., Matthews, J. V, Gonsalves, M., and Amiel, D.,** *Calcif. Tissue Res.*, 19, 27, 1975.
3. **Tonino, A. J., Davidson, C. L., Klopper, P. J., and Linclau, L. A.,** *J. Bone Jt. Surg.*, 58b, 108, 1976.
4. **Meyrueis, J. P., Bonnet, G., De Bazelaire, E., and Zimmermann, R.,** *Rev. Chir. Orthop.*, 64(Suppl. 2, 108, 1978.
5. **Woo, S. L.-Y., Akeson, W. H., Coutis, R. P., Gomez, M. A., Cobb, N. G., Gonzalves, M., and Amiel, D.,** Trans. 7th Annu. Meet. Soc. Biomaterials, Troy, N.Y., 1981, 58.
6. **Levin, M. P., Getter, L., Cutright, D. E., and Bhaskar, S. N.,** *Oral Surg.*, 38, 344, 1974.
7. **Guillemin, G., Fournie, J., Patat, J.-L., and Chetail, M.,** *C.R. Acad. Sci.*, 293, 371, 1981.
8. **Kronenthal, R. L.,** *Polymer in Medicine and Surgery*, Kronenthal, R. L., Oser, Z., and Martin, E., Eds., Plenum Press, New York, 1975, 119.
9. **Graham, N. B.,** *Br. Polym. J.*, 10, 260, 1978.
10. **Leray, J. and Vert, M.,** *Plastics in Medicine and Surgery*, 3rd Int. Conf. The Plastics and Rubber Institute, London, 1979, preprint #2.1.
11. **Dorre, E., Geduldig, D., Happel, M., Lade, R., Prussner, P., Willert, H. G., and Zichner, L.,** *J. Biomed. Mater. Res. Symp.*, 7, 493, 1976.
12. **Kulkarni, R. K., Pani, K. C., Newman, C., and Leonard, F.,** *Arch. Surg.*, 93, 839, 1966.
13. **Kulkarni, R. K., Moore, E. G., Hegyeli, A. F., and Leonard, F.,** *J. Biomed. Mater. Res.*, 5, 169, 1971.
14. **Schmitt, E. E. and Polistina, R. A.,** U.S. Patent 3,463,158, 1969.
15. **Cutright, D. E., Hunsuck, E. E., and Beasley, J. D.,** *Oral Surg.*, 29, 393, 1971.
16. **Getter, L., Cutright, D. E., Bhaskar, S. N., and Ausburg, J. F.,** *Oral Surg.*, 30, 344, 1972.
17. **Cutright, D. E. and Hunsuck, E. E.,** *Oral Surg.*, 33, 28, 1972.
18. **Schmitt, E. E. and Polistina, R. A.,** U.S. Patents 3,739,773, 1973.
19. **Brady, J. M., Cutright, D. E., Miller, R. A., and Battistone, G. C.,** *J. Biomed. Mater. Res.*, 7, 155, 1973.
20. **Hegyeli, A. F.,** *Fed. Proc. Fed. Am. Soc. Exp. Biol.*, 29 (Abstr.), 744, 1971.
21. **Cutright, D. E., Perez, B., Beasley, J. D., Larson, W. T., and Posey, W. R.,** *Oral Surg.*, 37, 142, 1974.
22. **Miller, R. A., Brady, J. M., and Cutright, D. E.,** *J. Biomed. Mater. Res.*, 11, 711, 1977.

23. **Alexander, H., Weiss, A. B., Parsons, J. R., Stauchler, I. D., and Gona, O.**, Trans. 5th Annu. Meet. Soc. Biomaterials, Clemson, S.C., 1979, 72.
24. **Aragona, J., Parsons, J. R., Alexander, H., and Weiss, A. B.**, *Clin. Orthop.*, 160, 268, 1981.
25. **Alexander, H., Parsons, J. R., Stauchler, J. D., Corcoran, S. F., Gona, O., Mayott, C., and Weiss, A. B.**, *Orthop. Rev.*, 10, 41, 1981.
26. **Alexander, H.**, personal communication, 1981.
27. **Parsons, J. R., Alexander, H., Corcoran, S. F., Korolux, J. M., and Weiss, A. B.**, Trans. 25th Annu. Orthoped. Res. Soc., San Francisco, 1979, 162.
28. **Alexander, H., Corcoran, S., Parsons, J. R., and Weiss, A. B.**, Proc. 9th Northeast Bioeng. Conf., Rutgers University, New Brunswick, N.J., 1981, 115.
29. **Lewis, D.**, Tran. 7th Annu. Meet. Soc. Biomaterials, Troy, N.Y., 1981, 61.
30. **Schmitt, E. E. and Polistina, R. A.**, U.S. Patent 3, 297, 033, 1967.
31. **Frazza, E. J. and Schmitt, E. E.**, *J. Biomed. Mater. Res.*, 1, 43, 1971.
32. **Wasserman, D.**, U.S. Patent 3, 375, 008, 1971.
33. **Manton, C. G. F., Phillips, C. I., Martin, B., Bartholomew, R. S., and Capperauld, I.**, *Br. J. Ophthalmol.*, 58, 941, 1974.
34. **Ruderman, R. J., Hegyeli, A. F., Hattler, B. G., and Leonard, F.**, *Trans. Am. Soc. Artif. Organs*, 18, 30, 1972.
35. **Boswell, G. A. and Schriber, R. M.**, U.S. Patent 3,773,919, 1973.
36. **Wise, D. L., McCormick, G. J., Willet, G. P., and Anderson, L. C.**, *Life Sci.*, 19, 867, 1976.
37. **Jacanicz, I. M., Nash, H. A., Wise, D. L., and Gregory, J. B.**, *Contraception*, 8, 227, 1973.
38. **Sinclair, R. G.**, *Am. Chem. Soc. Div. Org. Coat. Plast. Chem. Pap.*, 34, 684, 1974.
39. **Beck, L. R., Cowsar, D. R., Lewis, D. H., Gigson, J. W., and Flowers, C. E., Jr.**, *Am. J. Obstet. Gynecol.*, 135, 419, 1979.
40. **Reed, A. M.**, Ph.D. thesis, University of Liverpool, England, 1978, 269.
41. **Gilding, O. K. and Reed, A. M.**, *Polymer*, 20, 1459, 1979.
42. **Gilding, D. K.**, in *Fundamental Aspects of Biocompatibility*, Vol. 1, Williams, F. F., Ed., CRC Press, Boca Raton Fla., 1981, 43.
43. **Vert, M., Chabot, F., Leray, J., and Christel, P.**, *Makromol. Chem.*, Suppl. 5, 30, 1981.
44. **Carothers, W. H., Dorough, G. L., and Van Natta, F. J.**, *J. Am. Chem. Soc.*, 54, 761, 1932.
45. **Schobel, A. and Wiehler, G.**, *Liebigs Ann. Chem.*, 595, 112, 1955.
46. **Higgins, N. A.**, U.S. Patent 2,676,945, 1954.
47. **Lowe, C. E.**, U.S. Patent 2,668,162, 1954.
48. **Kleine, J. and Kleine, J. J.**, *Makromol. Chem.*, 30, 23, 1959.
49. **Dittrich, W. and Schulz, R. C.**, Angew. Makromol. Chem., 15, 109, 1971.
50. **Holten, C. H.**, *Lactic Acid*, Verlag Chemie, Weinheim, 1971, 221.
51. **Lillie, E. and Schulz, R. C.**, *Makromol. Chem.*, 176, 1901, 1975.
52. **Schindler, A. and Harper, D.**, *J. Polym. Sci. Polym. Lett. Ed.*, 14, 729, 1976.
53. **Chabot, F., Vert, M., Chapelle, S., and Granger, P.**, *Polymer*, 24, 53, 1983.
54. **Schulz, R. C.**, in *Kinetics and Mechanism of Polyreactions*, IUPAC Int. Symp., Budapest, 1969, Académiai Kiado, Budapest, 1971, 185.
55. **Vert, M. and Chabot, F.**, Unpublished data.
56. **Leray, J. L., Chabot, F., de Charentenay, F. X., Christel, P., and Sedel, L.**, Trans. 3rd Annu. Meet. Soc. Biomaterials, New Orleans, La., 1977.
57. **Sedel, L., Chabot, F., Christel, P., de Charentenay, F. X., Leray, J., and Vert, M.**, *Rev. Chir. Orthop.*, 64(Suppl. 2), 92, 1978.
58. **Vert, M., Chabot, F., Leray, J., and Christel, P.**, French Patent Appl., 78-29978, 1978.
59. **Sevitt, S.**, *Bone Repair and Fracture Healing in Man*, Churchill Livingstone, Edinburg, 1981.
60. **Leray, J. L., Chabot, F., Christel, P., and Vert, M.**, Trans. 7th Annu. Meet. Soc. Biomaterials, Troy, N.Y., 1981, 37.
61. **Fischer, E. W., Sterzel, M. J., and Wegner, G.**, *Kolloid Z. Polym.*, 252, 980, 1973.
62. **Hastings, G. W.**, *Composite*, 193, 1978.
63. **Christel, P., Chabot, F., Leray, J. L., Morin, C., and Vert, M.**, in *Biomaterials 1980*, Winter, G. D., Gibbons, D. F., and Plenk, H., Jr., Eds., John Wiley & Sons, New York, 1981, 271.
64. **Audion, M., Michelet, M., Chabot, F., and Vert, M.**, to be published.

Chapter 7

# POLYMER MEMBRANES

**J. M. Courtney, J. D. S. Gaylor, H. Klinkmann, and M. Holtz**

## TABLE OF CONTENTS

I. Introduction .......... 144

II. Extracorporeal Blood Purification — Membrane Separation .......... 144

III. Extracorporeal Blood Purification — Membrane Selection .......... 149
- A. Permeability .......... 150
- B. Mechanical Properties .......... 151
- C. Blood Compatibility .......... 152
- D. Membrane Design .......... 152
- E. Membrane Systems .......... 153
  - 1. Cellulose Membranes .......... 153
  - 2. Cellulose Acetate Membranes .......... 159
  - 3. Cellulose Nitrate Membranes .......... 160
  - 4. Collagen Membranes .......... 160
  - 5. Polyether Block Copolymer Membranes .......... 161
  - 6. Polyacrylonitrile-Based Membranes .......... 162
  - 7. Polycarbonate-Based Membranes .......... 163
  - 8. Methacrylate-Based Membranes .......... 165
  - 9. Polystyrene-Based Membranes .......... 166
  - 10. Poly(vinyl pyrrolidone)-Based Membranes .......... 166
  - 11. Polyamide-Based Membranes .......... 166
  - 12. Polypeptide Membranes .......... 166
  - 13. Polysulfone Membranes .......... 167
  - 14. Polyelectrolyte Membranes .......... 167
  - 15. Poly(vinyl alcohol)-Based Membranes .......... 167
  - 16. Heparinized Membranes .......... 168
  - 17. Sorbent Membranes .......... 169
  - 18. Future Development .......... 170

References .......... 170

# I. INTRODUCTION

This article is concerned with polymer membranes used in clinical applications. The role of such membranes as biomaterials is to act as an imperfect barrier and thereby permit the transfer of solute, liquid, or gas in contact with blood during clinical conditions.

The clinical procedures dealt with in this article may be grouped under the heading extracorporeal blood purification, where extracorporeal blood purification denotes the contact of membranes with whole blood and the transfer of solute or liquid. The clinical procedures are

- Hemodialysis
- Hemofiltration
- Simultaneous hemofiltration-hemodialysis
- Sequential ultrafiltration-hemodialysis

The precise nature of the structure-property relationship of the polymer membranes is dependent ultimately on the membrane separation process envisaged. However, in the general sense, it is necessary to consider three properties. These properties are

- Permeability
- Mechanical strength
- Blood compatibility

Permeability is essential for the desired transfer, mechanical strength for membrane handling, and blood compatibility for clinical usage. To comply with clinical acceptability, the polymer membrane is required to achieve appropriate levels of these properties, while membrane development is directed towards an enhancement of these properties.

# II. EXTRACORPOREAL BLOOD PURIFICATION—MEMBRANE SEPARATION

Membrane separation processes[1-6] enable the uneven transport of solute or solvent under the driving force of a gradient based on concentration, pressure, electrical potential, or temperature. The separation processes relevant to the use of polymer membranes in extracorporeal blood purification are dialysis, osmosis, and ultrafiltration. These processes may be summarized (Table 1) in terms of driving force, mass flow, and principal result.

Dialysis[6] takes place when solutions of different concentration are separated by a membrane and a permeable solute passes from the more concentrated solution through the membrane under the action of a concentration gradient driving force, the process being one of diffusion. In a dilute multisolute system, each solute diffuses across the polymer membrane under the action of its individual concentration gradient driving force.

Osmosis occurs when a membrane separates a solvent from solution containing impermeable solute and solvent passes through the membrane into the solution under the action of a concentration gradient driving force. The osmotic pressure is the excess pressure which must be applied to the solution in order to prevent the entry of solvent, and being dependent on membrane and solute characteristics, it is not a simple solution property.

Ultrafiltration[7] takes place when solvent passes through a membrane under the action of a pressure gradient driving force. The passage of fluid may be associated with the transfer of solute (the convective component), with this movement of solute contributing to the overall solute transfer. The rate of ultrafiltration is dependent on the hydrostatic and osmotic pressure gradients across the membrane. Solutes up to a critical size, which is determined by the properties of the membrane, pass through the membrane at the same rate.

**Table 1**
**MEMBRANE SEPARATION PROCESSES**

| Process | Driving force | Mass flow | Principal result |
|---|---|---|---|
| Dialysis | Concentration gradient | Solute transfer | Solute leaves more concentrated solution |
| Osmosis | Concentration gradient | Solvent transfer | Solvent enters more concentrated solution |
| Ultrafiltration | Pressure gradient | Solvent transfer | Solvent leaves more concentrated solution |

From McMullan, J. T., Ed., *Physical Techniques in Medicine*, Vol. 1, Copyright 1977 by John Wiley & Sons. Reprinted by permission of John Wileys & Sons.

Dialysis, osmosis, and ultrafiltration are the separation processes which permit the utilization of polymer membranes in extracorporeal blood purification. Dialysis allows the passage of solute by diffusion, with the transport rate a function of solute size. Osmosis and ultrafiltration influence fluid removal and ultrafiltration may also be used for solute removal, with membrane porosity the controlling factor.

The clinical procedures using polymer membranes in extracorporeal blood purification have had as the primary objective the replacement of functions normally carried out by the human kidney. The role of the human kidney[6,8,9] is the regulation of the volume and composition of the internal fluid environment of the body. The kidney maintains the normal composition of blood, regulates the acid-base balance, and participates in the control of blood pressure. The stages in the operation of the kidney are glomerular filtration, tubular reabsorption, and tubular secretion, with the transfer processes in the reabsorption and secretion stages complex and not completely understood.

Extracorporeal blood purification utilizes polymer membranes and membrane separation processes in order to compensate for renal insufficiency. The principal objective is the controlled removal of solute and water.

The most widely used clinical procedure in extracorporeal blood purification is that of hemodialysis,[10] a procedure generally referred to as the artificial kidney. In hemodialysis (Figures 1 and 2), blood from a patient is spread out as a film over a membrane before returning to the patient. On the other side of the membrane from the blood, a dialysate fluid or balanced isotonic solution is passed. Permeable solutes present in higher concentration in the blood pass through the membrane by dialysis and enter the dialysate solution. The removal of certain blood constituents can be avoided by the addition of these constituents to the dialysate in approximately physiological concentrations. Ultrafiltration and dialysis take place simultaneously. The driving force for ultrafiltration is the hydrostatic pressure gradient between blood and dialysate with osmotic pressure gradients representing a secondary consideration. It is normal practice to avoid osmosis, but it can be used to achieve water removal by the addition of certain solutes to the dialysate.

Since the early hemodialysis studies,[11-13] there have been many proposals for dialyzer design aimed at an optimum utilization of the polymer membranes. Three basic design configurations have emerged.[14] These are the coil dialyzer (Figure 3), the flat bed or parallel plate dialyzer (Figure 4), and the capillary or hollow fiber dialyzer (Figure 5).

The coil dialyzer uses membrane in the form of lay-flat tubing sandwiched between mesh screening, with the assembly wound around a central core. Dialysate is circulated through the interstices of the mesh at right angles to the flow of blood.

In the parallel plate dialyzer, blood flows within a membrane envelope formed by two sheets of polymer, dialysate normally flowing countercurrent to the blood. The parallel plate concept has given rise to numerous versions, since possibilities exist for varying the number

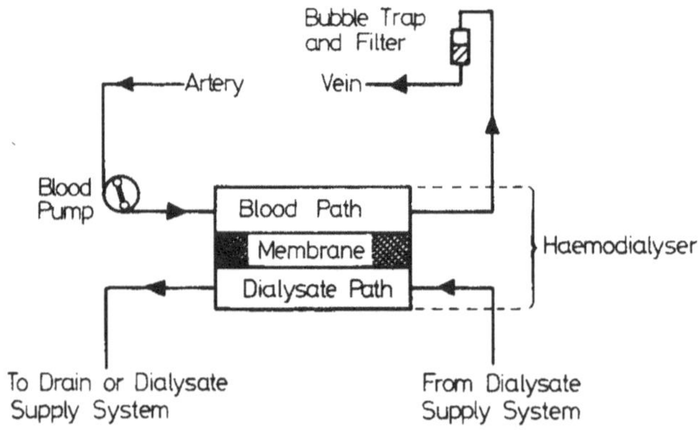


FIGURE 1. Schematic of extracorporeal hemodialysis technique. (From Gaylor, J. D. S., *A Textbook of Biomedical Engineering*, Kenedi, R. M., Ed., Blackie, Glasgow, 1980, 115. With permission.)

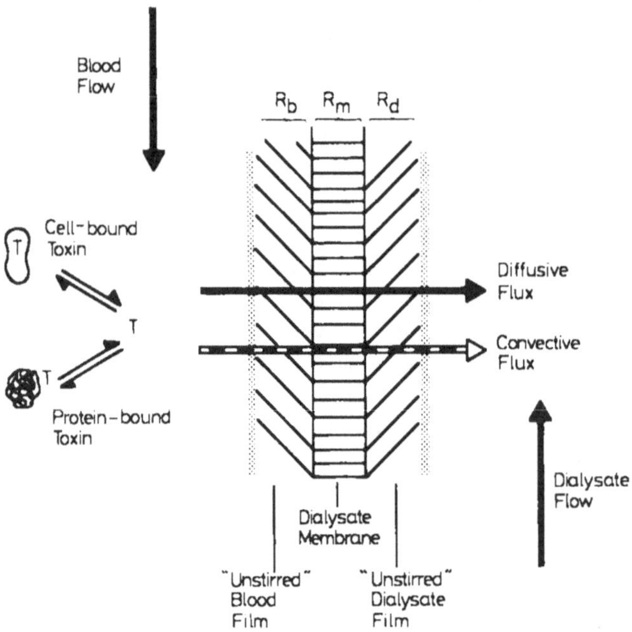


FIGURE 2. Schematic representation of the transport processes involved in clinical dialysis. The toxin, T, may be free in the bloodstream, contained in cells, or bound to macromolecules, but only the free toxin can cross the dialysis membrane. The three mass transfer resistances are the membrane resistance, $R_m$, and the liquid film resistances on the blood and dialysate sides, $R_b$ and $R_d$, respectively. Blood flow and dialysate flow are shown in the countercurrent mode. (From Gardner, C. R., *Membrane Separation Processes*, Meares, P., Ed., Elsevier, Amsterdam, 1976, 529. With permission.)

and dimensions of the membrane envelopes, the method of introducing blood into the envelope, and the configuration of membrane support structures.

Hollow fiber dialyzers are constructed on the shell and tube heat exchanger principle using fibers of about 200-μm internal diameter, a typical dialyzer containing >10,000 fibers. Hollow fiber dialyzers offer the highest ratio of effective membrane area to machine size

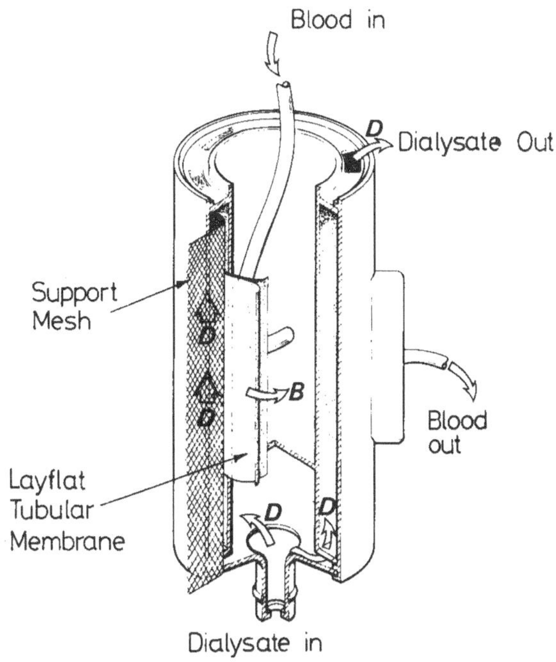


FIGURE 3. Sectioned view of a coil dialyzer showing direction of blood (B) and dialysate (D) streams.

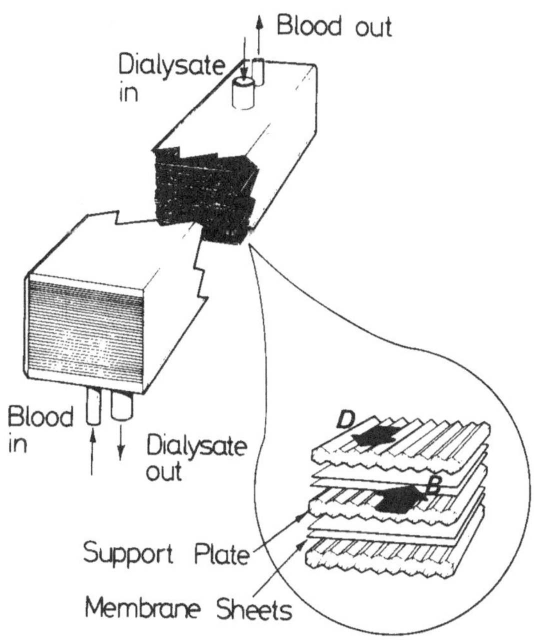


FIGURE 4. Sectioned view of a parallel plate dialyzer showing direction of blood (B) and dialysate (D) streams.

and provide a means of reducing blood priming volume for a given membrane area. With an appropriate ratio of fiber diameter to wall thickness, the need for membrane backing support is eliminated. Presentation of the fibers is important and this has led to the devel-

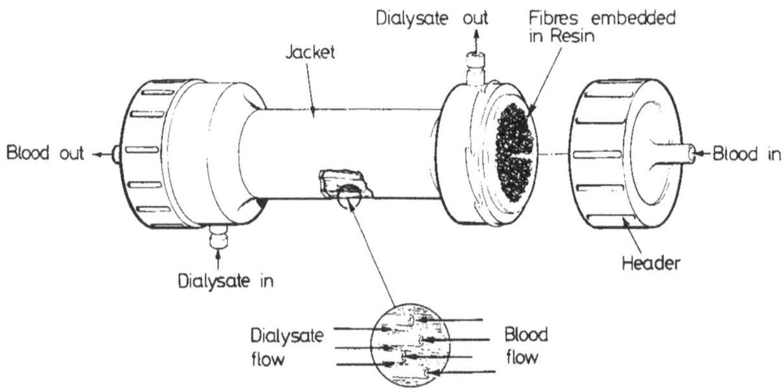


FIGURE 5. Sectioned view of a hollow fiber dialyzer. (From Gaylor, J. D. S., *A Textbook of Biomedical Engineering*, Kenedi, R. M., Ed., Blackie, Glasgow, 1980, 115. With permission.)

opment of technology[15-17] to ensure reproducible dialyzer manufacture. Entry of blood into the fibers is achieved by embedding in a suitable polymer and cutting to expose the fiber ends. The most widely used polymers for embedding are polyurethane, epoxide resin, and polysiloxane.

Hemodialysis remains the most important clinical procedure for extracorporeal blood purification. However, in an effort to improve performance in terms of solute and water removal and to enhance clinical efficiency, alternative procedures have been proposed. Basically, these procedures endeavor to make a more effective use of ultrafiltration,[18,19] either for the transport of water or for the convective transport of solute. Examination of these procedures is complicated by the fact that terminology has not been uniform and expressions such as diafiltration,[20-22] hemofiltration,[23-25] hemodiafiltration,[26,27] and hemoanadiafiltration[28] have been used. To reduce confusion, this chapter will attempt to follow the proposed guildelines for nomenclature.[29,30] Hence, the processes to be considered are hemofiltration, simultaneous hemofiltration-hemodialysis, and sequential ultrafiltration-hemodialysis.

The clinical use of hemofiltration stems from the work of Henderson and co-workers[20,31-34] and has received considerable attention[35-41] as an alternative to hemodialysis. Hemofiltration seeks to improve on hemodialysis by overcoming the problem of those solutes, which because of size are removed inefficiently from the blood by the diffusive process of dialysis. Hemofiltration is a procedure for extracorporeal blood purification in which solute removal is accomplished by convective transport using membranes of higher hydraulic permeability than that of conventional hemodialysis membranes (Table 2). Since hemofiltration involves a much stronger ultrafiltration than hemodialysis, it is necessary to compensate for excessive fluid removal. Thus, in hemofiltration, blood is treated by a combination of ultrafiltration and convective solute loss with a physiological saline solution added to replace fluid losses exceeding the planned removal of excess total body water. The physiological solution may be added before or after the ultrafiltration stage (Figure 6). Hemofiltration, being based on ultrafiltration and partial restoration of fluid, approximates more closely than hemodialysis to the operation of the human kidney.

Simultaneous hemofiltration-hemodialysis,[27] sometimes, like hemofiltration itself, referred to as hemodiafiltration, is a combination of hemofiltration and hemodialysis, where both convection and diffusion contribute significantly to solute removal. The objective is to obtain solute removal over a molecular weight range superior to the removal possible with hemodialysis or hemofiltration alone.

Sequential ultrafiltration-hemodialysis[42] involves the incorporation of an additional

**Table 2**
**COMPARISON BETWEEN HEMODIALYSIS AND HEMOFILTRATION MEMBRANES**

| Application polymer type | Cuprophan® 150 PM hemodialysis cellulose | Rhone-Poulenc® RP-AN69 hemodialysis/ hemofiltration polyacrylonitrile | Amicon® XM-50 hemofiltration polysulfone |
|---|---|---|---|
| Hydraulic permeability (mℓ/min/cm²/mmHg) × $10^6$ | 3.8 | 45.0 | 110.0 |
| Sieving coefficients[a] | | | |
| Urea (60 daltons) | 1.00 | 1.00 | 1.0 |
| Creatinine (113 daltons) | 0.98 | 1.00 | 1.0 |
| Vitamin $B_{12}$ (1,355 daltons) | 0.75 | 0.96 | 1.0 |
| Inulin (5,200 daltons) | 0.37 | 0.80 | 1.0 |
| Bovine albumin (60,000 daltons) | 0 | ~0 | ~0 |

[a] The sieving coefficient is defined as the ratio of solute concentration in ultrafiltrate to that in the bulk of the fluid being ultrafiltered.

From Keshaviah, P., *Chronic Replacement of Kidney Function*, Vol. 75, AIChE Symposium Series, No. 187, Villarroel, F. and Dedrick, R. L., Eds., American Institute of Chemical Engineers, New York, 1979, 24. With permission.

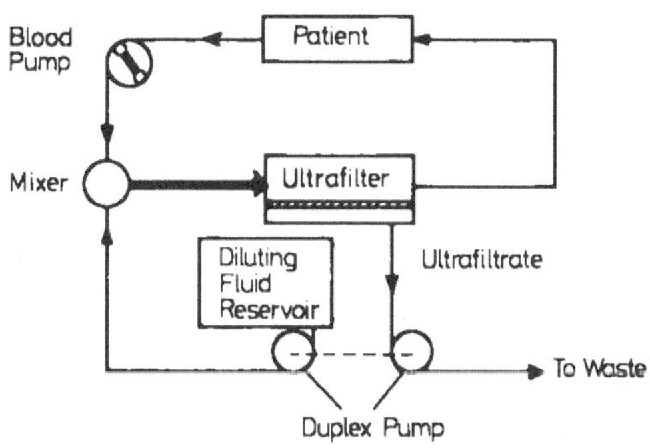


FIGURE 6. Schematic of hemofiltration technique with predilution configuration.

ultrafiltration stage into the chemical treatment. Such a stage can augment solute and water removal and overcome problems arising from too rapid removal of water during hemodialysis. Therefore, sequential ultrafiltration-hemodialysis[42,43] is a process for extracorporeal blood purification in which solute and fluid are removed by convective transport through conventional hemodialysis membranes. This ultrafiltration stage, which uses transmembrane pressures greater than those in normal hemodialysis, may be performed before, after, or independently of hemodialysis.

## III. EXTRACORPOREAL BLOOD PURIFICATION — MEMBRANE SELECTION

Polymer membranes used in extracorporeal blood purification must comply with requirements of permeability, mechanical strength, and blood compatibility[44,45] and it is important to consider each of these properties in turn prior to an examination of polymer membranes proposed for clinical use. However, it is appropriate to indicate at this stage that membrane selection has been strongly influenced by the fact that membranes of regenerated cellulose became established clinically at an early stage in the development of hemodialysis. The acceptance and widespread use of regenerated cellulose membranes meant the existence of a reference material with which to compare the properties of other polymer membranes with respect to permeability, mechanical strength, and blood compatibility.

**A. Permeability**

In a consideration of the permeability of polymer membranes used in the major application of extracorporeal blood purification, viz. the treatment of chronic renal failure, it is necessary to realize and accept that the permeability requirements cannot be stipulated exactly. This is because information is incomplete on the solutes occurring in uremia and which should be removed in a treatment involving polymer membranes.

Regarding hemodialysis, the net solute flux of species A, $J_A$, and the net volume flux, $J_V$, are given by[44,46]

$$J_A = P_m \Delta^\circ C_A + \overline{C}_A(1 - \sigma) J_v \quad (1)$$

$$J_V = L_p (\Delta P - \sigma \Delta \pi) \quad (2)$$

where $P_m$ is the membrane permeability factor, $\Delta^\circ C_A$ and $\Delta P$ are the respective solute concentration and hydrostatic pressure differences across the membrane, $\overline{C}_A$ is the average solute concentration in the membrane, $L_P$ is the hydraulic conductivity cofficient, $\sigma$ is the Staverman reflection coefficient, and $\Delta\pi$ is the osmotic pressure difference that would be exerted across the membrane if the membrane was completely impermeable to the solute.

An important feature of extracorporeal blood purification is that solute diffusion is determined by factors additional to membrane permeability. The presence of concentration boundary layers associated with low velocity regions on either side of the membrane gives rise to the following relationship between $P_m$ and the overall permeability $P_o$, in the absence of ultrafiltration

$$1/P_o = 1/P_m + 1/P_b + 1/P_d \quad (3)$$

where $P_b$ and $P_d$ denote permeability through the boundary layers on the blood and dialysate sides, respectively.

In terms of resistances, Equation (3) may be restated as

$$R_o = R_m + R_b + R_d \quad (4)$$

In the early years of hemodialysis, membrane permeability was assessed in terms of the ability of the membrane to remove solutes such as urea, creatinine, uric acid, and glucose, which with a molecular weight range of 60 to 180, are commonly referred to in extracorporeal blood purification as low molecular weight solutes.

With the low molecular weight solutes, $R_m$ is of the same order as $R_b + R_d$. As solute size increases, $R_m$ increases in comparison to $R_b + R_d$ and a stage is reached for higher molecular weight solutes when $R_o$ and $R_m$ are about equal. In 1971,[47] a hypothesis was put forward that the efficient removal of these higher molecular weight solutes should form the basis of clinical treatment. In this situation when $R_o \simeq R_m$, the solute removal from the patient can be shown to be proportional to the product of membrane area and dialysis time. This "square meter-hour" hypothesis has had important consequences.

1. Attention was focussed on these higher molecular weight solutes or "middle molecules" as they came to be termed and membrane development and membrane permeability assessment were forced to take account of this change in emphasis.

2. One consequence of the hypothesis, viz. that for a given polymer membrane it is possible to reduce treatment time and compensate by increasing membrane area, was adopted clinically.
3. The concept of middle molecules stimulated interest in alternative procedures to hemodialysis and the development of membranes with permeability characteristics different from those of regenerated cellulose.

Interest in middle molecules has continued to the present[48] with controversy remaining with respect to the separation and identification of such molecules and their possible role in uremia and other diseases. Because of the interest in middle molecules, it has become customary to evaluate membrane permeability over a solute molecular weight range of about 50 to 5000.

Polymer membrane evaluation has included the selection of test procedures and solutes,[49-55] the relationship between diffusive and hydraulic permeabilities,[56] and the comparison of experimental and calculated permeability values.[57] Evaluation has also taken into account developments such as hollow fiber membranes[58] and hemofiltration.[59-61]

Hollow fiber membranes require special test circuits,[58,62,63] while the in vitro assessment of sheet membranes[64] follows one of two procedures, viz. measurement of an overall resistance or measurement of the membrane diffusion resistance.

In the measurement of an overall resistance,[65-67] a test solution is circulated, but in order to simulate clinical conditions, the fluid film resistances are not completely eliminated. The results, which are normally expressed in terms of $t_{1/2}$, the time taken for the solute concentration to reach one half of its initial value, must be quoted with respect to results obtained for a reference membrane.

In measurement of the membrane diffusion resistance, resistances can be eliminated by high speed stirring[49,50] or by the use of open pore cellular foam membrane support structures.[52] Values of $R_m$ are reported in units of minutes per centimeter or seconds per centimeter.

For the assessment of membrane permeability in the clinical situation, the overall solute transport is generally expressed in terms of clearance by analogy to the clearance concept in renal physiology.

Clearance is the amount of solute removed from the blood phase per unit time divided by the incoming blood concentration and represents the volumetric rate at which blood is cleared of solute. Dialyzance is the amount of solute removed from the blood phase per unit time divided by the difference in concentration between incoming blood and incoming dialysate. Dialyzance is most commonly used with reference to recirculating dialysis procedures utilizing immersed coil dialyzers. For machines where dialysate is passed only once, clearance and dialyzance are equal. The assessment of ultrafiltration properties consists of measuring water transport at different pressure gradients, plotting $J_V$ against $\Delta P$, and obtaining $L_p$ from the slope of the linear plot.[64]

Permeability evaluation as a guide towards membrane selection is largely concerned with the suitability of the membrane for the treatment of uremia. The suitability of a membrane for other applications, such as drug and poison removal,[68] hepatic support,[69,70] psoriasis,[71] or schizophrenia[72] is determined by animal studies or clinical assessment.

**B. Mechanical Properties**

The procedures which have become established for the assessment of the mechanical properties of polymer membranes are directed towards providing an indication of the ability of the membrane to perform satisfactorily during extracorporeal blood purification. Tests are normally carried out with dry membrane and with membrane soaked in saline for 1 hr at 37°C. The properties measured include breaking load, elongation at break, and resistance to creep.

The resistance to creep is important, since excessive membrane distension during extracorporeal blood purification can cause blockage of the dialysate channels. In this respect, guidelines have been published for the determination of the minimum yield stress required of a membrane to avoid excessive distension under the normal pressure gradient in a standard dialyzer. As expected, the evaluation of hollow fibers is more complicated than that of membrane in sheet form.[44,45,58]

### C. Blood Compatibility

The relationship between polymer structure and blood compatibility, the assessment of blood compatibility, the relevance of in vitro and animal studies to the clinical situation, and, indeed, what constitutes blood compatibility form important topics in biomaterials. Theories on blood compatibility of polymers,[73] blood compatibility testing,[74] and the evaluation of surface properties for blood contacting polymers[75] are dealt with elsewhere in this book.

Regarding membranes for extracorporeal blood purification, blood compatibility assessment is influenced by the fact that the clinical situation generally involves contact of the membrane with anticoagulated blood and the availability of regenerated cellulose as a reference material. Membranes are evaluated by means of in vitro circuits in which human blood is contacted with the polymer in a test cell, animal studies using test cells or dialyzers, and clinical assessment.[46] In an evaluation of polymer membranes, information can be derived from studies such as the following:

1. Platelet and leucocyte counts
2. Platelet and leucocyte function
3. Release of platelet constituents, particularly β-thromboglobulin and 5-hydroxytryptamine
4. Platelet aggregation response
5. Platelet adhesion and aggregation
6. Erythrocyte adhesion
7. Adsorption of albumin and fibrinogen
8. Activation of complement system
9. Partial thromboplastin time
10. Fibrinopeptide A release

### D. Membrane Design

By adapting the definition of the perfect hemodialysis membrane given by Leonard and Bluemle,[76] the properties of a membrane for extracorporeal blood purification may be summarized as[6]

1. Permeability to the desired solute or solutes
2. Impermeability to the proteins in the blood
3. Impermeability to bacteria or viruses which may be present in nonsterile dialysate
4. Adequate water permeability
5. Sufficient mechanical strength for clinical use
6. Satisfactory blood compatibility
7. Absence of pyrogens

As previously observed, knowledge of the solutes to be removed is often limited and membrane permeability is normally directed towards a molecular weight range. Transport of solute and water can be accomplished by a combination of diffusion and convection, as in hemodialysis, or by convection alone, as in hemofiltration.

Various approaches are available to the designer of membranes for extracorporeal blood purification. These approaches, which can be used in combination, are

1. Preparation of membranes from polymer structures providing a hydrophilic/hydrophobic balance (such membranes swell in contact with blood and dialysate, thus permitting solute and water transport, while retaining insolubility)
2. Stretching of wet membranes
3. Leaching of substances incorporated into the membrane
4. Manufacture of membranes with an asymmetric structure

The hydrophilic/hydrophobic balance can be achieved by cellulose regeneration, insolubilization of water-soluble polymers, copolymerization of selected monomers, and chemical modification of membranes. With regard to membrane manufacture, membranes based on regenerated cellulose are obtained by processes established for the production of packaging films,[77,78] while other polymer membranes are generally produced by solvent casting techniques.[3]

Membranes in hollow fiber form[79] are prepared by techniques developed from the technology used in the manufacture of hollow textile fibers. Hollow fiber membranes are produced by extruding or spinning the polymer through a spinneret or small annular die. Production techniques[80] include wet spinning, in which a polymer solution is spun and coagulated in a nonsolvent, dry spinning, in which a polymer in a volatile solvent is spun into an evaporative column, and melt spinning, in which molten polymer is spun and cooled.

Asymmetric or anisotropic membranes comprise a dense ultrathin skin with an underlying porous substructure. Such membranes originated with the development of cellulose acetate membranes for desalination by reverse osmosis.[81] The membranes are solvent cast, often from a mixed solvent system, with the skin forming during a brief drying period and the porous substructure during gelation, which may be induced by immersion in a nonsolvent.

Asymmetric membranes have a characteristic structure. This structure[7,82] represents a combination of a dense skin for selectivity and a highly permeable porous substructure for support and permits high rates of ultrafiltration for a given solute molecular weight cut off. Membrane properties can be altered by varying casting conditions and by leaching. Asymmetric membranes can be produced in the form of flat sheet or hollow fibers.[79,83]

### E. Membrane Systems

In considering polymer structures proposed as the basis of membrane systems for extracorporeal blood purification,[6,84-87] it is important to bear in mind the following points:

1. Numerous membrane systems have been suggested, but few have reached the stage of clinical evaluation.
2. Research after 1971 has been influenced by the "square meter-hour" hypothesis.
3. The dominant role of regenerated cellulose membranes has greatly affected membrane development and assessment.

#### *1. Cellulose Membranes*

Regenerated cellulose membranes form the logical starting point in a consideration of membrane systems. Cellulose membranes have a long association with hemodialysis[12] and remain the most widely used hemodialysis material. Indeed, the availability of cellulose membranes contributed greatly to the establishment of hemodialysis as a routine clinical procedure.

The structure of cellulose (Figure 7) consists of a large number of anhydroglucose units joined together by β-glucosidic linkages.[88] A feature of cellulose of great importance to membrane formation and utilization is the strong intermolecular attraction resulting from hydrogen bonding. This bonding, which is promoted by the regular arrangement of hydroxyl groups, ensures that cellulose is insoluble in solvents, and, since thermal degradation occurs

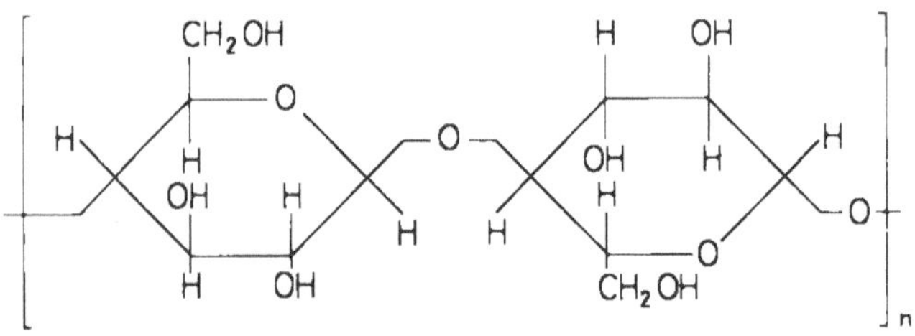


FIGURE 7. Cellulose structure.

before flow, cellulose cannot be melted. Therefore, the manufacture of cellulose membranes[77,78] requires the production of a soluble or thermoplastic derivative as an intermediate. The other important consequence of the intermolecular bonding is that cellulose, while swelling in an aqueous medium, remains insoluble and is suitable for use in contact with blood.

The manufacture of cellulose membranes involves three basic steps.

1. Chemical modification to produce the soluble or thermoplastic cellulose derivative
2. Membrane formation by solvent casting or melting
3. Treatment of the cellulose derivative to achieve regeneration into cellulose

In the strict sense, regeneration is incomplete, since there is a reduction in molecular weight. Production of the cellulose derivative is a key factor and three techniques are used.

1. The cuprammonium process forms a soluble complex by dissolution of cellulose in cuprammonium solution. Regeneration is by reaction with acid.
2. The viscose process forms a xanthate by reaction of cellulose with alkali and carbon disulphide. Regeneration is by reaction with acid.
3. Thermoplastic cellulose acetate is produced and hydrolyzed to cellulose by reaction with alkali.

Cellulose films generally can be identified by a system of figures and letters derived from a code used in the manufacture of regenerated cellulose films as wrapping materials.[78] The figures refer to the nominal weight in grams of 10 $m^2$ of film and, thus, are indicative of thickness. The letters denote the condition of the film by indicating whether it is coated or uncoated, since cellulose films may be coated with cellulose nitrate in order to reduce moisture sensitivity. Regenerated cellulose films used as membranes for extracorporeal blood purification do not require treatment for reducing moisture sensitivity and were formerly described by the letters PT, signifying that the membranes were uncoated or "plain, transparent". However, it is now more common to use the letters PM, denoting that the membrane is uncoated and suitable for medical application.

Regenerated cellulose membranes are too brittle to be handled satisfactorily in the dry state. The addition of a plasticizer is essential and glycerol is normally used. Plasticization and the moisture sensitivity of the membranes make careful storage a necessity. Contact of the cellulose membrane with an aqueous environment leads to displacement of the plasticizer, an increase in thickness and a marked change in mechanical properties, with a fall in the breaking load and a rise in the elongation at break.

With respect to solute transport characteristics, cellulose membranes appear to act as sieves or microporous barriers.[66,89] Thus, solute permeability, for a given membrane, is principally in inverse proportion to the molecular volume of the diffusing solute.[53] Since the influence of charge or adsorption properties on solute transport can be ignored, it follows that a cellulose membrane is selective to closely related solutes only when the molecular volumes of the solutes approximate to the pore size of the membrane.

## Table 3
## TYPICAL PROPERTIES OF CUPROPHAN® FLAT SHEET MEMBRANES

| Type | Thickness (μm) | | Breaking[a] load, (N) | Elongation[a] at break, (%) | Ultrafiltration (mℓ/h/m²/mmHg) | Permeability at 37°C × 10³(cm/min) | |
|---|---|---|---|---|---|---|---|
| | Dry | Wet | | | | Urea | Vitamin $B_{12}$ |
| 80 PM | 8.0 ± 0.5 | ~16 | 21.58 | 11 | 4.9 ± 0.5 | 70 | 8.3 |
| 100 PM | 10.5 ± 0.5 | ~22 | 28.45 | 11 | 3.7 ± 0.5 | 60 | 7.2 |
| 150 PM | 11.5 ± 0.5 | ~23 | 29.43 | 10 | 3.5 ± 0.5 | 53 | 5.8 |
| 200 PM | 13.5 ± 0.5 | ~28 | 34.34 | 13 | 2.9 ± 0.3 | 52 | 4.9 |
| 250 PM | 17.5 ± 0.5 | ~37 | 41.20 | 17 | 1.9 ± 0.3 | 40 | 3.0 |

[a] Values obtained for 15-mm-wide samples.

Reproduced by permission of Membrana GmbH, Wuppertal, West Germany.

## Table 4
## TYPICAL PROPERTIES OF CUPROPHAN® HOLLOW FIBER MEMBRANES

| Type | 8/200 UF4.2 | 8/200 UF5.5 | 11/200 UF2.9 | 11/200 UF4.0 | 11/200 UF4.2 | 11/200 UF4.4 | 16/215 UF3.0 |
|---|---|---|---|---|---|---|---|
| Name | FI IM | FI IM | CI IM | CI IM | CI IM | CI IM | D2 IM |
| Inner diameter — dry (μm) | 200 ± 15 | 200 ± 15 | 200 ± 15 | 200 ± 15 | 200 ± 15 | 200 ± 15 | 215 ± 15 |
| Wall thickness — dry (μm) | 8 ± 1.1 | 8 ± 1.1 | 11 ± 1.1 | 11 ± 1.1 | 11 ± 1.1 | 11 ± 1.1 | 16 ± 1.6 |
| Outer diameter—wet ca. (μm) | 260 | 260 | 270 | 270 | 270 | 270 | 290 |
| Tensile strength — dry (*N*) | 0.785 | 0.785 | 1.472 | 1.472 | 1.472 | 1.472 | 1.962 |
| Elongation at break — dry (%) | 15 | 15 | 15 | 15 | 15 | 15 | 20 |
| Ultrafiltration mℓ/m²/hr/mmHg | 4.2 ±0.4 | 5.5 ±0.5 | 2.9 ±0.3 | 4.0 ±0.4 | 4.2 ±0.4 | 4.4 ±0.4 | 3.0 ±0.3 |
| Vitamin $B_{12}$; permeability at 37°C × 10³ cm/min | 4.8 | 6.6 | 3.3 | 4.6 | 4.8 | 5.1 | 3.4 |

Reproduced by permission of Membrana GmbH, Wuppertal, West Germany.

The most important regenerated cellulose membranes are those manufactured by Membrana GmbH under the tradename Cuprophan® often referred to in the literature as Cuprophane. These membranes are produced in flat sheet, tubular, and hollow fiber form using the cuprammonium process and are available generally to device manufacturers. Cuprophan® hollow fibers are filled with a water-soluble, alcohol-soluble liquid, such as isopropylmyristate, to minimize deformation during handling and dialyzer assembly. Typical properties of Cuprophan® membranes are listed in Tables 3 and 4. It is important to appreciate the significance for extracorporeal blood purification of the availability of Cuprophan® membranes and the willingness of the manufacturers to undertake membrane development.[90] Cuprophan® membranes are used clinically in a number of devices. These membranes also provide reference standards. In this respect, the most common reference standard membrane is Cuprophan® 150PM (150PT).

There is no range of membranes comparable to that of Cuprophan® produced by the viscose process. Since 300PT membranes regenerated from the xanthate have been assessed and British Cellophane Ltd. prepared a flat sheet 150PT membrane under the name Tec-

**Table 5**
**TYPICAL PROPERTIES OF NEPHROPHAN® FLAT SHEET MEMBRANES**

| Solute | Mol. Wt. | Permeability × $10^3$ (cm/min) | Permeability Nephrophan®/ permeability Cuprophan® (150PM) |
|---|---|---|---|
| Urea | 60 | 26.53 | 1.26 |
| Creatinine | 113 | 17.50 | 1.66 |
| Uric acid | 168 | 8.02 | 1.61 |
| Vitamin $B_{12}$ | 1355 | 3.59 | 1.48 |

*Note:* Values obtained at 37°C with test cell designed to measure an overall permeability.

tophane, the reasons for the domination of the cuprammonium process would appear to be commercial rather than technical.

Manufacture of cellulose membranes by deacetylation of cellulose acetate derives importance from the fact that it is used in the production of hollow fiber membranes by the Cordis Dow® Corporation.[15] In a typical preparation, cellulose acetate plasticized with tetramethylene sulfone is melt extruded to produce hollow fibers, which are hydrolyzed by reaction with sodium hydroxide. The tetramethylene sulfone is removed during the hydrolysis and is replaced with glycerol. The Cordis Dow® fibers pioneered the clinical utilization of hollow fiber hemodialysis systems.

Regenerated cellulose membranes are now fully established and there has been considerable assessment.[53,55,56,66,91-96] There have also been several approaches adopted in order to achieve an improvement in properties.

One approach set out to optimize membranes produced by the cuprammonium process,[97-100] the general objective being to increase permeability by decreasing polymer crystallinity. Examination was made of the influence on membrane properties of alterations to the molecular weight of cellulose, concentration of the casting medium, the addition of poly(propylene glycol) and dimethylsulfoxide, and treatment with zinc chloride and glycerol. Other aspects examined were regeneration with acetic acid instead of sulfuric acid[98] and the improvement of mechanical properties by cross-linking with formaldehyde.[100]

Another approach[101,102] was designed to obtain regeneration of cellulose with less degradation than that occurring in the cuprammonium and viscose processes. Solutions of cellulose in cupriethylene diamine and cadmiumethylene diamine were regenerated to give membranes termed cuenophane and cadophane, respectively.

Klinkmann and co-workers[67] reported on membranes produced by the viscose process, but using a special stretching technique and having glycerol and sorbitol as additives to the casting medium. These membranes are described generally as Nephrophan®. Assessment of Nephrophan® membranes[103] indicated possible advantages over Cuprophan® 150PT with respect to ultrafiltration. Interest in the Nephrophan® membranes has continued[104] and promising in vitro results have been obtained (Table 5).

Weber et al.[105] described two methods of strengthening cellulose membranes. Internal support was achieved with cellulose fiber strands and external support was achieved with polyamide mesh or cellulose nonwoven mesh. Such mechanical strengthening must be balanced against a loss in permeability.

Hoenich et al.[106,107] carried out the assessment of wet gel membranes. These were Cuprophan® membranes which were not allowed to dry out following regeneration. The wet

### Table 6
### WET GEL MEMBRANES (CUPROPHAN® PMW 200) IN VITRO MIDDLE MOLECULAR CLEARANCE (mℓ/min)[a]

| | $^{14}C$ EDTA (mol wt 380) | $^{3}H$ Vitamin $B_{12}$ (mol wt 1355) | $^{14}C$ Inulin (mol wt 5175) | UF correction (mℓ/min) |
|---|---|---|---|---|
| Cuprophan® PMW 200 | 76.3 ± 2.8 | 39.6 ± 2.9 | 13.5 ± 2.1 | 5.5 |
| Cuprophan® PM 150 | 67.4 ± 1.9 | 33.7 ± 1.1 | 8.1 ± 3.2 | 2.1 |

[a] Blood flow, 200 mℓ/min; dialysate flow, 530 mℓ/min; temperature, 38°C.

*Note:* $^{3}H$ Tritium labelled; mean ±SD shown.

From Hoenich, N. A., White, T., Conceicao, S., and Kerr, D. N. S., *Technical Aspects of Renal Dialysis*, Frost, T. H., Ed., Pitman, London, 1978, 107. With permission.

### Table 7
### WET GEL MEMBRANES (CUPROPHAN® ± PMW 200) IN VIVO ULTRAFILTRATION (mℓ/min)[a]

| | | Running pressure (mmHg) | |
|---|---|---|---|
| | | 100 | 200 |
| Cuprophan® PMW 200 | n = 7<br>r = 0.98 | 17.9 | 36.4 |
| Cuprophan® PM 150 | n = 40<br>r = 0.73 | 11.9 | 22.8 |

[a] Blood flow, 200 mℓ/min; dialysate flow, 530 mℓ/min; temperature, 38°C.

*Note:* Running pressure defined as $(P_{B\,Out} + P_{D\,Out})/2$. Results obtained with Meltec Multipoint Dialyser, surface area of 1.0 $m^2$.

From Hoenich, N. A., White, T., Conceicao, S., and Kerr, D. N. S., *Technical Aspects of Renal Dialysis*, Frost, T. H., Ed., Pitman, London, 1978, 107. With permission.

membrane offered possible advantages over Cuprophan® 150PM for middle molecule removal and ultrafiltration (Tables 6 and 7), but the application appeared limited to dialyzers of the nondisposable flat bed type. Cuprophan® membrane permeability has also been enhanced by reducing the cellulose content per unit volume.[108]

Another objective undertaken was that of producing ultrathin cellulose membranes of thickness 200 to 1500 nm. The approach[109,110] selected involved casting a solution of cellulose nitrate in cyclohexanone onto a water surface and converting to cellulose by reduction with ammonium hydroxide. Superiority over Cuprophan® 150PT for solute and water permeability was obtained, although support of the membrane by a woven or nonwoven fabric was necessary. Subsequent work included the examination of supporting materials[111] and the production of cellulose by ammoniacal deacetylation of cellulose acetate.[112] Variations in solvent and deacetylation were studied[113,114] with the aim of achieving a membrane thickness of 2.2 to 2.5 μm. Permeability enhancement and satisfactory blood compatibility were

**Table 8**
**ULTRATHIN CELLULOSE MEMBRANES CLEARANCE RATIOS:CLEARANCE WITH ULTRATHIN MEMBRANE DIVIDED BY CLEARANCE WITH CUPROPHANE MEMBRANE**

| $Q_B$ mℓ/min | BUN | $PO_4$ | Creatinine | Uric acid | PAH | BSP | Vitamin $B_{12}$ |
|---|---|---|---|---|---|---|---|
| 50 | 1.07 | 1.12 | 1.08 | 1.23 | 1.38 | — | — |
| 100 | 1.32 | 1.40 | 1.21 | 1.57 | 1.47 | 1.32 | 1.08 |
| 150 | — | — | — | — | — | 1.68 | — |
| 200 | 1.72 | 1.64 | 1.67 | 1.52 | 1.34 | 1.75 | 1.37 |

From Kjellstrand, C. M., Petersen, R. J., Evans, R. L., Shideman, J. R., Santiago, E. A., Busselmeier, T. J., and Rozelle, L. T., *Trans. Am. Soc. Artif. Intern. Organs*, 18, 106, 1972. With permission.

**Table 9**
**ULTRATHIN CELLULOSE MEMBRANES PLATELET AND WHITE BLOOD CELL COUNTS BEFORE AND AFTER DIALYSIS MEAN AND SD (NUMBER IN PARENTHESIS IS NUMBER OF STUDIES)**

| Membrane | Before dialysis | After dialysis | Change (%) |
|---|---|---|---|
| | Platelets/mm$^3$ × $10^{-3}$ | | |
| Ultrathin | 176 ± 83 (6) | 98 ± 52 (6) | −44 |
| Cuprophane | 129 ± 32 (4) | 94 ± 43 (4) | −28 |
| | WBC/mm$^3$ × $10^{-3}$ | | |
| Ultrathin | 7.8 ± 4.0 (9) | 9.3 ± 8.6 (9) | +19 |
| Cuprophane | 6.8 ± 2.3 (5) | 6.5 ± 1.6 (5) | −4 |

From Kjellstrand, C. M., Petersen, R. J., Evans, R. L., Shideman, J. R., Santiago, E. A., Busselmeier, T. J., and Rozelle, L. T., *Trans. Am. Soc. Artif. Intern. Organs*, 18, 106, 1972. With permission.

demonstrated by in vitro studies with dogs[115] (Tables 8 and 9). This approach also illustrates the problem of balancing increased permeability with decreased mechanical strength.

Other techniques reported for the enhancement of cellulose membrane properties include preswelling[116] and reinforcement with a fibrous web.[117] Regenerated cellulose membranes, in particular Cuprophan®, continue to dominate the hemodialysis field. Such membranes not only provide properties acceptable for clinical usage, but also offer scope for property improvement by alteration to production variables.

The search for alternative membrane systems to those of cellulose extends over 20 years and is still ongoing. At various times, this search has been influenced by attempts to achieve the following:

1. Improved urea permeability
2. Improved permeability to middle molecules
3. Improved water transport
4. Development of membranes suitable for hemofiltration

In this chapter, the membrane systems are examined in terms of the polymer structure on which they are based. Chronological order has not been maintained. Successful and unsuccessful approaches are described in order to ensure a better understanding of the possibilities and limitations.

*2. Cellulose Acetate Membranes*

Partial replacement of the hydroxyl groups of cellulose by acetyl groups reduces the effect of hydrogen bonding, increases interchain separation, and makes the polymer less polar. The properties of a cellulose acetate polymer depend on the degree of acetylation, the nature and proportion of plasticizer, and the chain length of the cellulose molecule.

Acetylation of cellulose introduces the possibilities of producing membranes by solvent casting or melting and the role of cellulose acetate in the preparation of cellulose membranes has been described. Interest in cellulose acetate membranes for extracorporeal blood purification was influenced by the success of cellulose acetate membranes in industrial ultrafiltration and reverse osmosis[118] and the availability of the technology for producing asymmetric membranes[82,83]

Envirogenics Company, a Division of Aerojet-General Corporation, initiated a program[119] for the manufacture of asymmetric cellulose acetate membranes, having a dense skin of 0.2 μm and a porous substructure of 50 to 100 μm. Variations in casting solution composition and gelation techniques produced membranes with the transport of water and solutes in the molecular weight range of middle molecules superior to that obtained with Cuprophan® 150PT. Clinical evaluation of the cellulose acetate membranes was carried out in a flat bed dialyzer,[120-122] and it was reported[123] that membranes could be produced in 40-mm width up to 40 m long. Another aspect of this program was the development of membranes produced from blends of cellulose acetate and diethylaminoethyl cellulose acetate. These membranes, suitable for the attachment of heparin, are considered later.

The role of cellulose acetate in the preparation of cellulose membranes in hollow fiber form has been described and it is, therefore, not surprising that the concept of membranes with enhanced water and solute permeability was extended to the production of cellulose acetate hollow fiber membranes.[124]

Reinforcement of cellulose acetate membranes has been studied by Sparks and co-workers.[125-127] Techniques developed in desalination research were adapted to produce cellulose acetate membranes reinforced with meshes and nonwoven materials. Faster solute transport and much greater ultrafiltration rates than those of Cuprophan® 150PT were obtained.[128,129]

To overcome the problem of controlling ultrafiltration with cellulose acetate membranes, the Celanese® Research Company, a Division of the Celanese® Corporation, examined various approaches for the reduction of ultrafiltration. These approaches[130-133] included water annealing, varying the acetyl content of the polymer, partial hydrolysis of the formed membrane, and varying the acetone to formamide ratio in the casting medium. The preferred membranes which emerged were designated CA-1 and CA-2. Both membranes were formed by casting cellulose acetate from a mixture of acetone and formamide, followed by coagulation in water. Desirable membrane properties required careful control of the casting solvent ratio and the drying conditions. The membranes offered superior ultrafiltration rates to those of Cuprophan® 150PT. Subsequent work[134,135] produced thinner membranes and membranes plasticized with glycerol. Alteration to the casting conditions used in the preparation of membrane CA-1 produced a membrane designated CA-3. The membrane CA-3 gave an ultrafiltration rate 20 times greater than that of Cuprophan® 150PT.

A problem in considering cellulose acetate membranes arises from the fact that the degree of substitution is often not stated. In fact, utilization of cellulose acetate ranges from cellulose triacetate membranes used in hemofiltration to polymers with a lower degree of substitution used as membranes in hemofiltration and hemodialysis. In the inevitable comparison with

**Table 10**
**COLLAGEN MEMBRANES PERMEABILITY MEASUREMENTS**

| | | P × 10⁴ (cm/sec) | |
|---|---|---|---|
| Solute | Mol. wt | Collagen (dry UV) | Cuprophane |
| Urea | 60 | 5.72 | 4.57 |
| Phenol | 94 | 4.42 | 3.35 |
| Creatinine | 113 | 3.37 | 2.71 |
| Uric acid | 168 | 2.23 | 2.15 |
| Sodium | — | 5.39 | 4.49 |
| Calcium | — | 3.33 | 2.45 |
| Phosphate | — | 1.82 | 1.78 |

From Stenzel, K. H., Rubin, A. L., Yamayoshi, W., Miyata, T., Suzuki, T., Sohde, T., and Nishizawa, M., *Trans. Am. Soc. Artif. Intern. Organs*, 17, 293, 1971. With permission.

regenerated cellulose membranes, interest has been focused not only on permeability, but also on membrane blood compatibility with studies based on factors such as platelet activity in chronic renal failure, using β-thromboglobulin as a marker.[136,137]

The low cost of cellulose acetate polymers and the existence of techniques for altering solute and water permeability offer scope to the membrane developer and seem likely to ensure continuing application for cellulose acetate membranes in extracorporeal blood purification.

*3. Cellulose Nitrate Membranes*

Nitration of cellulose presents another means of obtaining soluble film-forming polymers and the use of cellulose nitrate in the preparation of ultrathin cellulose membranes[109,110] has been considered. However, although cellulose nitrate formed the membrane material in the first hemodialysis experiments[11] and asymmetric cellulose nitrate membranes have found application in hemofiltration,[138] cellulose nitrate has not attained the importance of cellulose and cellulose acetate.

*4. Collagen Membranes*

Cellulose and cellulose acetate membranes represent examples of the successful adaptation of a naturally occurring material. Another attempt to utilize a naturally occurring material was the development of collagen membranes.

One approach, conducted over a number of years,[139-147] involved the formation of hemodialysis membranes from collagen obtained by enzymatic degradation of calfskin. Following degradation, the collagen membranes were produced by casting from acetic acid solution and the water-soluble membranes were cross-linked by heating at 110°C or by UV irradiation. These collagen membranes were essentially amino acid polymers containing positively and negatively charged side groups. Thus, chemical blocking could provide acidic, basic, or neutral membranes and the water sensitivity of the membranes could be controlled. Good permeability to urea and creatinine was reported (Table 10) and the membranes could be prepared in flat sheet, tubular, or hollow fiber form. Clinical trials have been carried out and other medical applications for the collagen polymer have been proposed, but the membranes have never emerged as alternatives to cellulose and cellulose acetate.

Another approach[148,149] examined a commercially available collagen polymer. Urea and water permeability were good, but the membranes were reported to be weak.

**Table 11**
**COPOLYETHER-ESTER MEMBRANES HALF-TIME ESCAPE RATE RATIOS**

| Greater than 1.10 | Ca. 1.0 (0.94 to 1.07) | Less than 0.91 |
|---|---|---|
| Urea (1.73) | Glutamine (1.0) | Glycine (0.82) |
| Creatinine (1.48) | D,L-Phenylalanine (0.96) | L-Alanine (0.85) |
| D-Threonine (1.25) | Glucose (1.07) | B-Alanine (0.54) |
| L-Leucine (1.32) | Sucrose (0.94) | Sarcosine (0.72) |
| Histidine (1.32) | | D,L-Serine (0.91) |
| Uric acid (1.96) | | L-Proline (0.87) |
| Arginine (1.29) | | L-Valine (0.89) |
| Ascorbic acid (1.79) | | Hydroxy-L-proline (0.76) |
| Citric acid (1.14) | | Raffinose·$5H_2O$ (0.62) |
| Glucuronic acid (1.41) | | |
| Glucosamine·HCl (1.34) | | |
| Thiamine chloride (1.36) | | |
| Bacitracine (1.81) | | |

*Note:* Ratio is $t_{1/2}$ Cuprophane divided by $t_{1/2}$ copolyester.

From Lyman, D. J., Loo, B. H., and Muir, W. M., *Trans. Am. Soc. Artif. Intern. Organs*, 11, 91, 1965. With permission.

### 5. *Polyether Block Copolymer Membranes*

The study by Lyman and co-workers[65,150-154] provides an example of an attempt to produce a membrane for extracorporeal blood purification by preparing a series of polymers and using membrane assessment to select the polymer with optimal properties. An important objective in the study of polyether block copolymer membranes was the preparation of membranes with a different permeability mechanism from that of regenerated cellulose. The concept adopted was to replace the sieve-like cellulose membranes with partition membranes obtained from block copolymers consisting of hydrophilic and hydrophobic segments. The hydrophilic segment would act primarily to provide swelling and partitioning properties and the hydrophobic segment would act to impart wet strength.

Using a modified melt polymerization technique,[151] copolymers of poly(oxyethylene glycol) and poly(ethylene terephthalate) were prepared. Alteration to the molecular weight of the poly(oxyethylene glycol) and its proportion in the copolymer led to a series of membranes. Assessment of solute permeability indicated that these membranes, unlike regenerated cellulose membranes, demonstrated no obvious relationship between the size of the diffusing solute and the rate of transport (Table 11).

In subsequent work,[153] the poly(ethylene terephthalate) segments were replaced by polyurethane segments using a two-stage solution polymerization. The results obtained indicated the dependence of solute transport on the structure of both hydrophilic and hydrophobic segments.

Despite the interesting permeability results, the polyether block copolymer membranes never achieved clinical status. The importance of these membranes is based largely on the fact that they enable useful guidelines to be drawn for membrane development.

**Table 12**
**AN 69 MEMBRANE (ACRYLONITRILE-SODIUM METHALLYL SULFONATE COPOLYMER)**

| Solute | Mol wt | $R_m \times 10^{-2}$ (min/cm) | $R_m$ Cuprophan® 150PT/$R_m$ AN69 |
|---|---|---|---|
| Urea | 60 | 0.106 | 1.4 |
| Creatinine | 113 | 0.18 | 1.65 |
| Sucrose | 342 | 0.38 | 2.2 |
| Raffinose | 595 | 0.55 | 2.3 |
| Vitamin $B_{12}$ | 1,355 | 0.70 | 3.6 |
| Inulin | 5,200 | 2.58 | 9.7 |
| Myoglobin | 17,800 | 8.00 | — |
| Ovalbumin | 45,000 | 75.00 | — |
| Bovalbumin | 70,000 | 200.00 | — |

*Note:* (1) Measurements performed at 37°C; solutes dissolved in saline and circulated to avoid significant resistance from boundary layers;
(2) Ultrafiltration rate at 37°C is 0.34 ± 0.04 mℓ/min/m²/mmHg.
(3) Wet membrane thickness, 28.5 ± 2 μm; pore size, 2 to 3 nm.

Typical performance data of HOSPAL AN69 membrane with permission of HOSPAL Ltd., Basle, Switzerland.

*6. Polyacrylonitrile-Based Membranes*

The ability of polyacrylonitrile to form fibers and films is well-known and polyacrylonitrile-based systems have formed the basis of several studies. The characteristic properties of polyacrylonitrile result from the small size of the CN side group, the high polarity of this group, and hydrogen bond formation between polymer chains. In addition to the film-forming properties, polyacrylonitrile-based membranes for extracorporeal blood purification are favored by the fact that polymerization and polymer purification can benefit from the solubility of the acrylonitrile monomer in water and the incompatibility between monomer and polymer. A disadvantage is the restricted range of suitable solvents.

If a polyacrylonitrile-based membrane is to function satisfactorily in extracorporeal blood purification, the normal structure must be altered in order to introduce solute and water permeability. Techniques available include copolymerization, membrane stretching, chemical modification, and the formation of asymmetric membranes.

The principal interest in polyacrylonitrile-based systems is due to the fact that a polyacrylonitrile-based membrane is one of the few synthetic polymer membranes in current clinical use. This membrane was produced by Rhone-Poulenc S.A. from acrylonitrile and sodium methallyl sulfonate. A Rhone-Poulenc patent[155] describes membranes prepared from acrylonitrile copolymers or from a mixture of copolymers, with the copolymers synthesized by conventional procedures. Optimal membrane properties are obtained by treating the membrane with water or with an aqueous nonsolvent, followed, if desired, by submitting the wet membrane to controlled stretching. An example cited for the in vitro dialysis of human blood is a membrane prepared by casting from dimethylformamide a mixture of two copolymers of acrylonitrile and sodium methallyl sulfonate. The monomer weight ratios were 91:9 and 89:11 with the copolymer proportion in the mixture 1:2.52. The cast membrane was treated with water at 90°C, subjected to a uniaxial degree of stretching of 180%, treated with water at 75°C, and passed through a bath of glycerol.

The acrylonitrile copolymer membrane used clinically[156-158] is referred to as RP-AN69, AN69, or sometimes simply polyacrylonitrile. The membrane is produced in flat sheets for use in a dialyzer marketed by Hospal A.G. Typical membrane properties are listed in Table 12. An interesting feature of the AN69 is that it permits clinical application of a membrane with permeability properties markedly different from those of regenerated cellulose. This permeability difference is relevant to the treatment of chronic renal failure and in the use of hemodialysis as a form of hepatic support.

Regarding chronic renal failure, the AN69 has a significantly higher permeability to solutes in the 1000 to 2000 wt range than Cuprophan® 150PT and is, therefore, more suitable for the removal of middle molecules. The ultrafiltration rate is several times greater than that of Cuprophan® 150PT.

On the assumption that removal of solutes from the blood would prove beneficial, hemodialysis with regenerated cellulose membranes was performed as a treatment for hepatic failure.[159] This failed to find clinical acceptance and interest in hemodialysis for the treatment of hepatic failure was not stimulated until the availability of the AN69 membrane. Several studies on the applicability to hepatic support of hemodialysis with the AN69 membrane have been carried out.[160-162]

The acrylonitrile-sodium methallyl sulfonate membranes have had a considerable impact on hemodialysis and hemofiltration and seem likely to have a continuing importance in extracorporeal blood purification. Disadvantages arise from the limitation to membranes in flat sheet form, restriction of use to a particular dialyzer, and the need to control the ultrafiltration rate with specialized monitoring equipment.

Another approach to the preparation of polyacrylonitrile-based membranes utilizes chemical modification by reaction with ethylene oxide gas. Membranes are prepared by solvent casting of copolymers of acrylonitrile and dimethylaminoethyl methacrylate.[163] Following solvent evaporation, membranes are reacted with pure ethylene oxide under slight vacuum. The chemical modification reaction[164] involves ionic opening of the ethylene oxide ring, which induces two chemical reactions. One reaction produces hydroxyl groups and promotes water sensitivity, while the other reaction leads to -C = N- conjugation and contributes to wet strength. The net result is a hydrophilic ladder polymer with membrane solute permeability, ultrafiltration, mechanical properties, and blood compatibility dependent on reaction time and temperature.[165] The relationship between membrane mass transfer resistance and ultrafiltration rate and reaction conditions is illustrated in Figures 8 and 9. These acrylonitrile copolymer membranes have been assessed in vitro and with dogs, but have not undergone clinical evaluation.

Having examined a number of synthetic polymers as potential hemodialysis membranes,[166] the Monsanto Corporation selected polyacrylonitrile as the most promising membrane material for the production of a hollow fiber artificial kidney.[167,168] An amine-modified polyacrylonitrile was prepared and formed into hollow fibers with internal diameter 200 to 300 μm and wall thickness 30 to 100 μm. Prototype artificial kidney units were tested in vivo with patients suffering from chronic renal failure, but the membranes did not attain widespread clinical use.

Polyacrylonitrile hollow fiber membranes are produced by Asahi Medical Co. Ltd. for hemofiltration and simultaneous hemofiltration-hemodialysis. These are asymmetric membranes with 200-μm internal diameter and 50-μm wall thickness.

Other polyacrylonitrile-based systems may emerge such as that proposed by the Batelle Institute[169] which copolymerizes acrylonitrile with vinyl sulfonic acid. In addition, polyacrylonitrile offers a means of enhancing the properties of poly(vinyl alcohol) membranes and this is considered later.

### *7. Polycarbonate-Based Membranes*

Studies of polycarbonate-based membranes for extracorporeal blood purification generally utilize polycarbonates derived from bisphenol A. The most significant outcome of such studies has been the eventual production of membranes for clinical use by C. R. Bard Inc.[170]

The objective in developing polycarbonate-based membranes is to combine the excellent

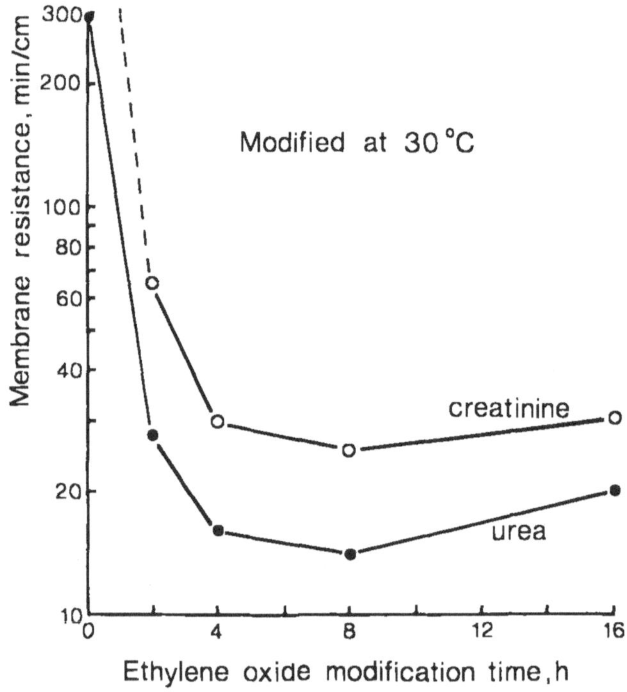


FIGURE 8. Acrylonitrile-dimethylaminoethyl methacrylate copolymer. Membrane mass transfer resistance as a function of ethylene oxide modification time. (From Courtney, J. M., Gaylor, J. D. S., Lindsay, R. M., and Martin, I., *Technical Aspects of Renal Dialysis*, Frost, T. H., Ed., Pitman, London, 1978, 126. With permission.)

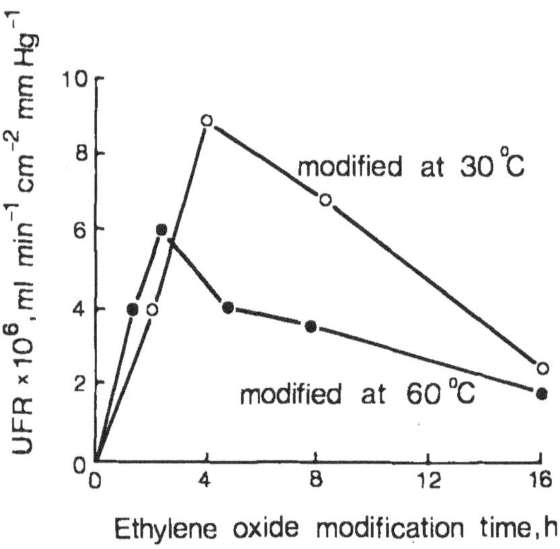


FIGURE 9. Acrylonitrile-dimethylaminoethyl methacrylate copolymer. Membrane ultrafiltration rate as a function of ethylene oxide modification temperature and time. (From Courtney, J. M., Gaylor, J. D. S., Lindsay, R. M., and Martin, I., *Technical Aspects of Renal Dialysis*, Frost, T. H., Ed., Pitman, London, 1978, 126. With permission.)

mechanical properties of aromatic polycarbonates with appropriate permeability to solutes and water. The approach, which led to membranes in clinical use, sought to achieve the balance of hydrophibic and hydrophilic properties by synthesizing polycarbonate-polyether block copolymers.[171] Block copolymers were synthesized by the action of phosgene on mixtures of bisphenol A and poly(ethylene glycol). Asymmetric membranes were produced[172] by casting from a solution of the block copolymer in chloroform and gelling with methanol after partial evaporation of solvent. A copolymer prepared from bisphenol A and the poly(ethylene glycol) Carbowax® 6000 (average molecular weight, 6700) in the ratio 75:25 gave an asymmetric membrane with a dense layer of 0.05 to 0.2 μm and a porous substructure of 25 to 30 μm. Assessment[173,174] indicated that, in comparison to Cuprophan® 150PT, permeabilities for urea, vitamin $B_{12}$, and water were, respectively, 1.09, 2.96, and 1.25 to 2.0 times greater. A problem with these membranes arose from the fact that the dense barrier layer was formed on the side of the membrane contacting the casting surface. This made reliable continuous casting difficult and necessitated alteration to the gelation technique.[175]

The preferred procedure for the continuous casting of the polycarbonate-polyether block copolymer membranes[170] involves dissolving the copolymer in a water-miscible solvent, such as 1,3 dioxolane, and using water as the gelling medium. In this way, the dense barrier layer is formed on the side facing the air during drying. This alteration in manufacturing procedure reduced the membrane thickness from 30 to 20 μm. Urea, creatinine, and vitamin $B_{12}$ permeability were increased, but, despite the reduction in membrane thickness, water permeability was decreased.

The polycarbonate-polyether block copolymer membranes offer potential advantages in extracorporeal blood purification.[176] Membranes can be produced in different forms[177] with the added advantage of heat sealability.[178] The mechanical properties of the membranes are reported to be good and unlike the situation with regenerated cellulose membranes, there is no pronounced change in properties on going from the dry to the wet state. High transmembrane pressures can be tolerated satisfactorily, membrane manufacture can be adapted to achieve a range of ultrafiltration rates, and protein adsorption is low. Thus, although originally developed for hemodialysis, the polycarbonate-polyether block copolymer membranes may find application in hemofiltration and sequential ultrafiltration-hemodialysis.

Kesting[179] has also reported on polycarbonate-polyether copolymer membranes. In this case, bisphenol A and Carbowax® 4000 were reacted with phosgene and asymmetric membranes were obtained by casting from a 50:50 mixture of dioxane and tetrahydrofuran and immersing in water. However, although other possibilities for producing polycarbonate-based membranes exist, only the membranes manufactured by C. R. Bard Inc. have achieved clinical status.

### *8. Methacrylate-Based Membranes*

Under the heading of methacrylate-based membranes, a number of approaches can be examined. These approaches range from the use of a hydrophobic polymer such as poly(methyl methacrylate) to impart wet strength to the use of a hydrophilic polymer such as poly(hydroxyethyl methacrylate) to impart permeability.

Ota et al.[180] reported on a dialyzer containing hollow fibers of poly(methyl methacrylate), 240-μm internal diameter and 50-μm wall thickness. Since the membrane was reported to be based on "pure" poly(methyl methacrylate), it must be assumed that membrane permeability was accomplished by introducing a degree of porosity. The poly(methyl methacrylate) hollow fiber membrane has been assessed in vitro and clinically. In comparison to regenerated cellulose membranes, improved permeability to middle molecules and to water has been claimed. The poly(methyl methacrylate) hollow fiber membrane has also been used clinically for simultaneous hemodialysis-hemofiltration.[181]

Muir et al.[182] described the potential of membranes prepared from copolymers of methyl or *n*-butyl methacrylate with acrylic or methacrylic acid. The preferred membranes were obtained from copolymers of *n*-butyl methacrylate and acrylic acid, and the transport characteristics could be altered by chemical modification of the membranes through reaction with ethylene oxide gas.[183] In addition to in vitro assessment, the blood compatibility of the butyl methacrylate-acrylic acid membranes was studied with dogs,[184] but the membranes did not come into clinical use.

In another exploitation of carboxylic acid groups,[185] methyl methacrylate has been copolymerized with acrylic acid and glycidyl methacrylate using solution polymerization. Membranes were cross-linked by successive treatments with heat, alkaline solution, and acidic solution.

There have been different attempts made to develop methacrylate hydrogel membranes.[186-189] The general approach involves the solution polymerization of hydroxyethyl methacrylate in the presence of a cross-linking agent such as tetraethylene glycol dimethacrylate. However, membranes based on methacrylate hydrogels have not achieved clinical success. The principal disadvantage is the unsatisfactory mechanical strength, often leading to suggestions for specialized support and reinforcement. Interest remains in the use of hydroxyethyl methacrylate in membranes based on poly(vinyl alcohol), which are considered later.

*9. Polystyrene-Based Membranes*

The copolymerization of acrylic acid with methacrylates has been previously considered. Styrene has also been copolymerized with this monomer and membranes have been subsequently assessed[190] for solutes of interest to the treatment of uremia. The transfer of solutes was comparable to that obtained with Cuprophan® 150PT.

*10. Poly(vinyl pyrrolidone)-Based Membranes*

The excellent film forming properties of poly(vinyl pyrrolidone) and the use of the polymer as a blood plasma substitute combined to promote interest in membranes based on vinyl pyrrolidone polymers and copolymers. One approach to producing vinyl pyrrolidone-based membranes is that of cross-linking the water-soluble poly(vinyl pyrrolidone). Markle et al.[191] described cross-linking with isocyanate. As expected, permeability decreased with increasing degree of cross-linking. Urea permeability superior to that of Cuprophan® 150PT was claimed.

Vinyl pyrrolidone has been copolymerized with ethyl acrylate and with styrene.[192] However, the blood compatibility was disappointing and membrane permeability was low, necessitating the leaching out of substances such as sucrose and dimethylsulfoxide.

*11. Polyamide-Based Membranes*

The approaches used to develop polyamide-based membranes illustrate the changing attitudes to extracorporeal blood purification. As a hemodialysis membrane, a polyamide-epoxide resin copolymer was studied.[193-196] The objective was to improve on Cuprophan® 150PT and, as in so many cases, this was not achieved. A more successful development has been the production of asymmetric hollow fiber membranes from an aromatic polyamide and the application of these membranes in hemofiltration.[197,198]

*12. Polypeptide Membranes*

In an interesting approach to hemodialysis membrane development, Klein and co-workers[199] set out to prepare synthetic polypeptides through the polymerization of *N*-carboxyanhydrides of amino acids and their derivatives, commencing with those of glutamic acid, lysine, and methionine. The investigation later narrowed[200] to representative cationic, anionic, and nonionic polypeptides with a group of copolymers formed from reactive leucine and methionine monomers. D,L-Leucine *N*-carboxyanhydride and D,L-methionine *N*-carboxyanhydride were

**Table 13**
**POLYSULFONE MEMBRANES, TYPICAL XM HOLLOW FIBER PROPERTIES**

| Ultrafiltration rate from saline (mℓ/cm²/min/mmHg) | Solute | Dialytic resistance (min/cm) |
|---|---|---|
| $9.0 \times 10^{-5}$ | NaCl | 21.6 |
| | Urea | 36.5 |
| | Creatinine | 45.5 |
| | Uric acid | 58.2 |
| | Bacitracin | 87.8 |
| | Inulin | 96.0 |

From Cross, R. A., Tyson, W. H., Jr., and Cleveland, D. S., *Trans. Am. Soc. Artif. Intern. Organs*, 17, 279, 1971. With permission.

copolymerized.[201] The methionine segments were oxidized to give both sulfoxide and sulfone functions. Permeability assessment indicated that solute transport was dependent on the degree of polymer hydration, which was itself dependent on the extent of methionine oxidation. Efforts were subsequently restricted to two polymer systems, an oxidized leucine-methionine copolymer and a commercially available poly(methyl-D-glutamate).[202] Advantages in physical properties, cost, and availability led to the neutral poly(methyl-D-glutamate) being selected for animal studies and clinical assessment.[203-205]

Other studies include those on L-leucine DL-methionine copolymers[206] and on copolymers of L-leucine, L-aspartic acid, and L-aspartic acide esters.[207] Polypeptide membranes are not in regular clinical use.

*13. Polysulfone Membranes*

The interest in polysulfone membranes centers on the asymmetric hollow fiber membranes developed by the Amicon Corporation for hemodialysis[208] and hemofiltration.[31] These fibers can be produced with an inside skin of <1-μm thickness and containing pores of 2- to 4-nm diameter.[209-211] Membrane structure can be altered to enhance solute or water transport. Exact details of polymer composition are not disclosed, but the membranes come in two material types, PM and XM (Tables 13 and 14), both of which are nonionic.[79]

*14. Polyelectrolyte Membranes*

In the area of biomedical polymers, the term polyelectrolyte has been used to denote complexes developed by the Amicon Corporation.[212-214] An example is the complex formed by reacting sodium polystyrene sulfonate with polyvinylbenzyltrimethyl ammonium chloride. The ionic charge or water sensitivity of the complex can be easily adjusted. Among the applications proposed for the polyelectrolyte complexes was that of membranes for extracorporeal blood purification with the ability to produce asymmetric membranes useful for both hemodialysis and hemofiltration. More recently, the term polyelectrolyte has been used to describe hemodialysis membranes prepared from poly(acrylic acid)-polyethylene-piperazine copolymers.[215]

*15. Poly(vinyl alcohol)-Based Membranes*

In this examination of polymer systems proposed as a basis for blood purification membranes, it is appropriate to conclude with poly(vinyl alcohol). With the possession of water-solubility and film-forming properties allied to a ready availability, poly(vinyl alcohol) is an obvious candidate material for suitable membrane development. Although the same is true of other water-soluble polymers, interest in poly(vinyl alcohol)-based membranes has been maintained.

**Table 14**
**POLYSULFONE MEMBRANES, TYPICAL PM HOLLOW FIBER PROPERTIES (AFTER DRYING AND AUTOCLAVING)**

| Ultrafiltration rate from saline (mℓ/cm²/min/mmHg) | Solute | Dialytic resistance (min/cm) |
|---|---|---|
| $3.4 \times 10^{-5}$ | NaCl | 20.4 |
| | Urea | 32.5 |
| | Creatinine | 29.2 |
| | Uric acid | 54.4 |
| | Bacitracin | 242 |
| | Inulin | 550 |

From Cross, R. A., Tyson, W. H., Jr., and Cleveland, D. S., *Trans. Am. Soc. Artif. Intern. Organs*, 17, 279, 1971. With permission.

Poly(vinyl alcohol) is normally prepared by alcoholysis of poly(vinyl acetate)[88] and products vary in molecular weight and in residual acetate content. The preparation of poly(vinyl alcohol)-based membranes can be considered in terms of two broad approaches. One approach is directed towards suitable cross-linking of the water-soluble polymer, while the other approach places emphasis on copolymerization prior to any cross-linking.

Markle et al.[191] described cross-linking of poly(vinyl alcohol) with formaldehyde and Odian and Leonard[216] described cross-linking with allyl methacrylate followed by irradiation. Neither of these procedures achieved clinical success and nor did that of Hills,[217] who proposed depositing poly(vinyl alcohol) onto a support scrim. However, investigation of cross-linked poly(vinyl alcohol) membranes has continued.[218]

Another important technique for producing poly(vinyl alcohol)-based membranes utilizes graft copolymerization.[219-222] Membranes have been prepared[219] by casting emulsions obtained from graft copolymerization of methyl acrylate and methyl methacrylate onto poly(vinyl alcohol) in aqueous solution and subsequently cross-linking. Other graft copolymer membranes reported include those with acrylonitrile[220] and with acrylonitrile-methyl acrylate-acrylamide[221] and acrylonitrile-hydroxyethyl methacrylate[222] combinations. Comonomer selection and treatment of cast membranes ensure that there is considerable scope for developing graft copolymer membranes.

Hemodialysis membranes prepared from ethylene-vinyl alcohol copolymers have been assessed in vitro.[223,224] Suitable membranes are produced by casting the copolymer from a solution in dimethylsulfoxide and contacting the membrane with water prior to solvent evaporation. Advantages over regenerated cellulose membranes were claimed with respect to permeability to water and solutes in the molecular weight range of middle molecules. The ethylene-vinyl alcohol copolymer membranes have been developed by the Kuraray Co. Ltd. This company has also described[225] the preparation of a poly(vinyl alcohol)-based membrane, where the poly(vinyl alcohol) is blended with poly(ethylene glycol) and cross-linked in alkaline solution. Asymmetric membranes in flat sheet or hollow fiber form are produced with a well-defined porosity.

### *16. Heparinized Membranes*

Extracorporeal blood purification generally involves the passage of anticoagulated blood and, at present, the only anticoagulant regarded as suitable is heparin. The widespread use of heparin has promoted consideration of the anticoagulant as a means of enhancing polymer blood compatibility and utilization of heparin by incorporation or attachment has formed the basis of several investigations.[226] Similarly, in membrane development, attempts have been made to utilize heparin in order to improve membrane blood compatibility or to reduce the requirement for anticoagulant addition.

The principal options available for the preparation of heparinized polymers are ionic or covalent attachment or prior reaction of heparin with a quaternary ammonium salt followed by incorporation of the reaction product into the polymer. A widely used reaction product is the complex formed by reacting heparin with tridodecylmethylammonium chloride. However, the incorporation approach with membranes appears to have the disadvantages of loss of permeability, deterioration of mechanical properties, and transient antithrombogenic properties, due to rapid elution of the complex from the membrane.[227]

Merrill et al.[228] described a technique for the heparinization of regenerated cellulose membranes. This technique involved the production of ethylene immonium ions using ethylene imine together with an acid catalyst. Exposure to heparin gave ionic attachment. With this approach, the improvement in blood compatibility was influenced by residual poly(ethylene imine) and cellulose abnormalities resulting from the strong acid catalyst. In subsequent work,[229] the problem of cellulose hydrolysis was eliminated by using nonpolar solvents. However, lengthy solvent exchange processes were necessary and residual poly(ethylene imine) remained in the membrane.

As part of a program for the development of asymmetric cellulose acetate membranes,[119] Martin et al.[230] prepared heparinized membranes from blends of cellulose acetate and diethylaminoethyl cellulose acetate. Ionic attachment of heparin was achieved by reacting the amino groups with hydrochloric acid followed by contact with heparin.

Approaches similar to that of Martin and co-workers have been used in the preparation of cellulose membranes[227] and acrylonitrile copolymer membranes.[231] The cellulose membranes consisted of regenerated cellulose containing 5% diethylaminoethyl cellulose. The acrylonitrile copolymer membranes contained dimethylaminoethyl methacrylate as comonomer.

Ionic attachment of heparin is a relatively straightforward method for producing heparinized membranes. However, such membranes have not yet achieved clinical status. Contact with blood or plasma may remove heparin which is ionically bound to a polymer,[232,233] making it difficult to ascertain if heparin release or heparin attachment is contributing to any improvement in blood compatibility. Techniques proposed for enhancing attachment include radiation grafting of heparin to amino groups[234] or cross-linking of the surface heparin with glutaraldehyde.[235]

Covalent attachment of heparin using cyanogen bromide was selected by Schmer and co-workers[236-239] as a procedure for the preparation of heparinized cellulose acetate and cellulose hollow fiber membranes. The technique, which gave a controlled leakage of small amounts of heparin, has been assessed with sheep[238] and clinically.[239] By analogy with the development of blood compatible polymers, interest in heparinized membranes seems certain to continue.

### *17. Sorbent Membranes*

The development of sorbent membranes was a consequence of the interest in hemoperfusion,[240-242] which may be regarded as the purification of blood by extracorporeal passage over an adsorbent material. Hemoperfusion, therefore, offers an adsorbent-based alternative to procedures based on dialysis or ultrafiltration in the areas of acute poisoning, uremia, and hepatic support. The adsorbent which has received most attention is activated carbon or charcoal.

Early studies demonstrated that hemoperfusion over uncoated activated carbon granules was unsatisfactory because of blood damage and the generation of fine carbon particles. Attempts to overcome these problems have included polymer coating of carbon granules, adsorbent immobilization, and the preparation of sorbent membranes.

In theory, adaptation of the membrane formation techniques previously considered would permit the incorporation of an adsorbent into various polymer systems. However, while a membrane based on activated carbon and cellulose nitrate has been reported,[243] the development of sorbent membranes has been dominated by regenerated cellulose-based materials prepared by Membrana GmbH and these membranes have received most assessment.[104,244-251]

The sorbent membranes of Membrana utilize Cuprophan® regenerated cellulose in different ways. One product is a hollow fiber consisting of a cellulose wall about 8 μm thick and an activated carbon core about 30 μm thick. Other membranes consist of an inner cellulose layer (8 to 10 μm thick) and an outer activated carbon layer (35 to 50 μm thick). These double layer membranes have been produced in the form of lay-flat tubing, sheet, or hollow fibers and permit the combined application of hemoperfusion and hemodialysis, with the ability to remove water particularly important for the treatment of uremia.

The use of activated carbon may be envisaged as a first stage in the development of sorbent membranes. Other adsorbents can extend the area of application. In this respect, Membrana GmbH have produced membranes containing aluminium oxide in order to enhance the removal of phosphate and the potential of a multisorbent system using filled cellulose acetate hollow fibers has been investigated.[252,253]

*18. Future Development*

Two broad approaches are available in the search to improve and extend the role of polymer membranes in extracorporeal blood purification: (1) development of new membrane systems and (2) modification of existing membrane systems. With respect to the latter approach, advantage is likely to be taken of the hollow fiber structure to enhance membrane properties by the culturing of cells,[254] the immobilization of enzymes,[255,256] or the attachment of platelet antiaggregating agents.[257]

## REFERENCES

1. **Tuwiner, S. B.,** *Diffusion and Membrane Technology*, Rheinhold, New York, 1962.
2. **Friedlander, H. Z. and Rickles, R. N.,** Membrane technology. II. Theory and development, *Anal. Chem.*, 37, 27a, 1965.
3. **Kesting, R. E.,** *Synthetic Polymer Membranes*, McGraw-Hill, New York, 1971.
4. **Cross, R. A. and Strathman, H.,** Barrier separation processes, in *An Introduction to Separation Science*, Karger, B. L., Snyder, L. R., and Horvath, C., Eds., John Wiley & Sons, New York, 1973, 469.
5. **Meares, P.,** The physical chemistry of transport and separation by membranes, in *Membrane Separation Processes*, Meares, P., Ed., Elsevier, Amsterdam, 1976, 1.
6. **Courtney, J. M., Gaylor, J. D. S., and Gilchrist, T.,** Dialysis, in *Physical Techniques in Medicine*, Vol. 1, McMullan, J. T., Ed., Wiley-Interscience, London, 1977, 195.
7. **Blatt, W. F.,** Principles and practice of ultrafiltration, in *Membrane Separation Processes*, Meares, P., Ed., Elsevier, Amsterdam, 1976, 81.
8. **Abbrecht, P. H.,** An outline of renal structure and function, in *The Artificial Kidney*, Vol. 64 Chemical Engineering Progress Symposium Series, No. 84, Dedrick, R. L., Bischoff, K. B., and Leonard, E. F., Eds., American Institute of Chemical Engineers, New York, 1968, 1.
9. **Pitts, R. F.,** *Physiology of the Kidney and Body Fluids*, 3rd ed., Year Book Medical, Chicago, 1974.
10. **Farrell, P. C.,** Hemodialysis, in *Chronic Replacement of Kidney Function*, Vol. 75, AIChE Symposium Series, No. 187, Villarroel, F. and Dedrick, R. L., Eds., American Institute of Chemical Engineers, New York, 1979, 10.
11. **Abel, J. J., Rowntree, L. G., and Turner, B. B.,** On the removal of diffusible substances from the circulating blood of living animals by dialysis, *J. Pharmacol. Exp. Ther.*, 5, 275, 1913.
12. **Thalhimer, W.,** Experimental exchange transfusion for reducing azotemia. Use of the artificial kidney for this purpose, *Proc. Soc. Exp. Biol. Med.*, 37, 641, 1937.

13. **Kolff, W. J., Berk, H. T. J., ter Welle, M., van der Ley, A. J. W., van Dijk, E. C., and van Noordwijk, J.**, The artificial kidney: a dialyser with great area, *Acta Med. Scand.*, 117, 121, 1944.
14. **Gaylor, J. D. S.**, Artificial organs, in *A Textbook of Biomedical Engineering*, Kenedi, R. M., Ed., Blackie, Glasgow, 1980, 115.
15. **Lipps, B. J., Jr.**, Permselective Membranes, U.K. Patent 1,245,053, 1971.
16. **Mahon, H. I., Doss, G. C., Vang, D. B., and Gavin, D. A.**, Method for Making a Hollow Fiber Separatory Element, U.K. Patent 1,366,615, 1974.
17. **Yamazaki, M. and Joh, Y.**, Setting an End Portion of a Bundle of Thread-Like Bodies, U.K. Patent 1,515,452, 1978.
18. **Henderson, L. W.**, Ultrafiltration, in *Replacement of Renal Function by Dialysis*, Drukker, W., Parsons, F. M., and Maher, J. F., Eds., Martinus Nijhoff, The Hague, The Netherlands, 1978, 135.
19. **Ing, T. S., Chen, W.-T., Daugirdas, J. T., Geis, W. P., and Hano, J. E.**, The old and the new in ultrafiltration techniques, *Artif. Organs*, 3, 185, 1979.
20. **Henderson, L. W., Besarab, A., Michaels, A., and Bluemle, L. W., Jr.**, Blood purification by ultrafiltration and fluid replacement (diafiltration), *Trans. Am. Soc. Artif. Intern. Organs*, 13, 216, 1967.
21. **Shinaberger, J. H., Miller, J. H., Rubini, M. E., Gardner, P. W., and Martin, F. E.**, Initial clinical evaluation of "diafiltration", *Trans. Am. Soc. Artif. Intern. Organs*, 15, 97, 1969.
22. **Mann, H., Heintz, R., and Brass, H.**, Diafiltration: Technik und Einsatzmöglichkeiten, *Biomed. Tech.*, 19, 170, 1974.
23. **Quellhorst, E., Schuenemann, B., and Doht, B.**, Hemofiltration — a new method for the purification of blood, *Artif. Organs*, 2, 83, 1978.
24. **Keshaviah, P.**, Hemofiltration, in *Chronic Replacement of Kidney Function*, Vol. 75, AIChE Symposium Series, No. 187, Villarroel, F. and Dedrick, R. L., Eds., American Institute of Chemical Engineers, New York, 1979, 24.
25. **Henderson, L. W.**, Historical overview and principles of hemofiltration, *Dialysis Transplant.*, 9, 220, 1980.
26. **Henderson, L. W., Lilley, J. J., Ford, C. A., and Stone, R. A.**, Hemodiafiltration, *J. Dialysis*, 1, 211, 1977.
27. **Leber, H.-W., Wizemann, V., Goubeaud, G., Rawer, P., and Schütterle, G.**, Hemodiafiltration: a new alternative to hemofiltration and conventional hemodialysis, *Artif. Organs*, 2, 150, 1978.
28. **Russo, L., Russo, A., Viola, M., and Russo, V.**, Artificial secretion of weakly bound middle molecules by hemoanadiafiltration, *Artif. Organs*, 4, 34, 1980.
29. Copenhagen Workshop, Sequential ultrafiltration-hemodialysis, *Dialysis Transplant.*, 7, 1074, 1978.
30. **Henderson, L. W., Colton, C., Leber, H., and Quellhorst, E.**, Hemofiltration, *Trans. Am. Soc. Artif. Intern. Organs*, 24, 788, 1978.
31. **Henderson, L. W., Ford, C., Colton, C. K., Bluemle, L. W., and Bixler, H. J.**, Uremic blood cleansing by diafiltration using a hollow fiber ultrafilter, *Trans. Am. Soc. Artif. Intern. Organs*, 16, 107, 1970.
32. **Colton, C. K., Henderson, L. W., Ford, C. A., and Lysaght, M. J.**, Kinetics of hemodiafiltration. I. In vitro transport characteristics of a hollow-fiber blood ultrafilter, *J. Lab. Clin. Med.*, 85, 355, 1975.
33. **Henderson, L. W., Colton, C. K., and Ford, C. A.**, Kinetics of hemodiafiltration. II. Clinical characteristics of a new blood cleansing modality, *J. Lab. Clin. Med.*, 85, 372, 1975.
34. **Henderson, L. W.**, Current status of hemofiltration, *Artif. Organs*, 2, 120, 1978.
35. **Quellhorst, E., Doht, B., and Schuenemann, B.**, Hemofiltration: treatment of renal failure by ultrafiltration and substitution, *J. Dialysis*, 1, 529, 1977.
36. **Quellhorst, E., Schuenemann, B., and Borghardt, J.**, Clinical and technical aspects of hemofiltration, *Artif. Organs*, 2, 334, 1978.
37. **von Herrath, D., Schaefer, K., Hufler, M., Asmus, G., and Hensel, A.**, Clinical aspects of hemofiltration, *J. Dialysis*, 1, 545, 1977.
38. **Schaefer, K., von Herrath, D., Gullberg, C. A., Asmus, G., Hufler, M., Offermann, G., Cremer, H., Heuck, C. C., and Ritz, E.**, Chronic hemofiltration — a critical evaluation of a new method for the treatment of blood, *Artif. Organs*, 2, 386, 1978.
39. **Schneider, H., Streicher, E., Hachmann, H., Chmiel, H., and von Mylius, U.**, Clinical experience with haemofiltration, *Proc. Eur. Dialysis Transplant Assoc.*, 14, 136, 1977.
40. **Streicher, E., Schneider, H., Knödler, U., Müller, H., Spohr, U., and Schmidt-Gayk, H.**, Chronic hemofiltration treatment, *Artif. Organs*, 4, 48, 1980.
41. **Bosch, J. P., Geronemus, R., Glabman, S., Lysaght, M., Kahn, T., and von Albertini, B.**, High flux hemofiltration, *Artif. Organs*, 2, 339, 1978.
42. **Bergström, J., Asaba, H., Fürst, P., and Oulès, R.**, Dialysis, ultrafiltration and blood pressure, *Proc. Eur. Dialysis Transplant Assoc.*, 13, 293, 1976.
43. **Bergström, J.**, Changes in blood pressure and hemodynamics during ultrafiltration and dialysis, *Dialysis Transplant.*, 7, 1087, 1978.

44. **Klein, E., Autian, J., Bower, J. D., Buffaloe, G., Centella, L. J., Colton, C. K., Darby, T. D., Farrell, P. C., Holland, F. F., Kennedy, R. S., Lipps, B., Jr., Mason, R., Nolph, K. D., Villarroel, F., and Wathen, R. L.,** Evaluation of hemodialyzers and dialysis membranes. Evaluation of hemodialysis membranes, *Artif. Organs*, 1, 21, 1977.
45. **Klein, E., Autian, J., Bower, J. D., Buffaloe, G., Centella, L. J., Colton, C. K., Darby, T. D., Farrell, P. C., Holland, F. F., Kennedy, R. S., Lipps, B., Jr., Mason, R., Nolph, K. D., Villarroel, F., and Wathen, R. L.,** Evaluation of hemodialyzers and dialysis membranes. In vitro characterisation of hemodialyzers, *Artif. Organs*, 1, 59, 1977.
46. **Courtney, J. M. and Gaylor, J. D. S.,** Membrane physics in medicine, in *Encyclopaedia of Physics and Its Applications*, Pergamon Press, Oxford, in press.
47. **Babb, A. L., Popovich, R. P., Christopher, T. G., and Scribner, B. H.,** The genesis of the square meter-hour hypothesis, *Trans. Am. Soc. Artif. Intern. Organs*, 17, 81, 1971.
48. **Klinkmann, H., Bergström, J., Dzúrik, R., and Funck-Brentano, J.-L., Eds.,** *Middle Molecules in Uremia and Other Diseases: Analytical Techniques, Metabolic Toxicity, and Clinical Aspects*, International Society for Artificial Organs, Cleveland, 1981.
49. **Kaufmann, T. G. and Leonard, E. F.,** Studies of intramembrane transport: a phenomenonological approach, *J. Am. Inst. Chem. Eng.*, 14, 110, 1968.
50. **Smith, K. A., Colton, C. K., Merrill, E. W., and Evans, L. B.,** Convective transport in a batch dialyser: determination of true membrane permeability from a single measurement, in *The Artificial Kidney*, Vol. 64, Chemical Engineering Progress Symposium Series, No. 84, Dedrick, R. L., Bischoff, K. B., and Leonard, E. F., Eds., American Institute of Chemical Engineers, New York, 1968, 45.
51. **Babb, A. L., Maurer, C. J., Fry, D. L., Popovich, R. P., and McKee, R. E.,** The determination of membrane permeabilities and solute diffusivities with applications to hemodialysis, in *The Artificial Kidney*, Vol. 64, Chemical Engineering Progress Symposium Series, No. 84, Dedrick, R. L., Bischoff, K. B., and Leonard, E. F., Eds., American Institute of Chemical Engineers, New York, 1968, 59.
52. **Babb, A. L., Maurer, C. J., Fry, D. L., Popovich, R. P., and Ramos, C. P.,** Methods for the in vivo determination of membrane permeabilities and solute diffusivities, *Trans. Am. Soc. Artif. Intern. Organs*, 14, 25, 1968.
53. **Colton, C. K., Smith, K. A., Merrill, E. W., and Farrell, P. C.,** Permeability studies with cellulosic membranes, *J. Biomed. Mater. Res.*, 5, 459, 1971.
54. **Wendt, R. P., Toups, R. J., Smith, J. K., Leger, N., and Klein, E.,** Measurements of membrane permeabilities using a rotating batch dialyzer, *I. & E. C. Fundam.*, 10, 406, 1971.
55. **Farrell, P. C. and Babb, A. L.,** Estimation of the permeability of cellulosic membranes from solute dimensions and diffusivities, *J. Biomed. Mater. Res.*, 7, 275, 1973.
56. **Klein, E., Holland, F. F., Donnaud, A., Lebeouf, A., and Eberle, K.,** Diffusive and hydraulic permeabilities of commercially available cellulosic hemodialysis films and hollow fibers, *J. Membr. Sci.*, 2, 349, 1977.
57. **Klein, E., Holland, F. F., and Eberle, K.,** Comparison of experimental and calculated permeability and rejection coefficients for hemodialysis membranes, *J. Membr. Sci.*, 5, 173, 1979.
58. **Klein, E., Holland, F., Lebeouf, A., Donnaud, A., and Smith, J. K.,** Transport and mechanical properties of hemodialysis hollow fibers, *J. Membr. Sci.*, 1, 371, 1976.
59. **Holland, F. F., Jr., Klein, E., Wendt, R. P., and Eberle, K.,** Rejection of solutes by hemodialysis/hemofiltration membranes, *Trans. Am. Soc. Artif. Intern. Organs*, 24, 662, 1978.
60. **Klein, E., Holland, F., and Eberle, K.,** Rejection of solutes by hemodialysis/hemofiltration membranes, *ASAIO J.*, 1, 15, 1978.
61. **Wendt, R. P., Klein, E., Bresler, E. H., Holland, F. F., Serino, R. M., and Villa, H.,** Sieving properties of hemodialysis membranes, *J. Membr. Sci.*, 5, 23, 1979.
62. **Stevenson, J. F., Parry, J. S., and Gupta, K. M.,** Hydraulic permeability of hollow-fiber membranes, *J. Biomed. Mater. Res.*, 12, 401, 1978.
63. **Naito, H., Shimizu, M., and Higashi, M.,** Relative performance test of hollow fiber dialyzers, *J. Dialysis*, 3, 173, 1979.
64. **Courtney, J. M.,** Preclinical evaluation of polymer membranes and coatings, *Br. Polym. J.*, 10, 288, 1978.
65. **Lyman, D. J.,** New synthetic membranes for the dialysis of blood, *Trans. Am. Soc. Artif. Intern. Organs*, 10, 17, 1964.
66. **Allen, W. T., Courtney, J. M., Gray, R. A., Klinkmann, H., and Muir, W. M.,** Comparative permeability measurements of cellulose-based membranes, *Proc. Eur. Dialysis Transplant Assoc.*, 5, 74, 1968.
67. **Klinkmann, H., Holtz, M., Willgerodt, W., Wilke, G., and Schoenfelder, D.,** "Nephrophane" — an improved variation of the cellulose membrane, *Proc. Eur. Dialysis Transplant Assoc.*, 5, 78, 1968.
68. **Winchester, J. F., Gelfand, M. C., Knepshield, J. H., and Schreiner, G. E.,** Dialysis and hemoperfusion of poisons and drugs — update, *Trans. Am. Soc. Artif. Intern. Organs*, 23, 762, 1977.

69. **Keynes, W. M.,** Haemodialysis in the treatment of liver failure, *Lancet*, 2, 1236, 1968.
70. **Opolon, P.,** High permeability membrane hemodialysis and hemofiltration in acute hepatic coma: experimental and clinical results, *Artif. Organs*, 3, 354, 1979.
71. **Anderson, P. C.,** Dialysis for psoriasis, *Artif. Organs*, 2, 202, 1978.
72. **Wagemaker, H.,** The effect of hemodialysis on fifteen, chronic, process schizophrenics, *Artif. Organs*, 2, 205, 1978.
73. **Bruck, S. D.,** Problems and artefacts in the evaluation of polymeric materials for medical uses, *Biomaterials*, 1, 103, 1980.
74. **Bruck, S. D.,** *Properties of Biomaterials in the Physiological Environment*, CRC Press, Boca Raton, Fla., 1980.
75. **Lindsay, R. M., Mason, R. G., Kim, S. W., Andrade, J. D., and Hakim, R. M.,** Blood surface interactions, *Trans. Am. Soc. Artif. Intern. Organs*, 26, 603, 1980.
76. **Leonard, E. F. and Bluemle, L. W., Jr.,** Evaluation of dialysis membranes, *Trans. Am. Soc. Artif. Intern. Organs*, 8, 182, 1962.
77. **Yarsley, V. E., Flavell, W., Adamson, P. S., and Perkins, N. C.,** *Cellulosic Plastics*, Iliffe, London, 1964.
78. **Briston, J. H. and Katan, L. L.,** *Plastics Films*, Iliffe, London, 1974.
79. **Baum, B., Holley, W., Jr., and White, R. A.,** Hollow fibres in reverse osmosis, dialysis, and ultrafiltration, in *Membrane Separation Processes*, Meares, P., Ed., Elsevier, Amsterdam, 1976, 187.
80. **Mahon, H. I., and Lipps, B.,** Hollow fiber membranes, in *Encyclopedia of Polymer Science and Technology*, John Wiley & Sons, New York, 1971, 258.
81. **Sourirajan, S.,** *Reverse Osmosis*, Logos, London, 1970.
82. **Kesting, R. E.,** Asymmetric cellulose acetate membranes, in *Reverse Osmosis and Synthetic Membranes, Theory — Technology — Engineering*, Sourirajan, S., Ed., National Research Council Canada, Ottawa, 1977, 89.
83. **Orofino, T. A.,** Technology of hollow fiber reverse osmosis systems, in *Reverse Osmosis and Synthetic Membranes. Theory — Technology — Engineering*, Sourirajan, S., Ed., National Research Council Canada, Ottawa, 1977, 313.
84. **Lyman, D. J. and Kim, S. W.,** Membranes in artificial kidney devices: past, present, and future, *Biomater. Med. Devices Artif. Organs*, 1, 431, 1973.
85. **Gardner, C. R.,** Biomedical applications of membrane processes, in *Membrane Separation Processes*, Meares, P., Ed., Elsevier, Amsterdam, 1976, 529.
86. **Klein, E.,** Progress in hemodialyzers and membranes — the last 20 years, *Clin. Nephrol.*, 9, 131, 1978.
87. **Lyman, D. J.,** Membranes, in *Replacement of Renal Function by Dialysis*, Drukker, W., Parsons, F. M., and Maher, J. F., Eds., Martinus Nijhoff, The Hague, 1978, 69.
88. **Brydson, J. A.,** *Plastics Materials*, 3rd ed., Newnes-Butterworths, London, 1975.
89. **Craig, L. C. and Pulley, A. O.,** Dialysis studies. IV. Preliminary experiments with sugars, *Biochemistry*, 1, 89, 1962.
90. **Henne, W. and von Sengbusch, G.,** A tribute to Dr. Werner Bandel: his contributions to the development of Cuprophan membranes, *Artif. Organs*, 3, 329, 1979.
91. **Alexander, R. W. and Galletti, P. M.,** The transfer of macromolecules in an artificial kidney, *Trans. Am. Soc. Artif. Intern. Organs*, 11, 95, 1965.
92. **Wilcox, C., Freeman, R. B., Maher, J. F., and Schreiner, G. E.,** Comparison of physical properties and permeability of six cellulose membranes, *Trans. Am. Soc. Artif. Intern. Organs*, 12, 44, 1966.
93. **Wezmar, J. and Leonard, E. F.,** In vitro assessment of six new hemodialysis membranes, *Trans. Am. Soc. Artif. Intern. Organs*, 16, 115, 1970.
94. **Fritzinger, B. K., Brauman, S. K., and Lyman, D. J.,** Membrane characteristics, permeability parameters, and frictional coefficients for Cuprophane, *J. Biomed. Mater. Res.*, 5, 3, 1971.
95. **Klinkmann, H. and Holtz, M.,** The permeation of essential amino acids through different dialysis membranes, *Proc. Eur. Dialysis Transplant Assoc.*, 9, 402, 1972.
96. **Klein, E., Holland, F. F., and Eberle, K.,** Comparison of solute permeability and rejection characteristics of normal and high flux cellulose hemodialysis membranes, *Proc. Eur. Dialysis Transplant Assoc.*, 16, 198, 1979.
97. **Meltzer, T. H., Gutfreund, K., Kulshrestha, V. K., and Stake, A. M.,** Optimized cellulose membranes for artificial kidney dialysis applications, *Trans. Am. Soc. Artif. Intern. Organs*, 14, 12, 1968.
98. **Staké, A. M.,** Optimized cellulose membranes for artificial kidney dialysis application, in *Proc. 2nd Annu. Contractors' Conf. of the Artificial Kidney Program of the National Institute of Arthritis, Metabolism, and Digestive Diseases*, Bethesda, Md., 1969, 80.
99. **Staké, A. M., Bock, F. C., Endres, A., and Meltzer, T. H.,** Statistical optimization of a cellulose membrane, *Appl. Polym. Symp.*, 13, 285, 1970.

100. **Staké, A., Bitten, J. F., Haas, W. R., Endres, A. C., and Bock, F. C.,** Optimized cellulose membranes for artificial kidney dialysis application, in *Proc. 3rd Annu. Contractors' Conf. of the Artificial Kidney Program of the National Institute of Arthritis, Metabolism, and Digestive Diseases,* Bethesda, Md., 1970, 57.
101. **Kulshrestha, V. K. and Kolff, W. J.,** Novel regenerated cellulose membranes for the artificial kidney, *Appl. Polym. Symp.,* 13, 297, 1970.
102. **Klinkmann, H., Lyman, D. J., Kirkham, R. L., Kulshrestha, V. K., and Kolff, W. J.,** Novel regenerated cellulosic membranes for hemodialysis, *Trans. Am. Soc. Artif. Intern. Organs,* 16, 121, 1970.
103. **Muir, W. M., Gray, R. A., and Courtney, J. M.,** Comparative membrane ultrafiltration rates, *Med. Biol. Eng.,* 8, 395, 1970.
104. **Klinkmann, H., Holtz, M., Falkenhagen, D., Ahrenholz, P., Bartsch, D., Bossin, F., Gensrich, H. J., Gröbe, V., Paul, D., and Schoner, H. W.,** Development and assessment of new cellulose membranes for blood purification, *Kidney Int.,* 18 (Suppl. 10), S-26, 1980.
105. **Weber, D. C., Kon, T., Mrava, G. L., and Nosé, Y.,** Cellulose membranes for hemodialysis, *Appl. Polym. Symp.,* 13, 197, 1970.
106. **Hoenich, N. A., Kerr, D. N. S., MacKenzie, J. C., and Parker, A. S.,** Undried cellulose gel film — a high permeability dialysis membrane, *Proc. Eur. Dialysis Transplant Assoc.,* 12, 565, 1976.
107. **Hoenich, N. A., White, T., Conceicao, S., and Kerr, D. N. S.,** Wet Cuprophan: in vivo and in vitro studies, in *Technical Aspects of Renal Dialysis,* Frost, T. H., Ed., Pitman, London, 1978, 107.
108. **Gullberg, C. A. and Goehl, H.,** Gel and pore membranes in blood purification, *Int. J. Artif. Organs,* 3, 189, 1980.
109. **Rozelle, L. T., Petersen, R. J., and Corneliussen, R. D.,** Development of a new concept in membrane structure for application in hemodialysis: ultrathin membranes, in *Proc. 2nd Annu. Contractors' Conf. of the Artificial Kidney Program of the National Institute of Arthritis, Metabolism, and Digestive Diseases,* Bethesda, Md., 1969, 84.
110. **Rozelle, L. T., Petersen, R. J., and Corneliussen, R. D.,** Ultrathin cellulose membranes for hemodialysis applications, *Appl. Polym. Symp.,* 13, 181, 1970.
111. **Rozelle, L. T., Petersen, R. J., and McClure, D. J.,** Development of a new concept in membrane structure for application in hemodialysis: ultrathin membranes, in *Proc. 4th Annu. Contractors' Conf. of the Artificial Kidney Program of the National Institute of Arthritis, Metabolism, and Digestive Diseases,* Bethesda, Md., 1971, 95.
112. **Rozelle, L. T. and Petersen, R. J.,** Development of a new concept in membrane structure for application in hemodialysis, in *Proc. 6th Annu. Contractors' Conf. of the Artificial Kidney Program of the National Institute of Arthritis, Metabolism, and Digestive Diseases,* Bethesda, Md., 1973, 165.
113. **Rozelle, L. T., Petersen, R. J., Shapiro, F. L., and Wathen, R. L.,** Development of a new concept in membrane structure for application in hemodialysis, in *Proc. 7th Annu. Contractors' Conf. of the Artificial Kidney Program of the National Institute of Arthritis, Metabolism, and Digestive Diseases,* Bethesda, Md., 1974, 70.
114. **Petersen, R. J., Keshaviah, P. R., Shapiro, F. L., and Wathen, R. L.,** Ultrathin cellulose membranes for application in hemodialysis, in *Proc. 8th Annu. Contractors' Conf. of the Artificial Kidney Program of the National Institute of Arthritis, Metabolism, and Digestive Diseases,* Bethesda, Md., 1975, 77.
115. **Kjellstrand, C. M., Petersen, R. J., Evans, R. L., Shideman, J. R., Santiago, E. A., Busselmeier, T. J., and Rozelle, L. T.,** In vivo studies of a new ultrathin membrane for hemodialysis, *Trans. Am. Soc. Artif. Intern. Organs,* 18, 106, 1972.
116. General Electric Company, Preswelled Cellulose Films and Process Therefor, U.K. Patent, 1,349,394, 1974.
117. Union Carbide Corporation, Reinforced Cellulosic Material and Method for Producing the Same, U.K. Patent, 1,358,314, 1974.
118. **Lonsdale, H. K. and Podall, H. E., Eds.,** *Reverse Osmosis Membrane Research,* Plenum Press, New York, 1972.
119. **Saltonstall, C. W., Jr., Martin, F. E., Shuey, H. F., and Mechalas, B. J.,** Improved membranes for hemodialysis, in *Proc. 2nd Annu. Contractors' Conf. of the Artificial Kidney Program of the National Institute of Arthritis, Metabolism, and Digestive Diseases,* Bethesda, Md., 1969, 69.
120. **Saltonstall, C. W., Jr., Martin, F. E., Shuey, H. F., and Shinaberger, J. H.,** Improved membranes for hemodialysis, in *Proc. 4th Annu. Contractors' Conf. of the Artificial Kidney Program of the National Institute of Arthritis, Metabolism, and Digestive Diseases,* Bethesda, Md., 1971, 91.
121. **Shinaberger, J. H., Miller, J. H., Rosenblatt, M. G., Gardner, P. W., Carpenter, G. W., and Martin, F. E.,** Clinical studies with membranes for hemodialysis, *Proc. Clin. Dialysis Transplant Forum,* 1, 36, 1971.
122. **Shinaberger, J. H., Miller, J. H., Rosenblatt, M. G., Gardner, P. W., Carpenter, G. W., and Martin, F. E.,** Clinical studies of 'low flow' dialysis with membranes highly permeable to middle weight molecules, *Trans. Am. Soc. Artif. Intern. Organs,* 18, 82, 1972.

123. **Saltonstall, C. W., Jr., Martin, F. E., Shinaberger, J. H., and Miller, J. H.,** Development of hemodialysis membranes, in *Proc. 5th Annu. Contractors' Conf. of the Artificial Kidney Program of the National Institute of Arthritis, Metabolism, and Digestive Diseases,* Bethesda, Md., 1972, 27.
124. **Sargent, J. A., Gotch, F. A., Lipps, B. J., Holmes, G. W., Keen, M. L., Evans, M. C., and Teisinger, G. L.,** Evaluation of experimental high flux artificial kidneys, in *Proc. 5th Annu. Contractors' Conf. of the Artificial Kidney Program of the National Institute of Arthritis, Metabolism, and Digestive Diseases,* Bethesda, Md., 1972, 12.
125. **Sparks, R. E., Lindan, O., and Mason, N. S.,** Research leading to an improved artificial kidney. Development of highly permeable membranes, in *Proc. 2nd Annu. Contractors' Conf. of the Artificial Kidney Program of the National Institute of Arthritis, Metabolism, and Digestive Diseases,* Bethesda, Md., 1969, 73.
126. **Sparks, R. E., Lindan, O., and Mason, N. S.,** Development of highly permeable membranes, in *Proc. 3rd Annu. Contractors' Conf. of the Artificial Kidney Program of the National Institute of Arthritis, Metabolism, and Digestive Diseases,* Bethesda, Md., 1970, 49.
127. **Sparks, R. E., Lindan, O., and Mason, N. S.,** Research leading to an improved artificial kidney. I. Cellulose acetate dialysis membranes, in *Proc. 4th Annu. Contractors' Conf. of the Artificial Kidney Program of the National Institute of Arthritis, Metabolism, and Digestive Diseases,* Bethesda, 1971, 93.
128. **Sparks, R. E., Lindan, O., Zimmerman, P., and Mason, N. S.,** Cellulose acetate dialysis membranes, in *Proc. 5th Annu. Contractors' Conf. of the Artificial Kidney Program of the National Institute of Arthritis, Metabolism, and Digestive Diseases,* Bethesda, Md., 1972, 41.
129. **Mason, N. S., Lindan, O., and Sparks, R. E.,** Reinforced cellulose acetate dialysis membranes, *Trans. Am. Soc. Artif. Intern. Organs,* 14, 31, 1968.
130. **Model, F. S., Plovan, S. G., Davis, H. J., and Sessa, P. A.,** To prepare and evaluate optimized hemodialysis membranes, in *Proc. 6th Annual Contractors' Conf. of the Artificial Kidney Program of the National Institute of Arthritis, Metabolism, and Digestive Diseases,* Bethesda, Md., 1973, 175.
131. **Model, F. S., Davis, H. J., and Gordon, M.,** Prepare and evaluate optimized hemodialysis membranes, in *Proc. 7th Annu. Contractors' Conf. of the Artificial Kidney Program of the National Institute of Arthritis, Metabolism, and Digestive Diseases,* Bethesda, Md., 1974, 72.
132. **Model, F. S., Davis, H. J., and Gordon, M.,** Prepare and evaluate optimized hemodialysis membranes, in *Proc. 8th Annu. Contractors' Conf. of the Artificial Kidney Program of the National Institute of Arthritis, Metabolism, and Digestive Diseases,* Bethesda, Md., 1975, 71.
133. **Model, F. S. and Davis, H. J.,** Preparation and evaluation of optimized hemodialysis membranes, in *Proc. 9th Annu. Contractors' Conf. of the Artificial Kidney Program of the National Institute of Arthritis, Metabolism, and Digestive Diseases,* Bethesda, Md., 1976, 70.
134. **Model, F. S. and Davis, H. J.,** Preparation and evaluation of optimized hemodialysis membranes, in *Proc. 10th Annu. Contractors' Conf. of the Artificial Kidney Program of the National Institute of Arthritis, Metabolism, and Digestive Diseases,* Bethesda, Md., 1977, 77.
135. **Rosenthal, A. J. and Davis, H. J.,** Preparation and evaluation of optimized hemodialysis membranes, *Proc. 11th Annu. Contractors' Conf. of the Artificial Kidney Program of the National Institute of Arthritis, Metabolism, and Digestive Diseases,* Bethesda, Md., 1978, 160.
136. **Adler, A. J., Lundin, A. P., Friedman, E. A., and Berlyne, G. M.,** Effect of hemodialysis on plasma beta thromboglobulin levels, *Trans. Am. Soc. Artif. Intern. Organs,* 25, 347, 1979.
137. **Azikawa, T., Nishiyama, H., and Koshikawa, S.,** Plasma β-thromboglobulin levels in chronic renal failure patients, *Artif. Organs,* 5, 54, 1981.
138. **Quellhorst, E., Rieger, J., Doht, B., Beckmann, H., Jacob, I., Kraft, B., Mietzsch, G., and Scheler, F.,** Treatment of chronic uraemia by an ultrafiltration kidney — first clinical experience, *Proc. Eur. Dialysis Transplant Assoc.,* 13, 314, 1976.
139. **Nishihara, T., Rubin, A. L., and Stenzel, K. H.,** Biologically derived collagen membranes, *Trans. Am. Soc. Artif. Intern. Organs,* 13, 243, 1967.
140. **Rubin, A. L., Riggio, R. R., Nachman, R. L., Schwartz, G. H., Miyata, T., and Stenzel, K. H.,** Collagen materials in dialysis and implantation, *Trans. Am. Soc. Artif. Intern. Organs,* 14, 169, 1968.
141. **Rubin, A. L., Miyata, T., and Stenzel, K. H.,** Collagen: medical and surgical applications, *J. Macromol. Sci. Chem.,* A3, 113, 1969.
142. **Stenzel, K. H., Sullivan, J. F., Miyata, T., and Rubin, A. L.,** Collagen dialysis membranes: initial clearance evaluation, *Trans. Am. Soc. Artif. Intern. Organs,* 15, 114, 1969.
143. **Rubin, A. L., Stenzel, K. H., Yamayoshi, W., and Miyata, T.,** Biologically derived collagen membranes and surfaces: their applications to problems of hemodialysis and chronic uremia, in *Proc. 3rd Annu. Contractors' Conf. of the Artificial Kidney Program of the National Institute of Arthritis, Metabolism, and Digestive Diseases,* Bethesda, Md., 1970, 65.
144. **Rubin, A. L., Stenzel, K. H., Yamayoshi, W., and Miyata, T.,** Collagen membranes and surfaces, in *Proc. 4th Annu. Contractors' Conf. of the Artificial Kidney Program of the National Institute of Arthritis, Metabolism, and Digestive Diseases,* Bethesda, Md., 1971, 86.

145. **Stenzel, K. H., Rubin, A. L., Yamayoshi, W., Miyata, T., Suzuki, T., Sohde, T., and Nishizawa, M.,** Optimization of collagen dialysis membranes, *Trans. Am. Soc. Artif. Intern. Organs*, 17, 293, 1971.
146. **Rubin, A. L., Stenzel, K. H., Miyata, T., Stubenbord, W. T., and Cheigh, J. S.,** Collagen membranes and surfaces, in *Proc. 5th Annu. Contractors' Conf. of the Artificial Kidney Program of the National Institute of Arthritis, Metabolism, and Digestive Diseases*, Bethesda, Md., 1972, 25.
147. **White, M. J., Kohno, I., Rubin, A. L., Stenzel, K. H., and Miyata, T.,** Collagen films: effect of crosslinking on physical and biological properties, *Biomater. Med. Devices Artif. Organs*, 1, 703, 1973.
148. **Kon, T., Mrava, G. L., Weber, D. C., and Nosé, Y.,** Collagen membrane for hemodialysis, *J. Biomed. Mater. Res.*, 4, 13, 1970.
149. **Mrava, G. L., Kon, T., Weber, D. C., and Nosé, Y.,** New membranes for hemodialysis, *Appl. Polym. Symp.*, 13, 235, 1970.
150. **Lyman, D. J.,** Investigation of synthetic membranes for hemodialysis, *Trans. Am. Soc. Artif. Intern. Organs*, 9, 92, 1963.
151. **Lyman, D. J., Loo, B. H., and Crawford, R. W.,** New synthetic membranes for dialysis. I. A copolyether-ester membrane system, *Biochemistry*, 3, 985, 1964.
152. **Lyman, D. J., Loo, B. H., and Muir, W. M.,** New synthetic membranes for the dialysis of blood. III. The concept of liquid partition of biological molecules by means of membranes, *Trans. Am. Soc. Artif. Intern. Organs*, 11, 91, 1965.
153. **Lyman, D. J. and Loo, B. H.,** New synthetic membranes for dialysis. IV. A copolyether-urethane membrane system, *J. Biomed. Mater. Res.*, 1, 17, 1967.
154. **Lyman, D. J.,** New synthetic membranes for extracorporeal hemodialysis, *Ann. N.Y. Acad. Sci.*, 146, 113, 1968.
155. **Rhone-Poulenc, S. A.,** Dialysis and Ultrafiltration of a Liquid, U.K. Patent 1,327,990, 1973.
156. **Man, N.-K., Terlain, B., Paris, J., Werner, G., Sausse, A., and Funck-Brentano, J.-L.,** An approach to "middle molecules" identification in artificial kidney dialysate, with reference to neuropathy prevention, *Trans. Am. Soc. Artif. Intern. Organs*, 19, 320, 1973.
157. **Man, N. K., Granger, A., Rondon-Nucete, M., Zingraff, J., Jungers, P., Sausse, A., and Funck-Brentano, J. L.,** One year follow-up of short dialysis with a membrane highly permeable to middle molecules, *Proc. Eur. Dialysis Transplant Assoc.*, 10, 236, 1973.
158. **Funck-Brentano, J. L. and Man, N. K.,** The polyacrylonitrile membrane and the Rhodial system: their practical application, in *Replacement of Renal Function by Dialysis*, Drukker, W., Parsons, F. M., and Maher, J. F., Eds., Martinus Nijhoff, The Hague, 1978, 125.
159. **Kiley, J. E., Pender, J. C., Welch, H. F., and Welch, C. S.,** Treatment of hepatic coma by hemodialysis, *Trans. Am. Soc. Artif. Intern. Organs*, 3, 86, 1957.
160. **Opolon, P., Rapin, J.-R., Huguet, C., Granger, A., Delorme, M.-L., Boschat, M., and Sausse, A.,** Hepatic failure coma (HFC) treated by polyacrylonitrile membrane (PAN) hemodialysis (HD), *Trans. Am. Soc. Artif. Intern. Organs*, 22, 701, 1976.
161. **Denis, J., Opolon, P., Nusinovici, V., Granger, A., and Darnis, F.,** Treatment of encephalopathy during fulminant hepatic failure by hemodialysis with high permeability membrane, *Gut*, 19, 787, 1978.
162. **Silk, D. B. A. and Williams, R.,** Experience in the treatment of fuliminant hepatic failure by conservative therapy, charcoal haemoperfusion and polyacrylonitrile haemodialysis, *Int. J. Artif. Organs*, 1, 29, 1978.
163. **Muir, W. M., Courtney, J. M., Gray, R. A., and Ritchie, P. D.,** Perm-selective dialysis membranes. I. Films based on dimethylaminoethyl methacrylate copolymers, *J. Biomed. Mater. Res.*, 5, 415, 1971.
164. **Courtney, J. M. and Gilchrist, T.,** Chemical modification of acrylonitrile copolymers: reaction of cationic copolymers with ethylene oxide, *J. Appl. Polymer Sci.*, 19, 1503, 1975.
165. **Courtney, J. M., Gaylor, J. D. S., Lindsay, R. M., and Martin, I.,** Acrylonitrile copolymer membranes: transport characteristics and blood compatibility, in *Technical Aspects of Renal Dialysis*, Frost, T. H., Ed., Pitman, London, 1978, 126.
166. **Salyer, I. O., Weesner, W. E., Bump, D. D., and Erbaugh, L. E.,** Membrane preparation and evaluation, in *Proc. 2nd Annu. Contractors' Conf. of the Artificial Kidney Program of the National Institute of Arthritis, Metabolism, and Digestive Diseases*, Bethesda, Md., 1969, 83.
167. **Salyer, I. O., Weesner, W. E., Tapp, J. S., Beemsterboer, G. L., Leonard, R. L., Bump, D. D., Erbaugh, L. E., and Ball, G. L., III,** Development of a polyacrylonitrile hollow fiber artificial kidney, in *Proc. 3rd Annu. Contractors' Conf. of the Artificial Kidney Program of the National Institute of Arthritis, Metabolism, and Digestive Diseases*, Bethesda, Md., 1970, 40.
168. **Salyer, I. O., Ball, G. L., III, Beemsterboer, G. L., Leonard, R. J., Merrill, J. T., Hampers, C. L., and Lowrie, E. G.,** The Monsanto polyacrylonitrile hollow-fiber artificial kidney, *J. Biomed. Mater. Res.*, 6, 59, 1972.
169. **Walch, A., Wollmann, K., Puchinger, H., and Pusch, W.,** Verwendung Einer Membran für Medizinisch - Biologische Zwecke, Offenlegungsschrift 2508416, Deutsches Patentamt, FRG, 1976.
170. **C. R. Bard, Inc.,** Thin Polycarbonate Membranes for use in Hemodialysis, U.K. Patent 1,556,897, 1979.

171. **Cantor, P.,** Modified polycarbonate membranes for hemodialysis, in *Proc. 3rd Annu. Contractors' Conf. of the Artificial Kidney Program of the National Institute of Arthritis, Metabolism, and Digestive Diseases,* Bethesda, Md., 1970, 63.
172. **Fisher, B. S. and Cantor, P.,** Modified polycarbonate membranes for hemodialysis, in *Proc. 5th Annu. Contractors' Conf. of the Artificial Kidney Program of the National Institute of Arthritis, Metabolism, and Digestive Diseases,* Bethesda, Md., 1972, 32.
173. **Fisher, B. S., Highley, W. S., Cantor, P. A., and Stone, W., Jr.,** Modified polycarbonate membranes for hemodialysis, *Trans. Am. Soc. Artif. Intern. Organs,* 19, 429, 1973.
174. **Barbour, B. H., Bernstein, M., Cantor, P. A., Fisher, B. S., and Stone, W., Jr.,** Clinical use of N1SR 440 polycarbonate membrane for hemodialysis, *Trans. Am. Soc. Artif. Intern. Organs,* 21, 144, 1975.
175. **Fisher, B. S., Highley, W. S., and Cantor, P. A.,** Modified polycarbonate membranes for hemodialysis, *Proc. 6th Annu. Contractors' Conf. of the Artificial Kidney Program of the National Institute of Arthritis, Metabolism, and Digestive Diseases,* Bethesda, Md., 1973, 170.
176. **Stone, W., Jr.,** Polycarbonate membrane for hemodialysis and hemofiltration, *Dialysis Transplant.,* 6 (4), 10, 1977.
177. **Fisher, B. S., Cantor, P. A., Highley, W. S., and Stone, W., Jr.,** Modified polycarbonate membrane for hemodialysis, in *Proc. 7th Annu. Contractors' Conf. of the Artificial Kidney Program of the National Institute of Arthritis, Metabolism, and Digestive Diseases,* Bethesda, Md., 1974, 78.
178. **Fisher, B. S., Cantor, P. A., Highley, W. S., Barbour, B., and Stone, W., Jr.,** Modified polycarbonate membrane for hemodialysis, in *Proc. 8th Annu. Contractors' Conf. of the Artificial Kidney Program of the National Institute of Arthritis, Metabolism, and Digestive Diseases,* Bethesda, Md., 1975, 79.
179. **Kesting, R. E.,** Control of polymer morphology for biomedical applications, I. Hydrophilic polycarbonate membranes for dialysis, *J. Macromol. Sci. Chem.,* A4, 655, 1970.
180. **Ota, K., Ozakawa, T., Kumagaya, E., Agishi, T., Sugino, N., Mitani, N., Fujii, Y., Kinura, M., Nagao, Y., Tsukamoto, H., Tanzawa, H., and Sakai, Y.,** Polymethylmethacrylate capillary kidney highly permeable to middle molecules, *Proc. Eur. Dialysis Transplant Assoc.,* 12, 559, 1976.
181. **Ota, K., Suzuki, T., Ozaku, Y., Hoshino, T., Agishi, T., and Sugino, N.,** Short-term hemodiafiltration using polymethylmethacrylate hemodiafilter, *Trans. Am. Soc. Artif. Intern. Organs,* 24, 454, 1978.
182. **Muir, W. M., Courtney, J. M., and Gray, R. A.,** Diffusion transport mechanisms in synthetic polymer membranes for haemodialysis, *Med. Biol. Eng.,* 8, 445, 1970.
183. **Muir, W. M., Gray, R. A., Courtney, J. M., and Ritchie, P. D.,** Perm-selective dialysis membranes. II. Films based on acrylic acid-*n*-butyl methacrylate copolymers — a critical comparison with cellulose films, *J. Biomed. Mater. Res.,* 7, 3, 1973.
184. **Muir, W. M., Martin, A. M., Courtney, J. M., Gray, R. A., and Gaylor, J. D. S.,** Clinical evaluation of new haemodialysis membranes, *Proc. Eur. Dialysis Transplant Assoc.,* 8, 445, 1971.
185. **Kin, U. Y. and Huang, Y. M.,** Hydrophilic polymer membranes containing carboxylic groups. I. Acrylic acid-methyl methacrylate-glycidyl methacrylate copolymers, *Eur. Polym. J.,* 15, 325, 1979.
186. **Ikenberry, L. D., Yasuda, H. K., and Clark, H. G.,** Characterization of membrane materials for hemodialysis, in *The Artificial Kidney,* Vol. 64, Chemical Engineering Progress Symposium Series, No. 85, Dedrick, R. L., Bischoff, K. B., and Leonard, E. F., Eds., American Institute of Chemical Engineers, New York, 1968, 69.
187. **Peterlin, A., Yasuda, H., and Ikenberry, L. D.,** Improved systems for hemodialysis, in *Proc. 2nd Annu. Contractors' Conf. of the Artificial Kidney Program of the National Institute of Arthritis, Metabolism, and Digestive Diseases,* Bethesda, Md., 1969, 86.
188. **Ratner, B. D. and Miller, J. F.,** Transport through crosslinked poly (2-hydroxyethyl methacrylate) hydrogel membranes, *J. Biomed. Mater. Res.,* 7, 353, 1973.
189. **Lee, K. H., Jee, J. G., Jhon, M. S., and Ree, T.,** Solute transport through crosslinked poly(2-hydroxyethyl methacrylate) membrane, *J. Bioeng.,* 2, 269, 1978.
190. **Huang, W. and Sparks, R. E.,** Mass transfer through hydrophilic-charged membranes, *Polym. Prepr., Am. Chem. Soc., Div. Polym. Chem.,* 17, 882, 1976.
191. **Markle, R. A., Falb, R. D., and Leininger, R. I.,** Development of improved membranes for artificial kidney dialysis, *Trans. Am. Soc. Artif. Intern. Organs,* 10, 22, 1964.
192. **Luttinger, M. and Cooper, C. W.,** Improved hemodialysis membranes for the artificial kidney, *J. Biomed. Mater. Res.,* 1, 67, 1967.
193. **Luttinger, M., Cooper, C. W., and Leininger, R. I.,** Preparation of novel hemodialysis membranes, *Trans. Am. Soc. Artif. Intern. Organs,* 14, 5, 1968.
194. **Luttinger, M., Cooper, C. W., and Leininger, R. I.,** Development of improved hemodialysis membranes, in *Proc. 2nd Annu. Contractors' Conf. of the Artificial Kidney Program of the National Institute of Arthritis, Metabolism, and Digestive Diseases,* Bethesda, Md., 1969, 71.

195. **Luttinger, M. and Cooper, C. W.,** Development of improved hemodialysis membranes, *Proc. 3rd Annu. Contractors' Conf. of the Artificial Kidney Program of the National Institute of Arthritis, Metabolism, and Digestive Diseases,* Bethesda, Md., 1970, 61.
196. **Luttinger, M. and Cooper, C. W.,** Parameters affecting preparation of nylon-based hemodialysis membranes, *J. Biomed. Mater. Res.,* 4, 281, 1970.
197. **Streicher, E. and Schneider, H.,** Asymmetric polyamide hollow-fiber filters in the hemofiltration system, *J. Dialysis,* 1, 727, 1977.
198. **Schneider, H. and Streicher, E.,** Clinical observations of the polyamide hollow-fiber hemofilter in hemofiltration systems, *J. Dialysis,* 1, 737, 1977.
199. **Klein, E., Perciaccante, V., Smith, J. K., Greer, W., Yatsu, B., and Boudreaux, G.,** Study of synthetic polypeptides as possible hemodialysis membranes, in *Proc. 2nd Annu. Contractors' Conf. of the Artificial Kidney Program of the National Institute of Arthritis, Metabolism, and Digestive Diseases,* Bethesda, Md., 1969, 78.
200. **Klein, E., Smith, J. K., May, P. D., and Leger, N. V.,** Study of synthetic polypeptides as possible hemodialysis membranes, in *Proc. 4th Annu. Contractors' Conf. of the Artificial Kidney Program of the National Institute of Arthritis, Metabolism, and Digestive Diseases,* Bethesda, Md., 1971, 89.
201. **Klein, E., May, P. D., Smith, J. K., and Leger, N.,** Permeability of synthetic polypeptide membranes. I. Poly-D,L-leucine-co-D,L-methionine, *Biopolymers,* 10, 647, 1971.
202. **Klein, E., Lindholm, D., Smith, J. K., and May, P. D.,** Study of synthetic polypeptides as possible hemodialysis membranes, in *Proc. 5th Annu. Contractors' Conf. of the Artificial Kidney Program of the National Institute of Arthritis, Metabolism, and Digestive Diseases,* Bethesda, Md., 1972, 29.
203. **Klein, E., Lindholm, D. D., Smith, J. K., and Holland, F. F., Jr.,** Study of synthetic polypeptides for hemodialysis membranes, in *Proc. 7th Annu. Contractors Conf. of the Artificial Kidney Program of the National Institute of Arthritis, Metabolism, and Digestive Diseases,* Bethesda, Md., 1974, 75.
204. **Dombeck, D. H., Holland, F. F., Klein, E., Lindholm, D. D., and Smith, J. K.,** Study of synthetic polypeptides for hemodialysis membranes, in *Proc. 8th Annu. Contractors' Conf. of the Artificial Kidney Program of the National Institute of Arthritis, Metabolism, and Digestive Diseases,* Bethesda, Md., 1975, 75.
205. **Klein, E., Bower, J., Lindholm, D., Holland, F., Holbert, R., May, P. D., Pearson, J., Smith, J. K., and Eberle, K.,** Study of synthetic polypeptides for hemodialysis membranes, in *Proc. 9th Annu. Contractors' Conf. of the Artificial Kidney Program of the National Institute of Arthritis, Metabolism, and Digestive Diseases,* Bethesda, Md., 1976, 78.
206. **Martin, E. C., May, P. D., and McMahon, W. A.,** Amino acid polymers for biomedical applications. I. Permeability properties of L-leucine DL-methionine copolymers, *J. Biomed. Mater. Res.,* 5, 53, 1971.
207. **Sederel, W. L., Bantjes, A., and Feijen, J.,** A route to anionic hydrophilic films of copolymers of L-leucine, L-aspartic acid and L-aspartic acid esters, *Polymer,* 16, 735, 1975.
208. **Cross, R. A., Tyson, W. H., Jr., and Cleveland, D. S.,** Asymmetric hollow fiber membranes for dialysis, *Trans. Am. Soc. Artif. Intern. Organs,* 17, 279, 1971.
209. **Cross, R. A., Cleveland, D. S., Kahn, P., Tyson, W. H., Thompson, C. R., Guzman, R. J., Hidalgo, J., Robinson, F. G., and Brown, T. L.,** An improved hollow fiber hemodialyzer, in *Proc. 5th Annu. Contractors' Conf. of the Artificial Kidney Program of the National Institute of Arthritis, Metabolism, and Digestive Diseases,* Bethesda, Md., 1972, 7.
210. **Lysaght, M. J., Thompson, C. R., Henderson, L. W., and Autian, J. W.,** An improved hollow fiber hemodialyzer, in *Proc. 6th Annu. Contractors' Conf. of the Artificial Kidney Program of the National Institute of Arthritis, Metabolism, and Digestive Diseases,* Bethesda, Md., 1973, 79.
211. **Lysaght, M. J., Ford, C. A., Henderson, L. W., and Lawrie, E. G.,** An improved hollow fiber hemodialyzer, in *Proc. 8th Annu. Contractors' Conf. of the Artificial Kidney Program of the National Institute of Arthritis, Metabolism, and Digestive Diseases,* Bethesda, Md., 1975, 136.
212. **Markley, L. L., Bixler, H. J., and Cross, R. A.,** Utilisation of polyelectrolyte complexes in biology and medicine, *J. Biomed. Mater. Res.,* 2, 145, 1968.
213. **Vogel, M. K., Cross, R. A., Bixler, H. J., and Guzman, R. J.,** Medical uses for polyelectrolyte complexes, *J. Macromol. Sci. Chem.,* A4, 675, 1970.
214. **Marshall, D. W., Cross, R. A., and Bixler, H. J.,** An evaluation of polyelectrolyte complexes as biomedical materials, *J. Biomed. Mater. Res.,* 4, 357, 1970.
215. **Rudman, A. R., Vengerova, N. A., Kalyuzhnaya, R. J., Eltsefon, B. S., and Zezin, A. B.,** Study of polyelectrolyte membrane permeability using low-molecular-weight metabolites, *Khim. Farm. Zh.,* 13(3), 82, 1979.
216. **Odian, M. and Leonard, E. F.,** Synthesis and evaluation of graded polyvinyl alcohol membranes, *Trans. Am. Soc. Artif. Intern. Organs,* 14, 19, 1968.
217. **Hills, B. A.,** Preparation of polymer films displaying some basic characteristics of natural membranes, *Trans. Am. Soc. Artif. Intern. Organs,* 14, 1, 1968.

218. **Peppas, N. A.,** Development of crosslinked poly(vinyl alcohol) biomembranes, *Polym. Prepr., Am. Chem. Soc., Div. Polym. Chem.*, 18, 794, 1977.
219. **Yamashita, S., Takakura, K., Imai, Y., and Masuhara, E.,** Polyvinyl alcohol based membranes for hemodialysis, *Kobunshi Ronbunshu*, 35, 283, 1978.
220. **Imai, Y., Masuhara, E., Takakura, K., Nakaji, O., and Yamashita, S.,** Modification of permeability of poly(vinyl alcohol)-acrylonitrile graft copolymer membranes by solvent treatment, *Kobunshi Ronbunshu*, 35, 765, 1978.
221. **Ohtsuka, Y., Hirabayashi, Y., Masubuchi, J., and Kaneko, N.,** Haemodialysis membranes prepared from latices of acrylonitrile-methyl acrylate-acrylamide graft copolymerized onto polyvinyl alcohol, *Kobunshi Ronbunshu*, 36, 257, 1979.
222. **Ohtsuka, Y., Hirabayashi, Y., and Takayama, S.,** Mechanical properties of membranes prepared from latices of acrylonitrile-2-hydroxyethyl methacrylate graft copolymerised onto poly(vinyl alcohol), *Kobunshi Ronbunshu*, 37, 449, 1980.
223. **Yamashita, S., Nagata, S., Takakura, K., and Yamada, K.,** Separation Membranes, U.K. Patent 1,503,270, 1978.
224. **Yamashita, S., Nagata, S., and Takakura, K.,** Studies on preparation conditions and properties of ethylene-vinyl alcohol copolymer membranes for haemodialysis, *Kobunshi Ronbunshu*, 36, 249, 1979.
225. **Yamauchi, K., Takakura, K., Kawai, S., and Ohmori, A.,** Membranes Based on Polyvinyl Alcohol, U.K. Patent 1,514,127, 1978.
226. **Gilchrist, T. and Courtney, J. M.,** The design of biocompatible polymers, in *Drug Design*, Vol. 10, Ariëns, E. J., Ed., Academic Press, New York, 1980, 251.
227. **Holland, F. F., Gidden, H. E., Mason, R. G., and Klein, E.,** Thrombogenicity of heparin-bound DEAE cellulose hemodialysis membranes, *ASAIO J.*, 1, 24, 1978.
228. **Merrill, E. W., Salzman, E. W., Lipps, B. J., Jr., Gilliland, E. R., Austen, W. G., and Joison, J.,** Antithrombogenic cellulose membranes for blood dialysis, *Trans. Am. Soc. Artif. Intern. Organs*, 12, 139, 1966.
229. **Britton, R. A., Merrill, E. W., Gilliland, E. R., Salzman, E. W., Austen, W. G., and Kemp, D. S.,** Antithrombogenic cellulose film, *J. Biomed Mater. Res.*, 2, 429, 1968.
230. **Martin, F. E., Shuey, H. F., and Saltonstall, C. W., Jr.,** Improved membranes for hemodialysis, *J. Macromol. Sci. Chem.*, A4, 635, 1970.
231. **Lindsay, R. M., Bourke, J. T. B., Reid, B. D., Linton, A. L., Gilchrist, T., Courtney, J. M., and Edwards, R. O.,** The role of heparin on platelet retention by acrylonitrile copolymer dialysis membranes, *J. Lab. Clin. Med.*, 89, 724, 1977.
232. **Falb, R. D., Takahashi, M. T., Grode, G. A., and Leininger, R. I.,** Studies on the stability and protein adsorption characteristics of heparinized polymer surfaces by radioisotope labeling techniques, *J. Biomed. Mater. Res.*, 1, 239, 1967.
233. **Gott, V. L., Whiffen, J. D., and Valiathan, S. M.,** Graphite-benzalkonium-heparin coatings on plastics and metals, *Ann. N.Y. Acad. Sci.*, 146, 21, 1968.
234. **Chawla, A. S. and Chang, T. M. S.,** Nonthrombogenic surface by radiation grafting of heparin: preparation, in vitro and in vivo studies, *Biomater. Med. Devices Artif. Org.*, 2, 157, 1974.
235. **Lagergren, H. R. and Eriksson, J. C.,** Plastics with a stable surface monolayer of cross-linked heparin: preparation and evaluation, *Trans. Am. Soc. Artif. Intern. Organs*, 17, 10, 1971.
236. **Schmer, G., Cole, J. Vizzo, J., and Krys, J.,** Anticoagulant-bound hemodialysis, in *Proc. 7th Annu. Contractors' Conf. of the Artificial Kidney Program of the National Institute of Arthritis, Metabolism, and Digestive Diseases*, Bethesda, Md., 1974, 69.
237. **Schmer, G., Cole, J. J., Vizzo, J. E., and Teng, L.-N.,** Anticoagulant-bound hemodialysis, in *Proc. 9th Annu. Contractors' Conf. of the Artificial Kidney Program of the National Institute of Arthritis, Metabolism, and Digestive Diseases*, Bethesda, Md., 1976, 68.
238. **Schmer, G., Teng, L. N. L., Cole, J. J., Vizzo, J. E., Francisco, M. M., and Scribner, B. H.,** Successful use of a totally heparin grafted hemodialysis system in sheep, *Trans. Am. Soc. Artif. Intern. Organs*, 22, 654, 1976.
239. **Schmer, G., Teng, L. N. L., Vizzo, J. E., Graefe, U., Militunovich, J., Cole, J. J., and Scribner, B. H.,** Clinical use of a totally heparin grafted hemodialysis system in uremic patients, *Trans. Am. Soc. Artif. Intern. Organs*, 23, 177, 1977.
240. **Kenedi, R. M., Courtney, J. M., Gaylor, J. D. S., and Gilchrist, T., Eds.,** *Artificial Organs*, Macmillan, London, 1977.
241. **Chang, T. M. S., Ed.,** *Artificial Kidney, Artificial Liver, and Artificial Cells*, Plenum Press, New York, 1978.
242. **Sideman, S. and Chang, T. M. S., Eds.,** *Hemoperfusion: Kidney and Liver Support and Detoxification*, Hemisphere, Washington, D.C., 1980.
243. **Rietema, K. and van Zutphen, P.,** The activated carbon loaded adsorber as an addition to the haemodialyser, *Proc. Eur. Soc. Artif. Organs*, 1, 100, 1975.

244. **Malchesky, P. S., Varnes, W., Nokoff, R., and Nosé, Y.,** The charcoal capillary haemoperfusion system, *Proc. Eur. Dialysis Transplant Assoc.*, 13, 242, 1976.
245. **Malchesky, P. S., Varnes, W., Piatkiewicz, W., and Nosé, Y.,** Membranes containing sorbents for blood detoxification, *Trans. Am. Soc. Artif. Intern. Organs*, 23, 659, 1977.
246. **Malchesky, P. S. and Nosé, Y.,** The use of membranes and sorbents for blood detoxification: Cuprophan sorbent membranes, in *Artificial Kidney, Artificial Liver, and Artificial Cells*, Chang, T. M. S., Ed., Plenum Press, New York, 1978, 185.
247. **Malchesky, P. S., Piatkiewicz, W., Varnes, W. G., Ondercin, L., and Nosé, Y.,** Sorbent membranes: device designs, evaluations and potential applications, *Artif. Organs*, 2, 367, 1978.
248. **Castro, L. A., Hampel, G., Gebhardt, R., Fateh, A., and Gurland, H.,** Combination of hemodialysis and hemoperfusion in a single hollow-fiber unit for the treatment of uremia, in *Artificial Kidney, Artificial Liver, and Artificial Cells*, Chang, T. M. S., Ed., Plenum Press, New York, 1978, 193.
249. **Gurland, H. J., Fernandez, J. C., Samtleben, W., and Castro, L. A.,** Sorbent membranes used in a conventional dialyzer format — in vitro and clinical evaluation, *Artif. Organs*, 2, 372, 1978.
250. **Gurland, H. J., Castro, L. A., Samtleben, W., and Fernandez, J. C.,** Combination of hemodialysis and hemoperfusion in a single unit for treatment of uremia, *Clin. Nephrol.*, 11, 167, 1979.
251. **Gurland, H. J., Castro, L. A., Hillebrand, G., and Schmidt, B.,** Sorbent membranes and their medical applications, in *Hemoperfusion: Kidney and Liver Support and Detoxification*, Sideman, S. and Chang, T. M. S., Eds., Hemisphere, Washington, D.C., 1980, 105.
252. **Klein, E. and Holland, F. F.,** Preparation of high clearance sorbent-filled hollow fibers, in *Proc. 11th Annu. Contractors' Conf. of the Artificial Kidney Program of the National Institute of Arthritis, Metabolism, and Digestive Diseases*, Bethesda, Md., 1978, 170.
253. **Klein, E., Holland, F. F., Eberle, K., Morton, F. C., and Cabasso, I.,** Sorbent-filled hollow fibers for hemopurification, *Trans. Am. Soc. Artif. Intern. Organs*, 24, 127, 1978.
254. **Wolf, C. F. W.,** Cells cultured on artificial capillaries and their use as a liver assist device, *Artif. Organs*, 4, 279, 1980.
255. **Mazzola, G., Cremonesi, P., Colombi, S., Vecchio, G., Agnellini, D., and Pietta, P. G.,** Application of immobilized enzymes in detoxification. II. Immobilization and characterization of enzymes on hollow fiber devices, *Minerva Nefrol.*, 26, 185, 1979.
256. **Callegaro, L., Denti, E., and Fontana, A.,** Immobilized hepatic enzymes as a tool in extracorporeal detoxification, in *Artificial Liver Support*, Brunner, G. and Schmidt, F. W., Eds., Springer-Verlag, Berlin, 1981, 198.
257. **Marconi, W., Bartoli, F., Mantovani, E., Pittalis, F., Settembri, L., Cordova, C., Musca, A., and Alessandri, C.,** Development of new antithrombogenic surfaces by employing platelet antiaggregating agents: preparation and characterization, *Trans. Am. Soc. Artif. Intern. Organs*, 25, 280, 1979.

Chapter 8

# POLYMERIC INSERTS AND IMPLANTS FOR THE CONTROLLED RELEASE OF DRUGS

**Neil B. Graham and David A. Wood**

## TABLE OF CONTENTS

I. Introduction .......... 182

II. Physical Forms of Drug-Polymer Composites as Inserts and Implants .......... 183

III. Release Mechanisms and Prediction of Release Profiles .......... 184
- A. Membrane Devices .......... 184
- B. Release from Monoliths with Dissolved Drug .......... 186
- C. Release from Matrices with Dispersed Drug .......... 186
- D. Release from Devices with Capillaries and Pores .......... 188
- E. Release from Hydrogels .......... 189
- F. Boundary-Layer Controlled Release .......... 189
- G. Release from Erodible Devices .......... 190

IV. Fundamentals of Polymer Chemistry for the Controlled Release of Drug from Polymeric Membranes and Matrices .......... 190
- A. Physical States and Structure of Polymers .......... 191
- B. The Swelling of Cross-Linked Polymers .......... 192
- C. Diffusivity of Drugs in Polymers .......... 193

V. Hydrophobic Nondegradable Devices .......... 194
- A. Silicones .......... 194
- B. Substituted Polyethylene .......... 195

VI. Hydrogel Systems .......... 197
- A. The Polymers Used in Hydrogels .......... 197

VII. Biodegradable and Bioerodible Systems .......... 202

VIII. Routes of Administration and Targeting .......... 205

IX. Polymer-Tissue Interactions and Biologically Active Polymers .......... 207

References .......... 208

# I. INTRODUCTION

Polymers are intrinsic to all living sytems, and life as we know it could not exist without the presence of organized large molecules. They fill a number of roles of which three may be clearly identified as follows.

1. Contributing mechanical strength to skeletal structures exemplified by bones, sinews, and muscles in humans and to cell membranes in all life at the cellular level
2. Providing very specific chemical action, as for example, their function as enzymes, RNA, DNA, and certain hormones
3. Controlling the concentration of small molecules in intra- and extracellular fluids by acting as membranes with precise diffusion characteristics or as adsorption centers to maintain a low, precisely controlled level of dissolved low molecular weight species

The increasing basic knowledge of the chemistry and biochemistry of cellular and body functions combined with an expanding analysis of the engineering operation of body parts has led in the past decade to an explosion of research into three areas related to (1) prosthetic devices, (2) polymeric drugs, and (3) drug carriers and devices for the controlled release of drugs. This chapter will discuss the latter class (3) with some additional brief reference to (2) as in reality overlap occurs between these classifications.

The principles of controlled release of biologically active compounds are found in a multitude of applications as varied as preventing barnacles from adhering to ships' bottoms, promoting the growth of trees, keeping flies out of houses and fleas off pets, to the particular focus of this work, that of the prophylaxis and treatment of human diseases. The usual method for drug administration is by taking tablets at intervals during the day. This technique is far from ideal as it provides wide variations in plasma levels at different times as shown in Figure 1 for a patient taking four tablets per day. For every drug there is a so-called therapeutic index which measures the ratio of the plasma level of the drug causing toxic side effects in a defined proportion of subjects to the level which is therapeutically ineffective. Obviously, drugs with high therapeutic indexes are the most desirable and those with low values will be suspect because of the probability of plasma levels becoming either too high or too low with undesirable consequences in both cases. A controlled release dosage form should achieve the "desired level" providing more consistent protection without toxic effects. The removal of the necessity for a patient to comply to a prescribed dosage regimen is also a major advantage of such systems.

Certain drugs, such as the prostaglandins, are rapidly metabolized to inactive compounds and effectively eliminated from the body. In this way the organism protects itself against continuation of the often dramatic effects induced by minute quantities of such active materials. For their application in therapy, a continuous adminstration is often necessary and a simple controlled release device provides an attractive alternative to an i.v. infusion.

Yet another benefit of controlled release is in reducing the patient exposure to a massive excess of drug over that required at the desired site of action. Normal therapy involves a gross overkill technique in which the entire body is subjected to drug when only a small local concentration is required.

The benefits of controlled release devices can thus be summarized as follows:

1. To improve the control of systemic blood levels of drug
2. To localize drug action at a particular site
3. To eliminate patient compliance problems

Major research effort is currently being made to apply the principles of physical chemistry,

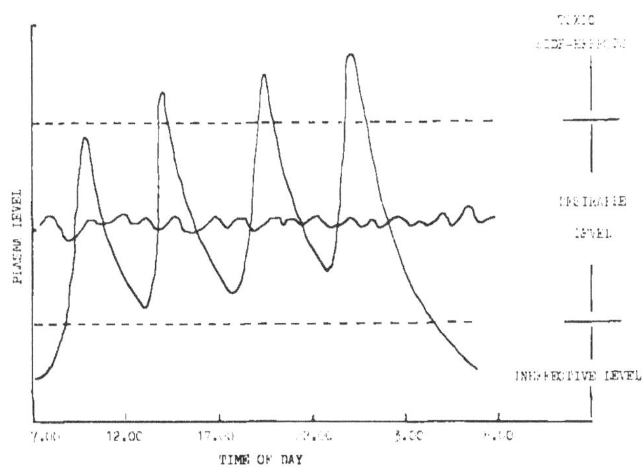


FIGURE 1. Hypothetical plasma level of drug with conventional multidosing schedule (e.g., four tablets daily) compared with controlled release device.

polymer science, and engineering to the design of controlled release devices for human and veterinary application. The pace of advance is rapid, and it is perhaps surprising the precision that has already been achieved in the release of drugs from quite complex polymeric devices and a high degree of precision promises to be obtained from even simpler polymer systems.

Conventional hydrophobic polymers, as rate controlling membranes or matrices, often will not provide a high enough release rate for short-term applications — particularly when water-soluble drugs are being used. They also have a significant disadvantage in that for surgical implantation (rather than insertion) in general two minor operations are required (i.e., for both implantation and removal of the device). Water-swollen matrices, commonly called hydrogels, will provide high permeation rates for water-soluble drugs while biodegradable polymer matrices offer a major potential improvement in the design of controlled release devices for implantation as they may decompose in vivo to nontoxic products which are readily eliminated from the body.

These sophisticated delivery devices can be conceived in many different shapes and sizes. Such devices could be utilized without surgery as inserts into accessible body cavities such as the conjunctival sac of the eye, the ear, mouth, GI tract, rectum, uterus, and vagina or merely placed in contact with the skin as in a transdermal therapeutic system.

An understanding of the physical and organic chemistry of the polymers used for these devices is most important if they are to be well-designed and this will be emphasized for the various types of controlled release devices. The principles involved are reasonably well-defined, but the precise predictions of the physical chemist can often be complicated or even confounded by the added complexity encountered when the device is placed in living tissue which may encapsulate, degrade, calcify, or swell it. An interdisciplinary approach is required for successful research in this field.

## II. PHYSICAL FORMS OF DRUG-POLYMER COMPOSITES AS INSERTS AND IMPLANTS

All physically controlled drug release devices can be classified into two distinct types as shown in Figure 2: (1) matrix devices in which the drug is dispersed throughout the polymer or can in some cases be completely soluble in the polymer and (2) membrane devices in which a core of drug is enveloped by a polymeric membrane. These two fundamental types form the basis for systems with a variety of geometrical configurations and topography.

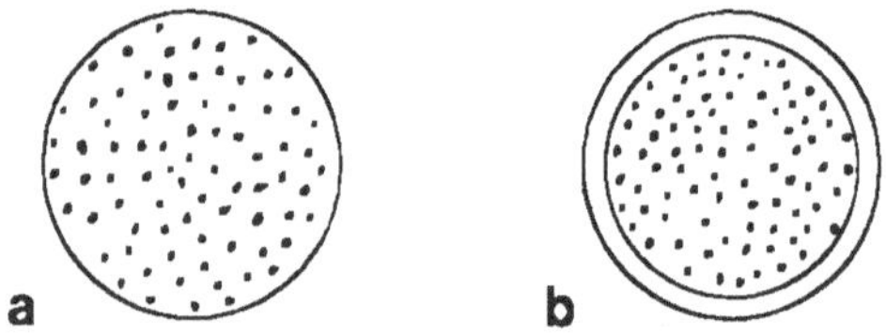


FIGURE 2. Schematic representation of controlled release devices. (a) Matrix device with dispersed drug; (b) membrane device with reservoir of drug.

Matrices are composed of solid polymer in which the drug is dissolved or dispersed. In most cases the loading of the drug exceeds its solubility in the polymer and the drug is present as a homogeneous dispersion of fine particles. These types are also referred to as monolithic devices and can be prepared by a variety of methods. With thermoplastic polymers, such as poly(lactic acid), a drug is dispersed in the melt which can be transfer molded as beads or extruded into rods.[1] Films can be prepared by casting from a common solvent system[1,2] and spray-drying from a solvent produces injectable powders.[3] When the polymer cannot be processed in this way, the polymerization reaction is often carried out in the presence of drug. Using this method, reactive liquids are polymerized by thermal curing[4] or irradiation[5] in molds to produce highly cross-linked matrices. In the case of hydrogels, it may be more convenient to prepare the polymeric device in a desired form prior to loading with a solution of the drug.

The membrane devices consist of a reservoir of drug, often dispersed in solvent, enveloped by a polymeric membrane which controls the rate of release of the drug to the surrounding medium. Microscopic devices take the form of microcapsules[6,7] or nanocapsules[8] while macroscopic devices can take a variety of shapes with the hollow cylinder being the most easily fabricated and most widely studied.[9,10] In general, a hollow tube of polymer is cut to the desired length and sealed at one end; the lumen is then filled with drug and the other end is sealed.

This chapter concentrates on the above types of formulation which involve the polymer as a "physical carrier". However, it should be noted that polymers are increasingly being used as "chemical carriers" in controlled release applications. In this case the drug molecules are linked to a polymeric backbone,[11,12] perhaps by spacer groups, or can be monomers or comonomers used to prepare polymeric molecules.[13,14] These polymers may be either soluble or insoluble in the body fluids at the time of administration. These approaches to drug therapy introduce the fields of polymeric drugs,[15,16] affinity labeled drugs,[17,18] and biologically active polymers.[19,20]

## III. RELEASE MECHANISMS AND PREDICTION OF RELEASE PROFILES

The release of drug from any of the different forms of drug-polymer composites, described above, must be predictable and often a constant release rate (zero order) is desired. For many years a considerable amount of mathematical analysis of the theoretical rates of diffusional[21,22] release from the various fixed geometrical configurations has been reported and correlated with experimental results. Three other basic rate determining mechanisms can control the release profile of drug, these being found in swelling, boundary-layer controlled, and erodible devices. The basic concepts involved in the treatment of each of these mechanisms are given below and are followed by a discussion of how these and polymer structure are related.

### A. Membrane Devices

Diffusion of a solute, such as a drug molecule, through a polymer occurs as random

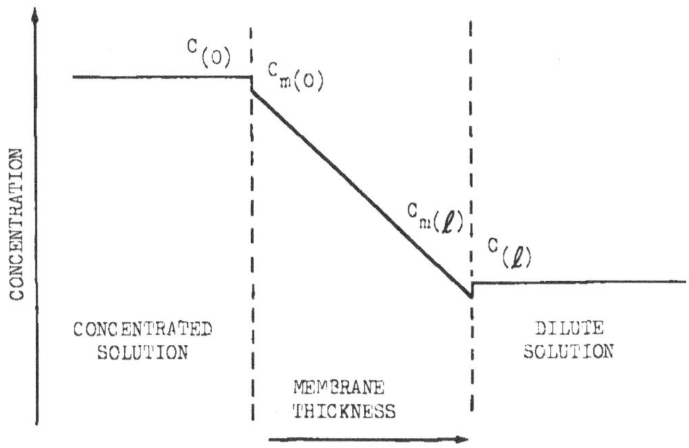


FIGURE 3. Diffusion of solute across a polymeric membrane showing concentration differences.

molecular movement and because a concentration gradient exists between the polymer phase and the external phase the drug tends to diffuse down the concentration gradient. It is generally assumed that the processes involved obey Fick's first law which for diffusion in a nonporous solid polymer membrane can be expressed as:

$$J = -D \frac{dC_m}{dx} \tag{1}$$

Where J is flux (g $cm^{-2}$ $sec^{-1}$)
$C_m$ is the concentration of the permeant in the membrane (g $cm^{-3}$)
$\frac{dC_m}{dx}$ is the concentration gradient, and
D is the diffusion coefficient ($cm^2$ $sec^{-1}$)

The negative sign reflects that the direction of flow is down the gradient.

Figure 3 shows the arrangement found for a membrane which separates two solutions of different concentration. It is assumed that the concentration of permeant in the surface layer is in equilibrium with either side. Therefore, if K is the partition (or distribution) coefficient of the solute between the two phases;

$$C_{m(0)} = K\,C_0 \text{ upstream } x = 0$$

$$C_{m(\ell)} = K\,C_\ell \text{ downstream } x = \ell$$

In the steady state

$$J = \frac{D\,(C_{m(0)} - C_{m(\ell)})}{\ell} \tag{2}$$

$$= \frac{D\,\Delta C_m}{\ell} \tag{3}$$

where $\ell$ is membrane thickness. Usually the concentration in the membrane is not known and $\Delta C$ is measured as the concentration difference between the two sides so that:

$$J = \frac{D\,K\,\Delta C}{\ell} \tag{4}$$

The terms D and K are not always easily measured and frequently values of permeability (P) are quoted where

$$P = D\,K \tag{5}$$

Therefore for a membrane device with constant activity, such as one with a saturated solution of the permeant over a large excess of insoluble material, the steady state release rate is expressed by:

$$\frac{dM_t}{dt} = \frac{A\,D\,K\,\Delta C}{\ell} \tag{6}$$

where $M_t$ is the mass released and A is the surface area of the device. When the concentration of drug in the surrounding body fluids is maintained at an extremely low level (i.e., sink conditions), the term $\Delta C$ in Equations (4) and (6) can be replaced by $C_s$, the saturated concentration of drug in the reservoir.

In membrane devices, particularly seen with steroid-silicone systems, "burst" and "lag" effects are found. Newly prepared devices require a period of time to establish a concentration gradient within the membrane and a "lag" effect is found if drug release from the device is assessed. When devices of this type have been stored for some time, the membrane becomes saturated with drug and a "burst" effect, observed as a higher release rate, is found until the concentration gradient associated with the steady state is established. Burst or lag effects are commonly encountered in most types of devices.[23] Inert[24] and biodegradable[25] matrices may often exhibit burst effects due to the presence of crystals of the drug at the surface[26] which readily dissolve when the device is placed in an aqueous environment.

**B. Release from Monoliths with Dissolved Drug**

The exact solutions for the desorption of solutes from a medium of given simple geometry (e.g., sphere, cylinder, and slab) are available in standard works.[27,28] However, these equations are clumsy for routine use and several approximate solutions can be used for early time and late time release (Table 1). An even simpler unifying equation for the simple geometric forms can be used if less rigourous treatment is acceptable; in this the fraction released at time, t, is given by:

$$\frac{M_t}{M_\infty} = 2\frac{S}{V}\left(\frac{Dt}{\pi}\right)^{1/2} \tag{7}$$

where $M_t$ is the amount of drug released after time, t, $M_\infty$ is the amount present initially, and S and V are the surface area and volume of the device. Practically "blocks", short cylinders, and disks are met more frequently and solutions for these are available in three dimensions where "end-effects" and "edge-effects" require to be taken into account.

**C. Release from Matrices with Dispersed Drug**

In many cases, the solubility of a particular drug in a given polymer is much lower than that required to provide an adequate amount of drug in a device of limited size. When this situation is encountered, the drug is assumed to be homogeneously dispersed as small particles throughout the polymer. The release kinetics of this type of system were derived by Higuchi,[29] using a model which assumes that the solid drug dissolves from the surface layer and that

**Table 1**
**APPROXIMATE SOLUTIONS FOR DIFFUSIONAL RELEASE OF SOLUTES FROM SLABS, CYLINDERS AND SPHERES (SEE TEXT FOR DEFINITIONS OF SYMBOLS)**

| Geometry | Early time | Late time |
|---|---|---|
| Slab<br>r = thickness | $\frac{M_t}{M_\infty} = 4\left(\frac{D\,t}{\pi\,r^2}\right)^{1/2}$ | $\frac{M_t}{M_\infty} = 1 - \frac{8}{\pi^2}\exp\left(\frac{-\pi^2\,D\,t}{r^2}\right)$ |
| Cylinder<br>r = radius | $\frac{M_t}{M_\infty} = 4\left(\frac{D\,t}{\pi\,r^2}\right)^{1/2} - \frac{D\,t}{r^2}$ | $\frac{M_t}{M_\infty} = 1 - \frac{4}{(2.405)^2}\exp\left(\frac{-(2.405)^2\,D\,t}{r^2}\right)$ |
| Sphere<br>r = radius | $\frac{M_t}{M_\infty} = 6\left(\frac{D\,t}{\pi\,r^2}\right)^{1/2} - 3\,\frac{D\,t}{r^2}$ | $\frac{M_t}{M_\infty} = 1 - \frac{6}{\pi^2}\exp\left(\frac{-\pi^2\,D\,t}{r^2}\right)$ |

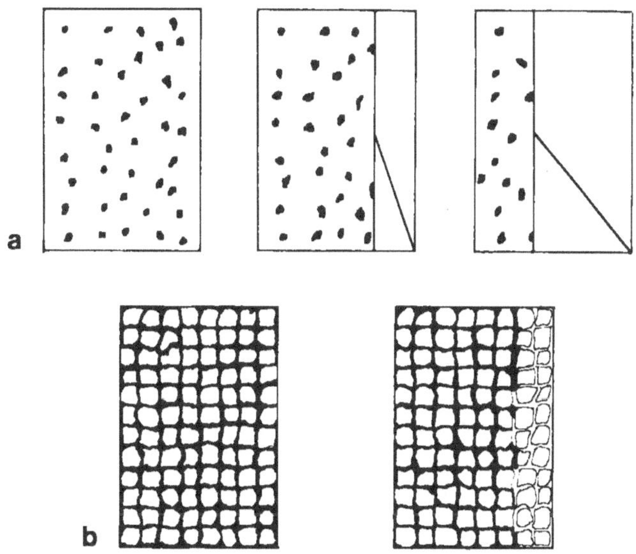


FIGURE 4. Models of drug release from polymeric devices. (a) Drug dispersed in continuous polymer; (b) drug dispersed in capillary channels of a porous polymer matrix.

this layer becomes exhausted of dispersed drug particles, as shown in Figure 4a. From the figure the dependence of concentration gradient with time is apparent. The validity of this matrix-controlled model has been demonstrated in several experimental studies.[30,31] In particular, the visual appearance of cylinders containing dispersed particles of steroid has shown a gradual depletion of the steroid from the surface layer which progressed towards the core with time.[32] The derivation of the Higuchi Equation (8) relies on Fick's first law and assumes:

1. A Pseudosteady state exists.
2. The drug particles are small compared to the average distance of diffusion in the device.
3. The diffusion coefficient is constant.
4. Perfect sink conditions exist in the external medium.

$$Q = (C_s\,D_s\,(2A - C_s)\,t)^{1/2} \tag{8}$$

and when

$$A \gg C_s \qquad Q = (2\, C_s\, D_s\, A\, t)^{1/2} \tag{9}$$

where Q is the amount of drug released at time, t, $C_s$ is the solubility of drug in the polymer, and A is the amount of drug present initially.

The most important point to note from these equations is that the amount of drug released is related to the square root of time ($Q$-$t^{1/2}$ relationship) essentially throughout the lifetime of the device. Whereas with monoliths with dissolved drug a $Q$-$t^{1/2}$ relationship is found initially and the second half of the release decays exponentially (see Table 1). Higuchi suggested that the Equation (8) would be valid only for systems in which A is greater than $C_s$ by a factor of three or four. More recently, the model has been developed and the newer equations can be used with more accuracy over a wider range of conditions.[33,34]

**D. Release from Devices with Capillaries and Pores**

All of the above models were based on simple diffusion of a permeant molecule in a homogeneous polymer phase. If, however, the polymer phase is discontinuous, such as in the situation where a considerable volume fraction of drug has been compressed into an infinite slab with a polymer powder, the situation is more complex. The small extent of diffusion through the polymer is neglected and it is assumed that diffusion occurs predominantly in the capillaries between polymer particles and through the pores left behind as drug particles dissolve and are replaced by the permeating liquid (Figure 4b). Account must be taken of the increase in the length of the diffusional path around the inert polymer particles and the effective cross-section of diffusion must be reduced to allow for the fact that the cross-section available for diffusion is that of the permeant-filled holes left behind by the drug particles after dissolution and not the total volume of the inert polymer-drug-permeant combination.

Higuchi[35] developed a similar equation to that above Equation (8) for release of drug from one face of an inert granular matrix containing a dispersion of drug. The increased path length in the capillaries using a tortuosity factor ($\tau \simeq 3$) and the cross section of pores left by the dissolved drug by a porosity factor ($\epsilon$) were taken into account. This provides an equation,

$$Q = \left(C_a\, D_a\, \frac{\epsilon}{\tau}\,(2A - \epsilon C_a)\, t\right)^{1/2} \tag{10}$$

where $C_a$ and $D_a$ are the solubility and diffusion coefficient of the drug in the permeating fluid. The terms Q, A, and t have similar definitions as in Equation 8 and again a $Q$-$t^{1/2}$ relationship is predicted. Higuchi deduced a related equation for the release from spheres of the same composite structure. More recently, the basic Higuchi models have been developed for more practical shapes — disks, biconvex tablets, and cylinders.[36-38]

Since it was first introduced, the Higuchi Equation (10) has frequently been applied to experimental data, particularly inert granular matrix tablets of the sustained-action type for oral administration. In some cases, the fundamental parameters have been studied in great detail. Values for $C_a$ and $D_a$ for a particular drug can be determined readily using conventional techniques. Desai et al.[39-42] studied the release of various drugs and model compounds from polyethylene and poly(vinyl chloride) matrices and described the components involved in the porosity term, $\epsilon$, and how to evaluate them. When the matrix tablets were subjected to vacuum treatment prior to study, tortuosity values obtained experimentally for poly(vinyl chloride) were in the range 1.5 to 4, which is in good agreement with theoretical values.[42] Similar studies with polyethylene matrices produced tortuosity values in the range 7 to 10 when a surfactant was added to the elution medium to wet the polymer.[41] Both types of polymer required some pretreatment before ideal release behavior was found and the differences in tortuosity values was attributed to the different particle size and density of the polymer powders. Apparent tortuosity values of the order of a thousand were found with polyethylene matrices when surfactant was not used to aid the removal of air from within the pores. In further extensions of these studies, the theoretical equations for the release of two noninteracting[43] and two interacting drugs[44] were presented and compared with experimental data.

Increasingly, other polymers (generally hydrophilic) are being included as additives to formulations in an attempt to modify the release rate from the major polymer component. In this way, the addition of gelatin or sodium alginate to silicone cylinders was found to increase the release rate of morphine sulfate (hydrophilic drug.)[45] When 20% alginate was added to the formulation, the devices swelled with water forming microscopic pores and channels and about 90% of the drug was released after 10 days compared to 9% from devices without alginate. The release of morphine sulfate was found to follow second-order kinetics and a loss of alginate was observed throughout most of the period of drug release. The release of macromolecules, such as proteins, from ethylene-vinyl acetate copolymers[46] may occur by a similar mechanism involving the uptake of water by osmotic pressure resulting in the formation of microchannels.

### E. Release from Hydrogels

With hydrogels a considerable proportion of the device is composed of water which may be several times the dry weight of the polymer. When the hydrogel is initially swollen and contains water-soluble drug, the release equations given previously (i.e., Table 1) can be applied. Davis[47] deduced the following empirical expression to calculate the apparent diffusion coefficient of any soluble drug in any hydrogel,

$$D_p = D_0 \exp -(0.05 + 10^{-6}M)P \quad (11)$$

where $D_p$ is the diffusion coefficient of the solute in the swollen polymer gel containing P% (by weight) of polymer; M is the molecular weight of the solute; and $D_0$ is the diffusion coefficient of the solute in water. The study involved both cross-linked poly(acrylamide) and poly(vinylpyrrolidone) hydrogels and solutes with a wide range of molecular weight (125,000 to 150,000) and included radio-labeled rabbit immunoglobulin, bovine serum albumin, insulin, a prostaglandin, and sodium iodide.

Various forms of the Higuchi Equations (8) and (10) have been used to describe the release of drugs from hydrogels,[48] and in a study by Chien and Lau,[49] the diffusion coefficient in the gel $D_m$ ($D_m = D_0\frac{\epsilon}{\tau}$) was related to the degree of cross-linking.

The release kinetics from initially dry hydrogels, as would be expected, are complicated by the added consideration of the diffusion of solvent into the polymer. Good[50] has derived equations where the diffusion coefficient is a time dependent variable and fitted experimental data based on the release of a water-soluble drug (tripelenamine·HCl) from a poly(2-hydroxyethyl methacrylate) hydrogel. This hydrogel had a low degree of swelling and the water uptake was approximately balanced by the loss of drug, and the dimensions of the device were assumed not to change during release. Recently, the much more complex situation of diffusion from a device which is simultaneously swelling,[51,52] and undergoing dimensional change, has been analyzed.

### F. Boundary-Layer Controlled Release

With drugs which are very poorly water soluble, the external medium in contact with the polymer phase may become saturated with drug and the release from the device is effectively stopped until drug has diffused out of the unstirred boundary layer. As an example of this phenomenon, also termed partition controlled release, Chien et al.[53] studied the release of estradiol diacetate from silicone devices and found a desirable zero-order (Q-t relationship) release when the steroid was poorly soluble in the eluting medium (water). However, when increasing proportions of poly(ethylene glycol) were added to increase the solubility of the steroid in the eluting medium, the release profile changed to a typical matrix-controlled type (Q-$t^{1/2}$ relationship).

The boundary-layer effect has considerable implications in vivo[54] where the composition and the movement of the eluting medium may vary from site to site. The effect should be considered when in vitro release kinetics are compared with in vivo release[49,55-57] and the in vitro methodology designed to favor reasonable correlation. If the release characteristics of a device change with the rate of stirring in an in vitro test, it is a strong indication of some measure of boundary control.

### G. Release from Erodible Devices

Hopfenberg[58] considered controlled release from erodible slabs, cylinders, and spheres. Where a single zero-order process controls erosion, the theoretical equations can be rearranged in the form

$$\frac{M_t}{M_\infty} = 1 - \left(1 - \frac{k_0\,t}{C_0\,a}\right)^n \qquad (12)$$

where $k_0$ is the single zero-order rate constant for the erosion process, $C_0$ is the uniform initial concentration of drug, $M_t$ is the amount of drug released at time, t, and $M_\infty$ is the total amount of drug present initially. For the infinite slab, n = 1 and a is the half-thickness; for the cylinder, n = 2 and a is the radius; for the sphere, n = 3 and a is the radius. From the analysis, it is evident that zero order is only obtained from the infinite slab and that delivery rates from cylinders and spheres should decrease with time. It is implicit in the overall concept that a boundary exists between unaffected polymer and the previously degraded material and that the rate determining step occurs at the boundary . Specifically, water sorption by the polymer and the diffusion of the drug out of the matrix have negligible effect on release rate. Cooney[59] discussed the effect of geometry on the dissolution of pharmaceutical tablets presenting the common shapes and suggesting that spheres and cylinders with internal bores cross and clover leaf-shapes may offer better release profiles.

Devices which undergo surface erosion have been reported,[60] however, at this time, most biodegradable systems have release profiles complicated by water sorption, degradation in the bulk, and diffusion of the drug from bulk polymer.

## IV. FUNDAMENTALS OF POLYMER CHEMISTRY FOR THE CONTROLLED RELEASE OF DRUG FROM POLYMERIC MEMBRANES AND MATRICES

In controlled release applications, the drug molecule is chosen for its biological action in a particular therapy and therefore is generally not considered to be a variable. Furthermore, in any proposed application the amount (dosage level) released over a chosen time scale from a particular dosage form are the desired objectives at the outset.

The physicochemical properties of the drug can have significant effect on its release from any polymeric system and molecular size, concentration, and solubility both in body fluids and the polymer are of extreme importance.

Perhaps the most important parameter in any controlled release system is the diffusion coefficient of the drug in the polymeric device. The rate of diffusion of one molecule through a medium depends mainly upon:

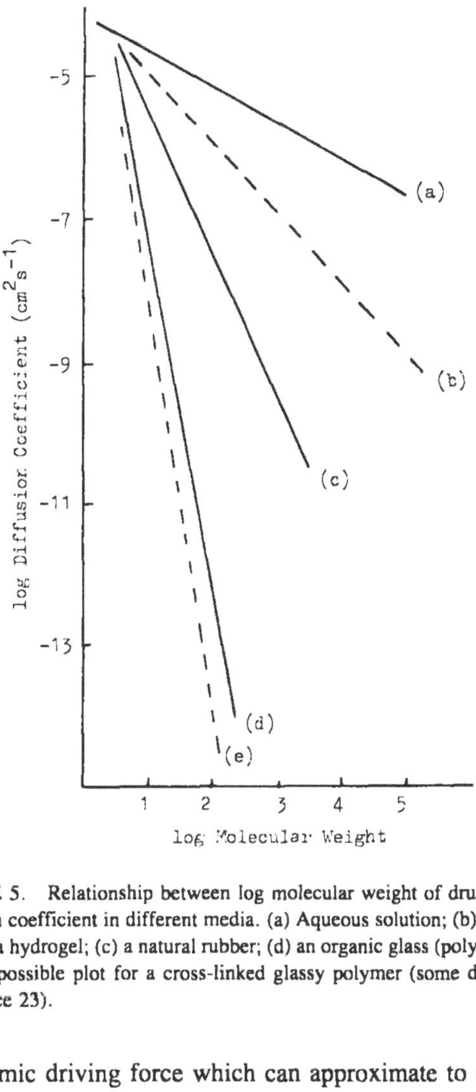


FIGURE 5. Relationship between log molecular weight of drug and its diffusion coefficient in different media. (a) Aqueous solution; (b) possible plot for a hydrogel; (c) a natural rubber; (d) an organic glass (polystyrene); and (e) possible plot for a cross-linked glassy polymer (some data from Reference 23).

1. The thermodynamic driving force which can approximate to a function of a concentration difference
2. The size of the diffusing molecule
3. The resistance to molecular movement presented by the medium

### A. Physical States and Structure of Polymers

The density of a polymer, indeed any particular material, will usually decrease in the order crystalline state, glassy state, rubbery state. This change in density is related to the molecular holes or free volume present in each state. This free volume increases from lowest in the crystalline state to highest in the rubbery state. As the free volume represents the holes into which diffusing molecules jump as they move down a concentration (or chemical potential) gradient, it is easy to postulate in general terms how diffusion coefficients increase as one goes from the crystalline to the glassy and then to the rubbery state. This is illustrated in Figure 5 in which the difference in both diffusion coefficient in different states can be seen and also how the different states show markedly different changes in their response to the diffusion of larger molecules. This plot demonstrates clearly why water-swollen hydrogels provide such appropriate systems for the short-term release of water-soluble drugs by providing the highest available diffusion coefficients amongst polymer matrices.

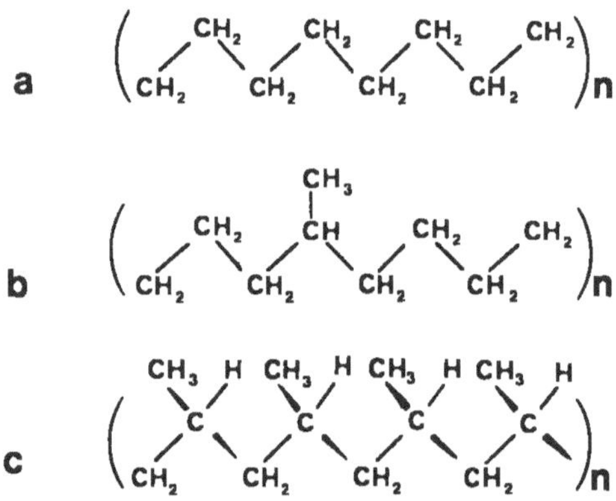

FIGURE 6. Polyethylene and ethylene/propylene structures producing rubber or crystalline polymers. (a) Crystalline; (b) rubbery; and (c) crystalline.

In the glassy state, polymers are hard, rigid, and often brittle.[61] There is a low level of molecular movement and the rate of diffusion of large molecules is very low. On heating at a defined rate, the glassy solid changes to a more flexible rubbery solid over a given small temperature range. Above this so-called glass transition temperature ($T_g$), the polymer is in the rubbery state and, if not constrained by some molecular interaction or bonding, will exhibit fluid flow. The fluid flow is prevented in many useful plastics by the presence of phase separation, crystallites, or cross-linking.

Crystallites can form in polymers when long sequences of the polymer chain have a stereoregular structure. The simplest example of this is polyethylene as shown in Figure 6. An approximate relationship $T_g = (0.5 \text{ to } 0.66)\ T_c$ is found to hold for many polymers (where $T_c$ in degrees Kelvin is the temperature at which the crystallites melt). The main point to note is that $T_c > T_g$ and as polymers are never 100% crystalline such polymers above their $T_g$ contain rubbery portions bound together with varying proportions of crystalline domains through which many chains pass.

Block or graft copolymers as represented in Figure 7 are comprised of long sequences of two or more different polymers. As high molecular weight chains rarely dissolve in each other such block or graft copolymers form a mixture of domains in which polymer separation has occurred and the units of each species have aggregated. If one type of these separate domains is glassy or crystalline while the other is rubbery, the effect is to prevent flow of the chains and hence bulk polymer.

The need for polymers with a range of precise requirements, as in controlled release applications, has led increasingly to the study and use of copolymers. Three broad groups of polymers used in controlled release can be divided into (1) inert hydrophobic, (2) inert hydrophilic, and (3) bioerodible. Examples of these three classes are given later. However, it should be noted that classification is far from simple and many of the devices which have been studied to date have been relatively complex. Classification of groups of polymers should be treated with caution because, for example, the "acrylates" can be

1. Inert hydrophobic, e.g., poly(methyl methacrylate)
2. Inert hydrophilic, e.g., poly(hydroxyethyl methacrylate)
3. Bioerodible, e.g., poly(methyl 2-cyanoacrylate).

### B. The Swelling of Cross-Linked Polymers

The ability of a polymer to swell with a given solvent is governed by the free energy of mixing of the solvent with the polymer and by the density of the cross-linking.[62,63] The

```
a)     AAAAAAAAAAAAAAAAAAAAAAABBBBBBBBBAAAAAAAAAAAAAAAAABBBBBBBBBAAAAAAAAAA



b)     AAAAAAAAAAAAAAAAAAAAAAAAAAAAAAAAAAAAAAAAAAAAAAAAAAAAAAAAAAAAAAAAAAAAAAAA
                           B            B                       B
                           B            B                       B
                           B            B                       B
                           B            B                       B
                           B            B                       B
                           B            B                       B
                           B            B                       B
                           B                                    B
                                                                B
```

FIGURE 7. Simple representation of (a) block and (b) graft copolymers, where A and B represent units derived from different monomers.

**Table 2**
**DIFFUSION OF $^3$H-LABELED PROGESTERONE THROUGH VARIOUS POLYMER MEMBRANES**

| Polymer | Progesterone diffused (%) |
|---|---|
| Polydimethylsiloxane (Silastic®) | 100 |
| Polyamide (nylon) | 1.0 |
| Acetylcellulose (cellophane) | 0.1 |
| Fluorethylene | 0.1 |
| Polyester (Mylar®) | 0.1 |
| Polycarbonate (Lexan®) | 0.1 |
| Polyethylene | 0.1 |
| Polystyrene copolymer (Cr 37) | 0.1 |

From Kincl, F. A., Benagiano, G., and Angee, I., *Steroids*, 11, 673, 1968. With permission.

theoretical analysis of the swelling is complex and will not be discussed here. The combination of a polymer with a swelling solvent which might be a plasticizer in the case of a commonly used membrane such as ethylene/vinyl acetate copolymers or water in the case of hydrogels is commonly used to moderate and control the diffusion of drugs through polymers.

The degree of swelling is increased by reducing the cross-linking density. As diffusion coefficients also increase with increasing swelling this can be used as a means of controlling the rate of release of incorporated soluble drugs.

### C. Diffusivity of Drugs in Polymers

Table 2 shows the relative permeabilities of $^3$H-labeled progesterone in a range of different polymeric membranes[64] and the difference between silicone rubber, with its very flexible backbone, and the other common polymers is readily apparent. The effect of soaking membranes in plasma prior to diffusion experiments with steroid was examined in another study.[65]

In addition to molecular size of the diffusing molecule other physicochemical properties are extremely important. The permeation of a molecule through a membrane is not only dependent upon its diffusion coefficient, but also its solubility in the polymer and its dis-

**Table 3**
**AVERAGE DIFFUSION RATE OF VARIOUS STEROIDS ACROSS SILICONE MEMBRANES**

| Steroid | Diffusion rate[a] |
|---|---|
| 19-Norprogesterone | 1353 |
| Progesterone | 469 |
| Testosterone | 317 |
| Megestrol acetate | 236 |
| Norethisterone | 73 |
| Estradiol | 61 |
| Mestranol | 43 |
| Corticosterone | 21 |
| Cortisol | 6 |

[a] $\mu g/(100\ mm^2)$ (0.1 mm) (24 hr).

From Kincl, F. A., Benagiano, G., and Angee, I., *Steroids*, 11, 673, 1968. With permission.

tribution coefficient between the polymer and the external medium. The effect of small changes in the structure of steroid molecules on their release from silicone[64] is shown in Table 3. Detailed studies[65-67] have since been carried out to relate the effect of these slight changes which have negligible effect on molecular size, but considerable effect on solubility. Similar studies have been reported for the release of steroid from hydrogels.[68]

Other additives apart from the drug may produce significant effects and can be used to modify the release profile. The effects of fillers[67,69] and plasticizers have been extensively studied while copermeant enhancement[70] and the addition of hydrophilic polymers[45] to produce a heterogeneous polymer matrix with microchannels appear to add further design variables. In the latter approach, as discussed previously, high molecular weight water-soluble materials (e.g., proteins) can be released from a hydrophobic polymer by the formation of microchannels.

## V. HYDROPHOBIC NONDEGRADABLE DEVICES

The release of biologically active compounds from "inert" synthetic polymers has been the subject of a considerable amount of research for many years and probably most classes of polymer have been screened in some controlled release application or another. Table 2 showed some which were studied as potential membranes for the release of steroids. Inert synthetic polymers have also been investigated as the wall material of microcapsules and many processes and applications have been reported. It is clear that in many cases the release rate of drug from simple macroscopic devices would be inadequate for most applications. However, two groups of polymers which showed some potential, the silicones and the substituted polyethylenes discussed in some detail below, have been the subject of concentrated study and their advantages and limitations have become apparent.

### A. Silicones

Silicone rubber[71] has been used in a variety of biomedical applications due to its inertness and good biocompatibility. In 1964, Folkman and Long[72] reported the use of this polymer in sustained release formulations.[73]

The backbone of the silicone polymer is made up of alternating silicon and oxygen atoms and the most widely used member of the family is poly(dimethylsiloxane). This material can be prepared as a fluid elastomer or a resin and is available commercially as Silastic®.

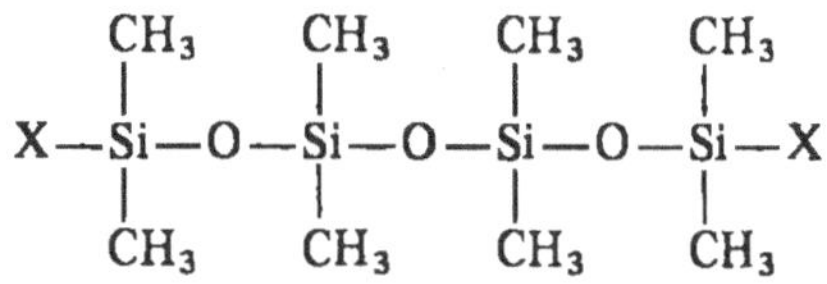

Silicone polymers end-blocked (group denoted X above) with $CH_3$, are unreactive and generally used as fluids. Alternatively, chains can have reactive hydroxyl end groups which may be cross-linked by curing at room temperature and these are used to prepare molded or cast devices. This type of polymer has been widely studied in implants and inserts for the controlled release of steroids,[66,67,74] chloroquine,[75] pyrimethamine, indomethacin,[76] atropine, and histamine.[77] The devices are generally macroscopic implants of both the membrane and the matrix type. Chien and Lambert[78] developed a system with microsealed compartments (10 to 200 μm) containing drug in a hydrophilic solvent enclosed within a matrix of silicone polymer. By careful choice of the composition of the hydrophilic solvent [e.g., aqueous solutions of poly(ethylene glycol)], the release rates of various drugs could be controlled over a wide range. A microsealed device of this type which delivered deoxycorticosterone acetate, was used to study the production of metacorticoid hypertension in rats.[79]

While the flexibility of the backbone of the silicone polymer in combination with the saturated solubility level of hydrophobic drugs in the polymer allows adequate release rates to be obtained, the release of hydrophilic drugs will be considerably slower due to their lower solubility. Attempts to increase the permeability of the silicone polymer to hydrophilic drugs employed the addition of a considerable proportion of hydrophilic polymer (e.g., sodium alginate) to produce microchannels, as described above. Devices modified in this way have been used to study morphine induced tolerance and physical dependence in animals.[80]

### B. Substituted Polyethylene

High density polyethylene is used in a variety of biomedical applications and the low density material has been used in the fabrication of a contraceptive intrauterine device which also releases progesterone.[81] The local release of a steroid, such as progesterone, has a synergistic effect on the contraceptive action of the device which is provided by its physical presence. This local direct drug action in the uterus reduces systemic side effects to undetectable levels and normal menstrual cycles occur. For controlled release devices, it is usual to add other monomers to provide a range of copolymers with the desired characteristics (see Figure 8). Thus, the Progestasert® device (Alza Corportion) uses an ethylene-vinyl acetate copolymer (9% vinyl acetate) as a rate controlling membrane. In this way a constant release of 65 μg/day of progesterone is obtained for a period of 1 year.[82,83] Another device developed by the Alza Corporation, and approved by the FDA in the U.S. is the Ocusert Pilo. The device, composed of two membranes of ethylene-vinyl acetate copolymer, is inserted into the cul-de-sac of the eye and provides a constant release of pilocaprine for the reduction of intraocular pressure in glaucoma. The two devices available, Pilo 20 and Pilo 40, release at a rate of 20 or 40 μg/hr for 1 week and offer much better control than eye drops instilled several times daily and remove problems with patient noncompliance.

The effect of the comonomer ratio of ethylene and vinyl acetate on the in vitro release of hydrocortisone from molded devices was studied by Fu et al.[84] It was found that increasing the proportion of vinyl acetate, thereby changing crystallinity, increased the release rate. In another study, an ethylene-vinyl acetate copolymer (40% vinyl acetate) was used to prepare

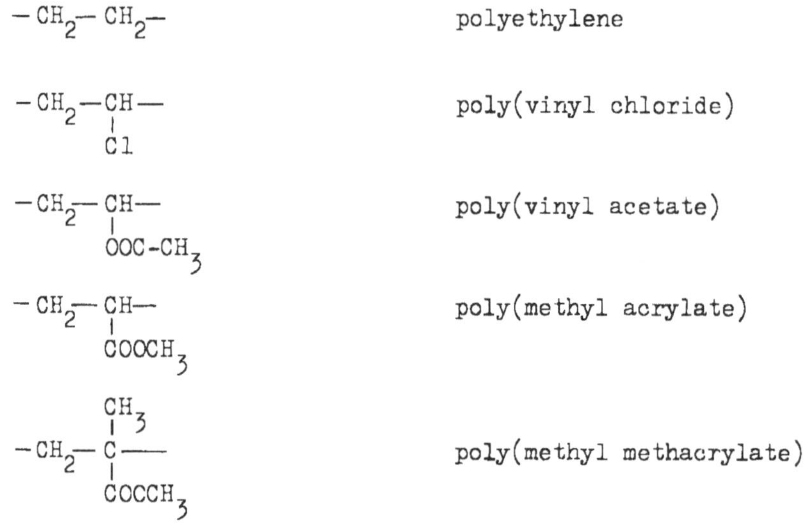


FIGURE 8. Some examples of the repeating units of substituted polyethylene polymers which have been used in controlled release devices.

pellets which released biologically active macromolecules (e.g., soy bean trypsin inhibitor, lysozyme, alkaline phosphatase, catalase, insulin, heparin, and DNA) for times in excess of 100 days at relatively constant rates depending on the particular preparation.[46] It was suggested that copolymers of this type could be used to prepare carriers for a wide range of enzymes and other biologically active macromolecules which would release at desired rates.[85] While it seems likely (see before) that the hydrophilic macromolecules would draw water into the matrix and swell it with the formation of microchannels, it was noted in that study that once the macromolecule had been released the membrane was not permeable to molecules of similar size.

Both polyethylene and poly(vinyl chloride) have been used to prepare matrix tablets by compression of the powdered polymer with hydrophilic drug and under certain conditions were found to obey the Higuchi Equation (10). Similar studies have been carried out with copolymers of methyl acrylate and methyl methacrylate.[86,87] Recently, methyl methacrylate membranes have been applied to a core of poly(hydroxyethyl methacrylate) and the resulting trilaminate device has provided zero order release of tetracycline for periods in excess of 100 days.[88] The trilaminate devices, when implanted intraperitoneally in rats, were used in a pharmacokinetic study of tetracycline.[89]

Surgeons using methyl methacrylate as a bone cement in total replacement of joints, especially hip joints, have found that the addition of antibiotics to the cement prior to use significantly reduced the incidence of infection. Several studies have been carried out[90,91] and the effect of several antibiotics on the mechanical properties has been reported.[92,93] The concept led to the evaluation of implantable beads of poly(methyl methacrylate) containing gentamycin for the treatment of chronic bone infections.[94] Such beads have been found in vitro and in vivo to provide adequate release locally for several months while serum concentrations were low thereby minimizing the risk of problems with side effects.

Devices composed of cross-linked diethylene glycol dimethacrylate have been prepared by $\gamma$-irradiation of the frozen monomer at $-78°C$.[95,96] Mixtures of the monomer with a variety of anticancer drugs were molded, frozen, and irradiated.[5] The effect of the incorporation of other polymers (e.g., poly(ethylene glycol), poly(methyl methacrylate), and poly(vinyl acetate) at 10% of the total weight) on the release profiles of the drugs were briefly discussed. It was found that the rate of release could be greatly enhanced by some of these incompatible additives which formed microchannels in the glassy matrix.

It is readily evident that a very diverse range of devices can be prepared from simple polymers, such as the two groups discussed above. Various constraints are placed by the solubility of the drug and the desired duration of action and in general each case has to be studied in some detail.

## VI. HYDROGEL SYSTEMS

Hydrogels comprise a large group of polymers which swell to a considerable degree with water. They have attracted considerable interest in prosthetic applications such as contact lenses or promising nonthrombogenic surfaces and in the context of this chapter as matrices for the controlled release of drugs. Their most interesting attribute is their high permeability to water-soluble drugs which is in contrast to the low permeability of these materials often obtained with hydrophobic polymers such as silicones. Their physical strength and durability in the dry and swollen states are important factors in their practical application and some indications of how these can be optimized are discussed later. They obey the release equations described earlier and can be obtained in monolithic slab, cylinder, or spherical configurations and also in envelope or matrix forms, all with the kinetics predicted by the well-developed mathematical models. Powders of hydrogels can provide injectable formulations which may be further developed.[97] The kinetics of release from devices which are changing their dimensions during their effective lifetime have recently been modeled, while other studies on systems which both lose crystallinity and swell simultaneously have been shown to provide constant rates of drug release from a monolithic device — something which is very useful, but not to date predicted by theory. Polymers can be designed which are hydrophobic at one pH and change to hydrophilic hydrogels at another. These can be used to design matrices which only release their contained drug at a particular pH. These points of design will be discussed in more detail below.

### A. The Polymers Used in Hydrogels

Any water-soluble polymer which can be rendered insoluble by cross-linking or incorporation into a block or graft copolymer (Figure 7) in which the other component is a hydrophobic unit can in principle be used as a hydrogel for the delayed or controlled diffusion of molecules which may or may not be drugs. Probably the best-known and most widely used hydrogel for such purposes is cross-linked collagen which is used as the basis of photographic films as well as in the pharmaceutical industry. While cross-linked collagen has been used for microencapsulation, it is not a very suitable polymer for preparing monolithic hydrogel devices with consistent performance properties especially if low degrees of swelling are desired. For this consistency with ease of preparation, research has favored the synthetic polymers of hydroxyethylmethacrylate,[50,98] acrylamide[47] and its *N*-sugar substituted derivatives,[99] *N*-vinylpyrrolidone[47,48] and poly(ethylene oxides),[100-104] as shown in Figure 9. Poly(glutamic acid)[102] and cross-linked dextrans and starches[97] have also been used. The most desirable hydrogels should be very strong and tough, but not brittle in the dry state, swell to a reproducible degree in water, buffer, or plasma, and be strong in the swollen state. Practically, the drug/polymer composite should be readily obtained in the desired geometry by a process capable of scaling up to production. It should not contain any toxic residues of monomer, initiators, stabilizer, or drug modified by the polymerization process. Although polyacrylamide has been the focus of some academic study, the risk of residual carcinogenic monomer might well rule it out in practice. The largest amount of work has been done on hydroxyethylmethacrylate which is unfortunately brittle as a homopolymer and has a relatively low degree of swelling (42%). In many studies it is polymerized with a cross-linking agent in aqueous solution containing drug and the residual initiation fragments and monomer are not removed.

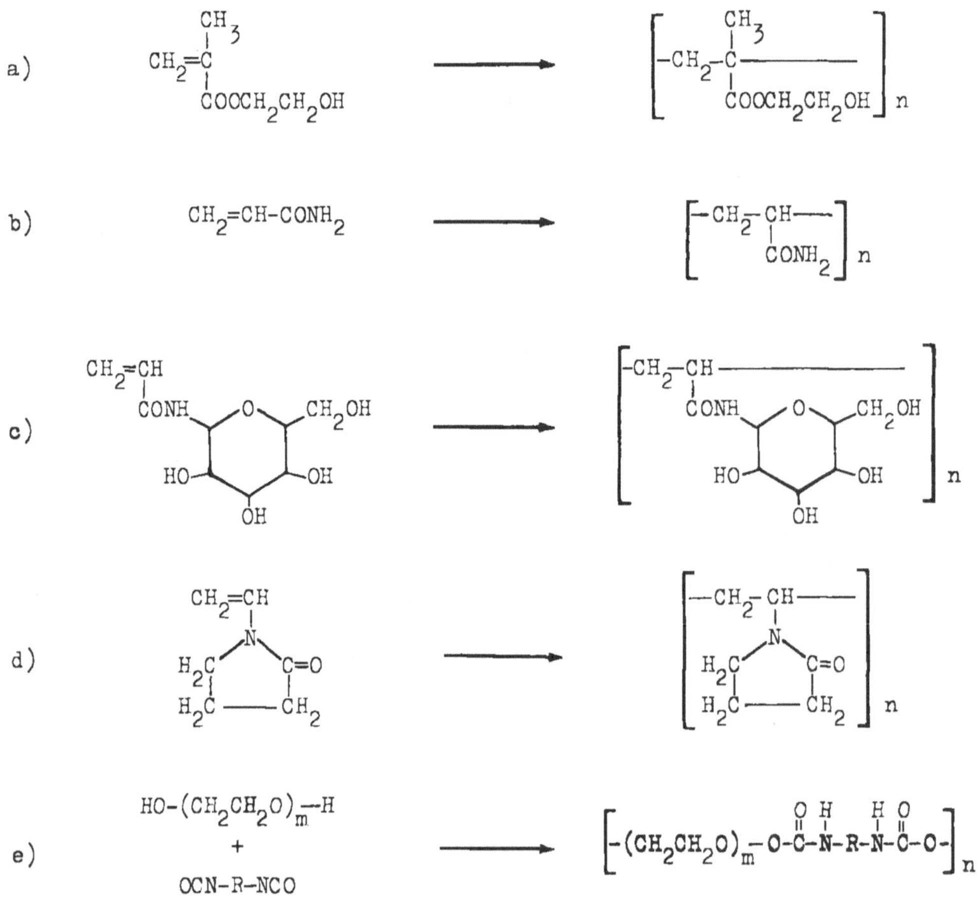

FIGURE 9. Structures of some commonly used hydrogels. (a) Poly(2-hydroxyethyl methacrylate) (HEMA); (b) polyacrylamide; (c) *N*-substituted derivative of polyacrylamide; (d) poly(vinyl-*N*-pyrrolid-2-one); (e) polyurethane prepared from poly(ethylene glycol) and a diisocyanate.

There is little difference in the release characteristics normally attributable to the chemical nature of the hydrogel for basically amorphous polymers.[47] The big differences arise from control of physical parameters such as the water content, degree of cross-linking, crystallinity, and morphology, i.e., if two or more phases are present. Thus, much recent research has been concerned with obtaining stronger hydrogels from what are essentially block or graft copolymers of either a thermoplastic or thermoset nature. Ciba-Geigy[101] has obtained cross-linked hydrogels of poly(ethylene oxide) capped with double bonds and subsequently co-polymerized with hydrophobic monomers to obtain polyphase systems which are claimed to have improved physical strength. ICI[104] achieved a similar multiphase system which also has the distinct advantage of retained polymer solubility allowing washing and purification of the polymer combined with the thermo-forming properties of a linear block copolymer. This allowed thin films to be obtained by the application of pressure and heat.

Another promising development along these lines by Graham et al.[103] involves cross-linked polymers of poly(ethylene oxide) using diisocyanates and polyols to provide cross-linking. If poly(ethylene glycols) of molecular weights above 2000 were used dry gels containing up to 50% crystalinity could be obtained. The crystallites act to reinforce an already quite perfect polymer network to provide a tough polymer akin in its physical properties to low density polyethylene. By control of the degree of cross-linking, hydrogels

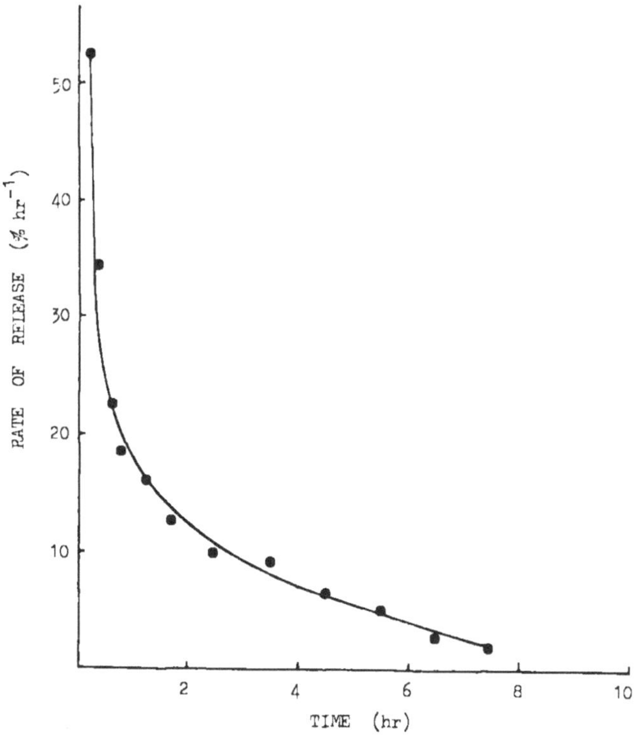


FIGURE 10. Release of prostaglandin $E_2$ from an initially fully swollen hydrogel based on poly(ethylene glycol) at 37°C and pH 7.4 (rate $\propto$ $t^{-1/2}$ relationship).

capable of swelling with water to typically five times their dry weight could be made. In the fully swollen state these polymers behaved like any other hydrogel, releasing their contained drug at a rate proportional to $(\text{time})^{-0.5}$ (Figure 10), but in the dry state slices of the polymer released a considerable proportion of the contained drug at a constant rate (Figure 11). This is because of a balance struck between decreasing drug release from an unswollen crystalline core and a physically coupled increasing permeability of the swelling outer layer. This constant release from a monolithic hydrogel is of obvious clinical importance and its first use has been in aiding the performance in delivery of pregnant women suffering from a so-called unripened cervix at full term.[105]

The above polymer in the form of a small vaginal pessary containing between 5 and 15 mg of prostaglandin $E_2$, depending on the assessed requirements of the individual patient, was inserted into the vaginal vault on the eve of the predicted delivery date. The uniform release of the prostaglandin $E_2$ over 12 to 24 hr induced cervical ripening and significantly reduced the need for a Caesarian operation as well as inducing normal and shortened labor with reduced pain in many patients. A major benefit of this device is that the prostaglandin $E_2$ which is normally quite unstable on storage was shown to be largely unchanged after 14 months storage at 4°C. This points to these drug-in-dry-hydrogel devices having enhanced practical application because of this ability to protect the contained drug from degradation or isomerization. This ability of hydrogels to stabilize prostaglandin $E_2$ has been confirmed in another briefly reported study using dry cross-linked starch gels.[106]

Other attempts to obtain constant rates of release from either fully swollen[88] or initially dried down[50] hydrogels have been reported in various studies. Lee et al.,[107] for example, have taken a drug containing monolithic cylinder prepared from derivatives of poly(methacrylate) and adsorbed and polymerized additional cross-linking monomers into

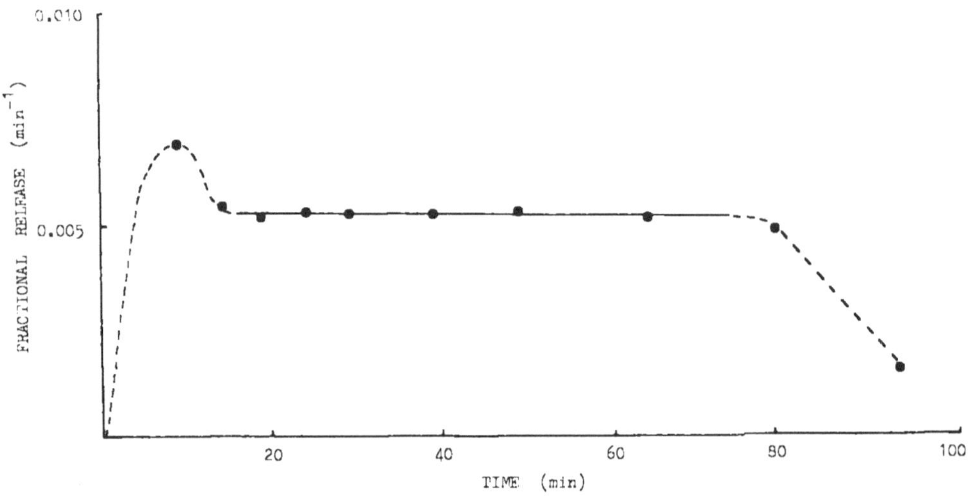


FIGURE 11. Release of promethazine hydrochloride from an initially dry, crystalline hydrogel based on poly(ethylene glycol) (rate ∝ t relationship for 37% of total release).

the external surface. This produces a skin around the device which has a lower degree of swelling than the core and as a result has a lower permeability to the contained drug. This skin becomes the rate controlling step of drug diffusion out of the device which is effectively converted into an "envelope" configuration for which the rate is predicted and found to be constant with time. Other related devices for the release of fluoride to teeth have also been reported.[108]

Cores of drug coated with rate controlling membranes have been used in even more sophisticated ways. Thus, by the incorporation of covalently bound carboxyl groups it is possible to obtain hydrogels which swell under basic conditions,[98] while swelling under acidic conditions can be obtained by the covalent incorporation of basic groups such as amine. Such groups have been incorporated into both vinyl and urethane polymers. An elegant example of such a use has been reported by Fildes of ICI,[109] who encapsulated a core tablet of quinoxaline di-*N*-oxide with a nylon copolymer containing basic groups in its backbone made by interfacial or condensation polymerization (Figure 12). These coatings were essentially unswollen at a pH of 7, but became highly swollen hydrogels at pH 4 and thus, having a high permeability to water-soluble drugs at this latter pH, released the contained drug in 4 hr. This arrangement was made for the purpose of protecting the drug in the rumen of cattle where the pH is around 7 and allowing release of the drug in the abomasum where the pH is around 4.

An example of systems of low swelling under neutral conditions, but releasing at high pH, is provided by the urethane polymers containing glucuronolactone.[110] The lactone group is not involved in the formation of a network by reaction of the hydrogens in this molecule with isocyanate. In high pH conditions the lactone ring opens and allows the polymer to dissolve if it is not cross-linked or to swell if it is cross-linked. These fundamental ideas were well-described in an early patent to the Czech Academy of Sciences[98] on the controlled release from vinyl hydrogels.

An even greater degree of sophistication has been developed by the Alza Corporation in their Oros® devices. In these, a solid tablet is coated with a hydrogel into which coating a very fine hole is introduced by a laser beam. In its simplest form, an elementary osmotic pump,[111] such as the Oros® device, consists of a membrane with an orifice in it (Figure 13a). In this type of device the membrane allows the passage of water, following an osmotic gradient, into the core which contains the drug which acts as an osmotic driving force for

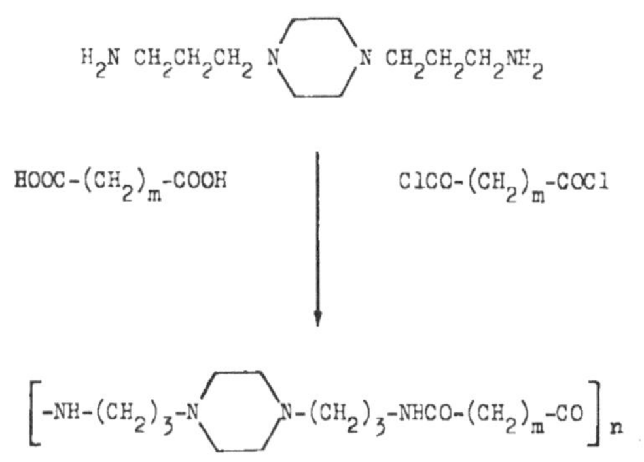

FIGURE 12. Structure of a polyamide polymer which responds to the pH of its environment (i.e., swells at about pH 4 while remaining unswollen at pH 7).

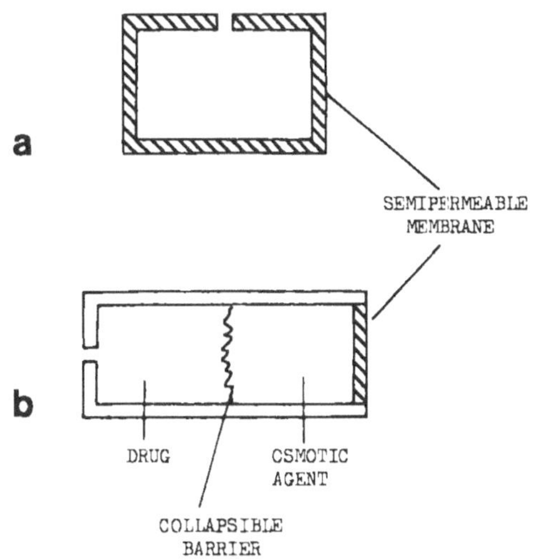


FIGURE 13. Schematic representation of osmotic pumps. (a) Elementary type; (b) pump with collapsible membrane separating drug compartment from osmotic agent.

water transport. The ingress of water is balanced by a saturated solution of drug being pumped out through the orifice. More sophisticated variations are available[112,113] (e.g., Figure 13b) in which the drug compartment is separated, by a movable barrier, from another compartment which contains an osmotic agent.

In principle there is no reason why any of the above devices cannot be utilized for use as inserts or implants in addition to their use orally. There has been a considerable amount of work done on devices for the controlled administration of drugs which come under a heading of bioengineering and are beyond the scope of this chapter. One should be mentioned, however, in the context of hydrogels as it is a commercial and very useful research tool for the study of the pharmacology and pharmacokinetics of drugs.[114] This is the Alzet® Osmotic Minipump. The implantable Alzet® device provides a typical delivery rate of 1 $\mu\ell$/hr for 1 week and is small enough for implantation in animals as tiny as mice. Before implantation the pump is primed by injecting the desired drug solution into the orifice. Several water-soluble drugs have been investigated using these devices, and poorly water-soluble steroids have been delivered using polyethylene glycol as solvent.[115]

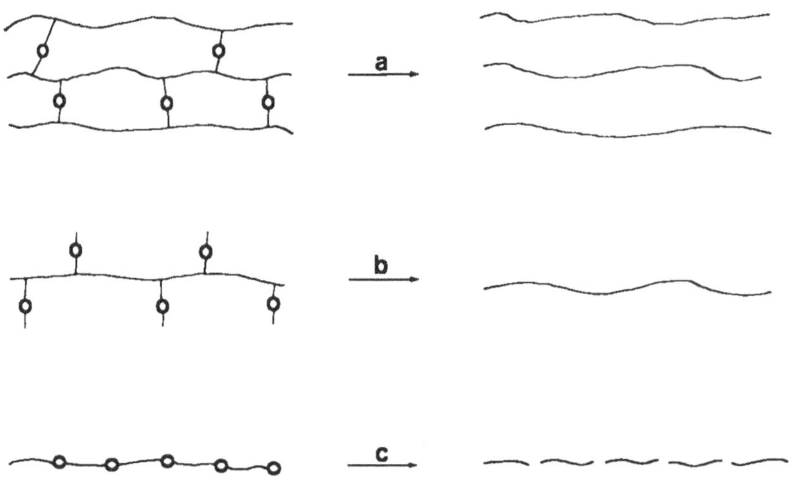


FIGURE 14. Schematic representation of three basic types of bioerodible polymers where -O- denotes a hydrolyzable linkage (see text for description).

## VII. BIODEGRADABLE AND BIOERODIBLE SYSTEMS

The use of biodegradable polymers in drug delivery systems[116] allows the implantation of the device without the need for its removal after the drug has been released. The terms biodegradable, bioadsorbable, and bioerodible are frequently used interchangeably. The term bioabsorbable has long been associated with surgical sutures which gradually dissolve in the body. Bioerodible polymers may erode by slow dissolution as a result of side-group hydrolysis or ionization while the polymer backbone remains undegraded — frequently the erosion occurs as a surface phenomenon. Biodegradable is a more general term which is applied to many different materials other than polymers[117] including drugs,[118] surfactants, insecticides, and synthetic compounds in general.[119,120]

Three basic classes of polymer are generally recognized in biodegradable systems (Figure 14):

1. Water-soluble polymers which are insolubilized by hydrolytically unstable cross-links
2. Linear polymers which are initially water insoluble which become solubilized by hydrolysis, ionization, or protonation of pendant groups, but which do not undergo backbone cleavage
3. Polymers which are water soluble and degrade to small soluble products by backbone cleavage

In general, biodegradation takes place by a hydrolysis reaction when the device is placed in the aqueous environment of the body. In some cases enzymes have been found to accelerate the degradation reaction considerably.[121,122] Perhaps the most important consideration with biodegradable implants is whether the degradation products are nontoxic and readily eliminated from the body. Standards have been proposed to allow a comparison between the cytotoxicity of novel biodegradable polymers and established ones.[123] Polymers which undergo erosion of the type (2) above would probably not be suitable for implantation because a high molecular weight polymer is not readily excreted from the body. However, this can be put to advantage by using surface eroding polymers as inserts, as opposed to implants, the degradation products of which would be less liable to be absorbed due to their high molecular weight. Heller and Trescony[124] have used partial esters of maleic anhydride copolymers to prepare devices which release drug by dissolution of the polymer. The rate

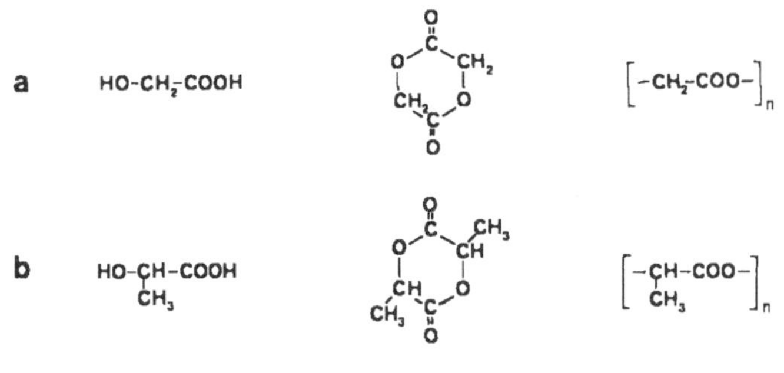

FIGURE 15. Structures of (a) glycolic acid (b) lactic acid, their cyclic dimers (glycolide and lactide), and homopolymers.

of surface erosion is controlled by the particular ester group and the pH of the environment. In a development of this system, the polymeric devices were coated with a hydrogel containing urease. The presence of urea resulted in the production of ammonium ions which accelerated the erosion, and the drug release, by a reversible mechanism. The Alza Corporation has patented an intrauterine device which bioerodes, and thereby releases drug, in the uterus.[125] A particular advantage of this type of device is that its shape can be designed for maximum retention in the uterus.

Polymers which undergo surface erosion are, almost by definition, hydrophobic while containing readily hydrolyzable groups. Workers at Alza have developed several novel polymers of this type which also undergo backbone cleavage to low molecular weight molecules which would therefore be of potential application in biodegradable implants.[126-128] The polymers have been given the general name Chronomer® and include hydrophobic poly(orthoesters) and poly(orthocarbonates). Heller,[129] one of the inventors, reviewed bioerodible drug delivery systems from a general point of view and included these as a possible group of polymers which could be used as implants. Choi,[130] another of the inventors, has described the development of these polymers for use in biodegradable systems. The Alza bioerodible polymers have been studied for controlled release of narcotic antagonists for the treatment of narcotic addiction[131] and of steroidal contraceptives for fertility control.[132]

Studies with polymers which undergo backbone cleavage have been restricted mainly to copolymers of lactic and glycolic acid[133,134] (Figure 15). Biodegradable drug delivery systems based on these polymers have shown the potential of this form of administration and have been reviewed by Wise et al.[135] The four main areas of application have used steroidal contraceptives,[25,136,137] narcotic antagonists,[1,138-140] antimalarials,[3,141,142] and anticancer drugs.[143,144] The forms in which these drug-polymer composites have been studied in vivo and in vitro include implantable cylinders,[136] spheres,[141] and films[139] and injectable microcapsules[137] and powdered formulations.[139] The polymers are generally prepared from the cyclic dimers of lactic and glycolic acid (i.e., lactide and glycolide) (Figure 15). Lactic acid has an asymmetric carbon atom, and therefore exists as two optical isomers. The different isomers and the racemic mixture produce polymers with significantly different properties. Poly(L-lactic acid), having steroregular sequences in the chain, has about 37% crystallinity while poly(DL-lactic acid) is totally amorphous.

Water uptake by a series of copolymers of L-lactic and glycolic acid has been studied recently by Gilding and Reed.[145] Poly(L-lactic acid) is the most hydrophobic and crystalline of the series and has the lowest equilibrium water content. With increasing the glycolic acid content, the degree of crystallinity and hydrophobicity decrease resulting in an increase in the equilibrium water content. A maximum of about 30% water uptake is found with the 30:70 copolymer (lactide to glycolide), thereafter with the onset of crystallinity the water content decreases.

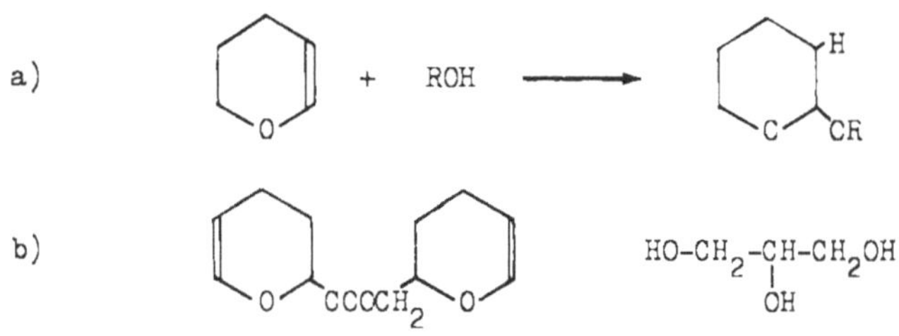

FIGURE 16. Biodegradable polymers based on dihydropyrans. (a) Reaction of dihydropyran with an alcohol; (b) typical monomers used to prepare cross-linked biodegradable polymers for drug delivery systems.

The presence of water in this type of polymer allows degradation to occur in the bulk amorphous region of the polymer matrix. Moiseev et al.[146] have considered the role of water, trace elements, and enzymes in the macrokinetics of polymer degradation in general and discussed the hydrolysis of poly(glycolic acid) in some detail. While the degradation of these polymers in the bulk complicates the prediction of the release profiles, it can have the fortuitous advantage of compensating for the decrease in release rate found with cylindrical or spherical devices which undergo surface erosion. Thus, a crudely constant release of $^{14}C$-norgestrel from cylinders of poly(lactic/glycolic acid), in the rat, has been followed for 2 years.[136] Attempts to correlate release with the molecular weight of the polymer have shown that these systems can be difficult to design. Woodland et al.[138] found no significant differences in the release rate of cyclazocine from composites prepared from poly(lactic acid) with molecular weight of 45,000 and 70,000. However, when poly(lactic acid) with molecular weight of about 150,000 and 450,000 was used to prepare implantable beads containing sulfadiazine, the amount of drug released after 90 days was about 40% from beads prepared from the higher molecular weight material compared to 80% with the lower weight material.[141] Further, small changes in the structure of the drug molecule may have profound effects on release rates. The in vivo release from particles of poly(lactic acid) of naltrexone, cyclazocine, and naloxone after 60 days was 68, 38, and 26%, respectively, while in vitro release studies showed considerably faster rates.[139]

Sterilization of poly(lactic/glycolic acid) copolymers by irradiation is known to affect their properties and is found to alter the molecular weight distribution.[145] Ethylene oxide sterilization has been used prior to in vivo studies with these devices.[147] Closely related polyesters may offer modified properties and poly(dioxanone) has been evaluated as a bioabsorbable suture material which can be sterilized by irradiation without significant loss of physical properties.[148] Poly(caprolactone) has recently been studied as a potential biodegradable polymer for controlled release of contraceptive steroids.[2,10]

Graham et al.[149] have developed a series of biodegradable polymers which possess both ester and glycosidic linkages and can be prepared using the well-known reaction of 3,4-dihydro-2H-pyran with alcohols to give a tetrahydropyranyl ether (Figure 16a). Monomers which contain two such dihydropyran groups, linked by an ester group, can form linear polymers with diols or can be used to prepare cross-linked matrices when multifunctional alcohols are used as comonomers[4] (Figure 16b). Drugs can be incorporated into the reaction mixture prior to molding and curing, and the resulting thermosetting polymers provide release over prolonged periods of time. Such formulations have been found to release $^{14}C$-labeled norethisterone in baboons for over 10 months. Related systems prepared by mixing preformed polymer with the antimalarial drug pyrimethanime have protected mice against *Plasmodium berghei* infection for at least 3 months. The range of possible monomers in addition to formulation variables has shown that devices have the potential to release drug over periods ranging from a few hours to in excess of 1 year.[149]

The identification of suitable biodegradable polymers for drug delivery systems has been relatively difficult, due to the number of properties required of the material for this application, and it is perhaps not surprising that polymers which have already been studied, and gained some acceptance, in other biomedical applications are attractive candidates. One such class is the poly(amino acids) and their derivatives which can be prepared with a wide range of properties and have been studied in several different disciplines. The in vivo degradation rate of copolymers of L-leucine and L-aspartic acid has been found to be dependent upon the degree of hydrophilicity of the polymer.[150] The resorption of sutures prepared from poly(glutamic acid) is delayed by partial esterification with lower alcohols,[151] while copolymers of glutamic acid and leucine form a biodegradable matrix which is well tolerated in rats.[152] Several studies have shown the potential of poly(glutamic acid) as a drug carrier in cancer chemotherapy where *p*-phenylene mustard[153] and cyclophosphamide[154] have been bound to the polymer.

Another class of biomedical polymer with potential use in biodegradable drug delivery systems is the poly (alkyl 2-cyanoacrylates). The alkyl 2-cyanoacrylate monomers have been used as biodegradable tissue adhesives in a variety of applications.[155] The butyl monomer is regarded as the most suitable due to a combination of its spreadability on biological fluids, rate of polymerization,[156] and its low toxicity.[157] The ability of the monomers to polymerize in the presence of water was exploited by Florence et al. to prepare microcapsules containing aqueous solutions of protein by *in situ* interfacial polymerization of butyl 2-cyanoacrylate in water-in-oil emulsions.[158] Microcapsules containing enzymes and other proteins have a variety of potential biomedical applications as demonstrated in many studies by Chang[159] and by other groups.[160] The possibility of using biodegradable polymers as the membranes of these "artificial cells" may encourage further study of these systems which have shown potential in therapy of substrate-dependent tumors and enzyme-deficiency diseases when nondegradable polymers have been used. Couvreur et al.[161,162] have prepared nanoparticles (200-nm diameter) from methyl and ethyl 2-cyanoacrylate monomers and studied the adsorption of several antineoplastic drugs. By virtue of their size, structure, and drug sorptive properties these nanoparticles have considerable potential as biodegradable lysosomotropic carriers and their tissue distribution, after i.v. injection in the rat, has shown their targeting potential.[163]

## VIII. ROUTES OF ADMINISTRATION AND TARGETING

As mentioned previously, polymeric inserts and implants can provide both prolonged release and a localized release of a drug at a particular site. With inserts the localized action can be obtained by placing the device in the vicinity of the receptor organ; thus a direct action on, for example, the eye or uterus is possible. Recently the Alza Corporation has described ocular inserts which can direct the release of drug to a particular region of the eye.[164] Other body cavities may favor a sustained release for systemic action. These more traditional routes, particularly the GI tract, are associated with a shorter duration of action with a time scale of hours, and a zero order release may not be optimal if the drug is absorbed from a particular region of the GI tract.[165] The vaginal and rectal routes may also have limited potential as sites for sustained release of certain drugs which are absorbed into the systemic circulation.[166,167] The use of transdermal delivery systems is rather new because the skin has traditionally been viewed as a barrier, but development in this area is probable. Synthetic polymers are also being investigated as wound dressings which deliver topical antimicrobial drugs.[168]

Traditionally, implants and depot injections have been designed to provide a prolonged release with a time scale of days, weeks, or even months. Considerable effort is being made in this area to extend the duration of action and to improve reproducibility of release. Implants

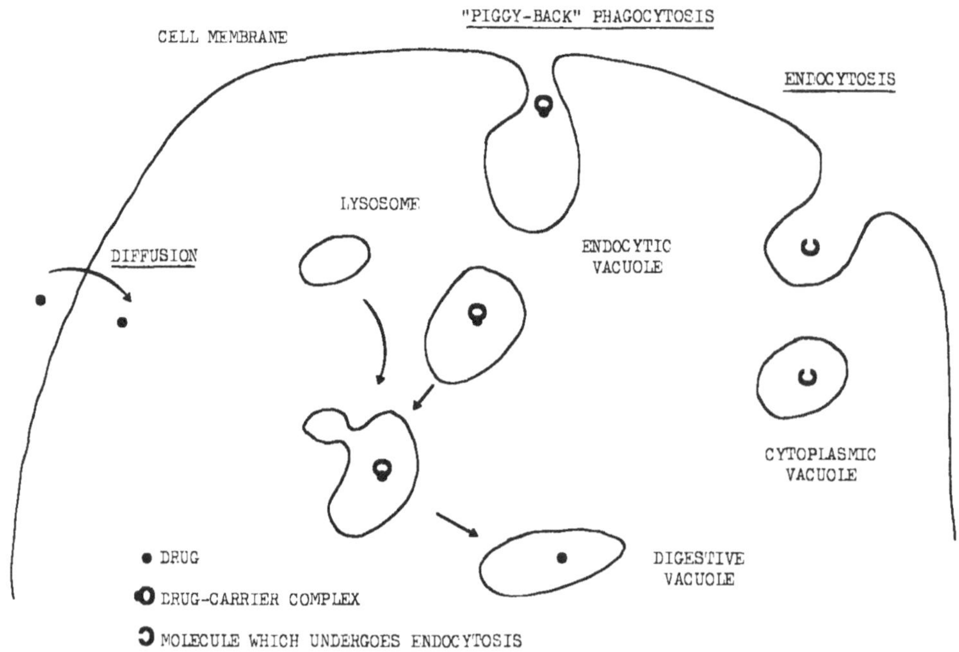


FIGURE 17. Diagramatic representation of entry of biologically active molecule into phagocytic cell.

also have some potential to localize action, as for example, the implantation of a device inside bone cavities for the treatment of chronic bone infections as described previously.[94]

More recently there has been considerable interest in the use of drug carriers in targeting therapy.[169] Many experimental approaches have been investigated particularly by the use of liposomes[170,171] and polymers, both natural and synthetic, as carriers. In many systems the targeting is accomplished by the selective uptake of the drug-carrier system by phagocytic cells following injection (Figure 17).

The macrophage[172] is a type of cell which is present throughout the body, in some cases as specialized cells (e.g., Kupffer cells of the liver). Characteristically, macrophages have the ability to take up large amounts of particulate matter of relatively large size (e.g., bacteria and protozoa) and have a highly developed ability to discriminate between particles of different types. These cells ingest solid or fluid matter by interiorizing part of their plasma membrane.[173,174] Where the material ingested is a particulate solid, the process is termed phagocytosis. The term pinocytosis is used where liquid droplets are engulfed, and the general term endocytosis encompases both cases. "Piggy-back" phagocytosis describes the way in which a substance, not normally ingested, gains entry to the cell by being phagocytosed along with a particle which is selectively taken up by the macrophage.[175] There is considerable interest in this type of approach for cancer chemotherapy and antimicrobial therapy for diseases in which the causative organism is localized within fixed macrophages.

Many attempts have been made to bind drugs and antibodies to a polymeric carrier which ideally will bind with, for example, tumor cells and subsequently release the active drug molecule. Poly(glutamic acid) was used as a biodegradable carrier for *p*-phenylene mustard and an antibody and the conjugate was found to suppress tumor growth in mice.[153] Dextran has been used as a carrier of daunomycin.[18] Unfortunately, a high degree of specificity for the target cells has not been achieved in most studies and systemic side effects still occur. In an attempt to avoid the use of biochemical differences between the target cell and normal cells to obtain specificity, the use of magnetic microspheres has been studied.[176,177] In this method magnetic particles are incorporated into the drug-carrier system and by the application of a magnetic field can be localized at a particular site in the body.

# IX. POLYMER-TISSUE INTERACTIONS AND BIOLOGICALLY ACTIVE POLYMERS

The possible interactions between implanted polymers (such as drug delivery systems) and body tissues are manifold, and over recent years a tremendous amount of research interest has revealed the complexities and provided fundamental knowledge. The complexities of these potential interactions are readily appreciated if one considers that living tissues consist of a wide variety of soluble low and high molecular weight compounds and insoluble structural components in dynamic equilibrium. Similarly, drug delivery systems can have present both low and high molecular weight material which is either soluble or insoluble. At present, little detailed information is available on the effects that these interactions may have on the release of drugs from delivery systems. "Medical grade" polymers and the control of additives have removed most of the problems associated with the leaching of toxic low molecular weight compounds. However, the uptake of lipid-soluble compounds from the body into poly(dimethylsiloxane) devices has been reported.[178]

Previously it was thought that "inert" polymers could be implanted without problem and bland or minimal responses with a wide range of polymers have been reported. However, the physical presence of an implanted polymer can cause foreign body tumorigenesis[179] and the shape and size of polymeric implants can produce different tissue responses.[180] Microporous hydrogels are encapsulated while the same polymer with a macroporous structure allows ingrowth of capillaries[181] or calcification.[182]

The adsorption of plasma proteins onto synthetic surfaces has been, and continues to be, the subject of a tremendous amount of research. The adsorption behavior of the particular proteins is dependent upon the hydrophilic/hydrophobic character of the surface. This field, in which the material in question is generally insoluble in the body fluids, still requires to be reviewed in terms of the "current understanding"[183] and our understanding of the interactions with soluble polymers is perhaps less clear.

Soluble polyelectrolytes of diverse nature have been found to exhibit biological activity. Synthetic heparinoid polymers have been designed and studied,[184] but many polymers with no apparent resemblance to natural macromolecules have been found to possess significant activity. The Pyran copolymer (divinyl ether-maleic anhydride copolymer, DIVEMA) has been the subject of many studies,[20,185] while its detailed structure is still uncertain.[186] DIVEMA has been found to exhibit antitumor, antiviral, antibacterial, antifungal, anticoagulant, and antiarthritic activity in addition to being an inducer of interferon and a macrophage activator. Both the biological activity and the toxicity of the material are related to its molecular weight. DIVEMA is by no means unique and other anionic polyelectrolytes display some biological activity. The antiviral activity of poly(acrylic acid) is related both to its molecular weight and tacticity.[187] However, atactic poly(methacrylic acid) has been reported to be devoid of antiviral action while the activity of the isotactic form is just detectable.

Even simple neutral water-soluble polymers can interact with living systems to a significant degree. Block copolymers of poly(oxyethylene)-poly(oxypropylene) (e.g., Pluronic®-F68) have been used in priming solutions for cardiopulmonary bypass and several advantages were found including an antisludging effect and a decrease in hemolysis.[188] However, poly(ethylene glycol) has been found to cause cell fusion, the mechanism of action being believed to be related to the reduction of the surface potential of the cells when the polymer adsorbed onto their surfaces.[189] Another study demonstrated that linking poly(ethylene glycol) to the enzyme catalase produced a conjugate which was nonimmunogenic, while experimental animals remained immune competent to the unmodified enzyme.[190] Poly(ethylene glycol) has also been used to study intestinal permeability in man because it is presumed to be nontoxic and nondegradable.[191]

Nevertheless, implanted polymeric prosthetic devices and controlled release implants and inserts have now been safely used in humans for many years. Increasingly, as the factors involved at the implant-tissue site are studied and the fact that even "inert" polymers have physicochemical properties, the trend is to design systems which interact with the living environment in a desired manner. With insight and imagination the complexities are becoming better understood. The current increase in cost of getting new drug entities to the market will guarantee that significant developments of controlled release systems will improve drug therapy for the treatment and prophylaxis of disease.

## REFERENCES


1. **Schwope, A. D., Wise, D. L., and Howes, J. F.,** Lactic/glycolic acid copolymers as narcotic antagonist delivery systems, *Life Sci.*, 17, 1877, 1975.
2. **Pitt, C. G., Jeffcoat, A. R., Zweidinger, R. A., and Schindler, A.,** Sustained drug delivery systems. I. The permeability of poly(ε-caprolactone), poly(DL-lactic acid) and their copolymers, *J. Biomed. Mater. Res.*, 13, 497, 1979.
3. **Wise, D. L., McCormick, G. J., Willet, G. P., and Anderson, L. D.,** Sustained release of an antimalarial drug using a copolymer of glycolic/lactic acid, *Life Sci.*, 19, 867, 1976.
4. **Graham, N. B.,** Injectable Medicinal Compositions, U.S. Patent 4,150,108, 1979.
5. **Kaetsu, I., Yoshida, M., and Yamada, A.,** Controlled slow release of chemotherapeutic drugs for cancer from matrices prepared by irradiation polymerization at low temperatures, *J. Biomed. Mater. Res.*, 14, 185, 1980.
6. **Luzzi, L. A.,** Microencapsulation, *J. Pharm. Sci.*, 59, 1367, 1970.
7. **Kondo, T.,** Microcapsules: their preparation and properties, in *Surface and Colloid Science*, Vol. 10, Matijevic, E., Ed., Plenum Press, New York, 1978, 1.
8. **Speiser, P.,** Microencapsulation by coacervation, spray encapsulation, and nanoencapsulation, in *Microencapsulation*, Nixon, J. R., Ed., Marcel Dekker, New York, 1976, 1.
9. **Kincl, F. A., Angee, I., Chang, C. C., and Rudel, H. W.,** Sustained release hormonal preparations. IX. Plasma levels and accumulation into various tissues of 6-methyl-17α-acetoxy-4,6-pregnadiene-3,20-dione after oral administration or absorption from polydimethylsiloxane implants, *Acta Endocrinol.*, 64, 508, 1970.
10. **Pitt, C. G., Gratzl, M. M., Jeffcoat, A. R., Zweidinger, R., and Schindler, A.,** Sustained drug delivery systems. II. Factors affecting release rates from poly(ε-caprolactone) and related biodegradable polyesters, *J. Pharm. Sci.*, 68, 1534, 1979.
11. **Ringsdorf, H.,** Structure and properties of pharmacologically active polymers, *J. Polym. Sci. Polym. Symp.*, 51, 135, 1975.
12. **Zaharko, D. S., Przybylski, M., and Oliverio, V. T.,** Binding of anticancer drugs to carrier molecules, in *Methods in Cancer Research*, Vol. 16, Part A, de Vita, V. T. and Busch, H., Eds., Academic Press, London, 1979, 347.
13. **Ferno, O., Fex, H., Hogberg, B., Linderot, T., Veige, S., and Diczfalusy, E.,** High molecular weight enzyme inhibitors. III. Polyestradiol phosphate (P.E.P.), a long-acting estrogen, *Acta Chem. Scand.*, 12, 1675, 1958.
14. **Rabadeux, J. C., Pleurdeau, A., and Pinazzi, C.,** Polycondensation of biologically active structures. IV. Polymers with potential pharmaceutical properties, *J. Polym. Sci. Polym. Chem. Ed.*, 18, 335, 1980.
15. **Batz, H. G.,** Polymeric drugs, *Adv. Polym. Sci.*, 23, 25, 1977.
16. **Samour, C. M.,** Polymeric drugs, *Chemtech*, 8, 494, 1978.
17. **Goldberg, E. P.,** Polymeric affinity drugs, in *Polymeric Delivery Systems*, Kostelnik, R. J., Ed., Gordon and Breach, New York, 1978, 227.
18. **Wilchek, M.,** Affinity therapy and polymer bound drugs, *Makromol. Chem. Suppl.*, 2, 207, 1979.
19. **Donaruma, L. G.,** Synthetic biologically active polymers, *Prog. Polym. Sci.*, 4, 1, 1974.
20. **Breslow, D. S.,** Biologically active synthetic polymers, *Pure Appl. Chem.*, 46, 103, 1976.
21. **Higuchi, W. I.,** Diffusional models useful in biopharmaceutics, *J. Pharm. Sci.*, 56, 315, 1967.

22. **Flynn, G. L., Yalkowsky, S. H., and Roseman, T. J.,** Mass transport phenomena and models: theoretical concepts, *J. Pharm. Sci.*, 63, 479, 1974.
23. **Baker, R. W. and Lonsdale, H. K.,** Controlled release: mechanisms and rates, in *Controlled Release of Biologically Active Agents*, Tanquary, A. C. and Lacey, R. E., Eds., Plenum Press, London, 1974, 15.
24. **Fu, J. C., Kale, A., and Moyer, D. L.,** Diffusion of pyrimethamine from silicone rubber and flexible epoxy drug capsules, *J. Biomed. Mater. Res.*, 7, 193, 1973.
25. **Anderson, L. C., Wise, D. L., and Howes, J. F.,** An injectable sustained fertility control system, *Contraception*, 13, 375, 1976.
26. **Baker, R. W., Michaelis, A. S., and Theeuwes, F.,** Process for Improving Release Kinetics of a Monolithic Drug Delivery System, U.S. Patent, 3,923,939, 1975.
27. **Crank, J.,** *The Mathematics of Diffusion*, 2nd ed., Clarendon Press, Oxford, 1975.
28. **Carslaw, H. S. and Jaeger, J. C.,** *Conduction of Heat in Solids*, Oxford University Press, London, 1959.
29. **Higuchi, T.,** Rate of release of medicaments from ointment bases containing drugs in suspension, *J. Pharm. Sci.*, 50, 875, 1961.
30. **Chien, Y. W., Lambert, H. J., and Lin, T. K.,** Solution-solubility dependency of controlled release of drug from polymer matrix: mathematical analysis, *J. Pharm. Sci.*, 64, 1643, 1975.
31. **Haleblian, J., Runkel, R., Mueller, N., Christopherson, J., and Ng, K.,** Steroid release from silicone elastomer containing excess drug in suspension, *J. Pharm. Sci.*, 60, 541, 1971.
32. **Roseman, T. J. and Higuchi, W. I.,** Release of medroxyprogesterone acetate from a silicone polymer, *J. Pharm. Sci.*, 59, 353, 1970.
33. **Paul, D. R. and McSpadden, S. K.,** Diffusional release of a solute form of polymer matrix, *J. Membr. Sci.*, 1, 33, 1976.
34. **Lee, P. I.,** Diffusional release of a solute from a polymeric matrix — approximate analytical solutions, *J. Membr. Sci.*, 7, 255, 1980.
35. **Higuchi, T.,** Mechanisms of sustained action medication: theoretical analysis of rate of release of solid drugs dispersed in solid matrices, *J. Pharm. Sci.*, 52, 1145, 1963.
36. **Cobby, J., Mayersohn, M., and Walker, G. C.,** Influence of shape factors on kinetics of drug release from matrix tablets. I. Theoretical, *J. Pharm. Sci.*, 63, 725, 1974.
37. **Fu, J. C., Hagemeir, C., Moyer, D. L., and Ng, E. W.,** A unified mathematical model for diffusion from drug-polymer composite tablets, *J. Biomed. Mater. Res.*, 10, 743, 1976.
38. **Cobby, J.,** Mathematical models for the release of drugs from matrix tablets, *J. Biomed. Mater. Res.*, 12, 627, 1978.
39. **Desai, S. J., Simonelli, A. P., and Higuchi, W. I.,** Investigation of factors influencing release of solid drug dispersed in inert matrices, *J. Pharm. Sci.*, 54, 1459, 1965.
40. **Desai, S. J., Singh, P., Simonelli, A. P., and Higuchi, W. I.,** Investigation of factors influencing release of solid drug dispersed in inert matrices. II. Quantitation of procedures, *J. Pharm. Sci.*, 55, 1224, 1966.
41. **Desai, S. J., Singh, P., Simonelli, A. P., and Higuchi, W. I.,** Investigation of factors influencing release of solid drug dispersed in inert matrices. III. Quantitative studies involving the polyethylene plastic matrix, *J. Pharm. Sci.*, 55, 1230, 1966.
42. **Desai, S. J., Singh, P., Simonelli, A. P., and Higuchi, W. I.,** Investigation of factors influencing release of solid drug dispersed in inert matrices. IV. Some studies involving the polyvinyl chloride matrix, *J. Pharm. Sci.*, 55, 1235, 1966.
43. **Singh, P., Desai, S. J., Simonelli, A. P., and Higuchi, W. I.,** Release rates of solid drug mixtures dispersed in inert matrices. I. Non-interacting drug mixtures, *J. Pharm. Sci.*, 56, 1542, 1967.
44. **Singh, P., Desai, S. J., Simonelli, A. P., and Higuchi, W. I.,** Release rates of solid drug mixtures dispersed in inert matrices. II. Mutually interacting drug mixtures, *J. Pharm. Sci.*, 56, 1548, 1967.
45. **McGinity, J. W., Hunke, L. A., and Combs, A. B.,** Effects of water-soluble carriers on morphine sulphate release from a silicone polymer, *J. Pharm. Sci.*, 68, 662, 1979.
46. **Langer, R. and Folkman, J.,** Sustained release of macromolecules from polymers, in *Polymeric Drug Delivery Systems*, Kostelnik, R. J., Ed., Gordon and Breach, New York, 1978, 175.
47. **Davis, B. K.,** Diffusion in polymer gel implants, *Proc. Natl. Acad. Sci. U.S.A.*, 71, 3120, 1974.
48. **Hosaka, S., Ozawa, H., and Tanazawa, H.,** Controlled release of drugs from hydrogel matrices, *J. Appl. Polym. Sci.*, 23, 2089, 1979.
49. **Chien, Y. W. and Lau, E. P. K.,** Controlled drug release from polymeric devices. IV. *In vitro-in vivo* correlation of the subcutaneous release of norgestomet from hydrophilic implants, *J. Pharm. Sci.*, 65, 488, 1976.
50. **Good, W. R.,** Diffusion of water soluble drugs from initially dry hydrogels, in *Polymeric Drug Delivery Systems*, Kostelnik, R. J., Ed., Gordon and Breach, New York, 1978, 139.
51. **Peterlin, A.,** Diffusion with discontinuous swelling. V. Type II diffusion into sheets and spheres, *J. Polym. Sci. Polym. Phys. Ed.*, 17, 1741, 1979.
52. **Peppas, N. A., Gurny, R., and Buri, P.,** Modelling of drug diffusion through swellable polymeric systems, *J. Membr. Sci.*, 7, 241, 1980.

53. **Chien, Y. W., Mares, S. E., Berg, J., Huber, S., Lambert, H. J., and King, K. F.**, Controlled drug release from polymeric delivery systems. III. *In vitro-in vivo* correlations for intra-vaginal release of ethynodiol diacetate from silicone devices in rabbits, *J. Pharm. Sci.*, 64, 1776, 1975.
54. **Flynn, G. L., Ho, N. F. H., Hwang, S., Owada, E., Molokhia, A., Behl, C. R., Higuchi, W. I., Yotsuyanagi, T., Shah, Y., and Park, J.**, Interfacing matrix release and membrane absorption — analysis of steroid absorption from vaginal device in the rabbit doe, in *Controlled Release Polymeric Formulations*, Paul, D. R. and Harris, F. W., Eds., Am. Chem. Soc. Symp. Ser. 33, American Chemical Society, Washington, D.C., 1976, 87.
55. **Kent, J. S.**, Controlled release of delmadinone acetate from silicone polymer tubing: *in vitro-in vivo* correlations to diffusion model, in *Controlled Release Polymeric Formulations*, Paul, D. R. and Harris, F. W., Eds., Am. Chem. Soc. Symp. Ser. 33, American Chemical Society, Washington, D.C., 1976, 157.
56. **Yolles, S., Elridge, J., Leafe, T., Woodland, J. H. R., Blake, D. R., and Meyer, F. J.**, Long-acting delivery systems for narcotic antagonists, in *Controlled Release of Biologically Active Agents*, Tanquary, A. C. and Lacey, R. E., Eds., Plenum Press, London, 1974, 177.
57. **Lovering, E. G. and Black, D. B.**, Diffusion layer effects on permeation of phenylbutazone through polydimethylsiloxane, *J. Pharm. Sci.*, 63, 1399, 1974.
58. **Hopfenberg, H. B.**, Controlled release from erodible slabs, cylinders, and spheres, in *Controlled Release Polymeric Formulations*, Paul, D. R. and Harris, F. W., Eds., Am. Chem. Soc. Symp. Ser. 33, American Chemical Society, Washington, D.C., 1976, 26.
59. **Cooney, D. O.**, Effect of geometry on the dissolution of pharmaceutical tablets and other solids: surface detachment kinetics controlling, *AIChEJ.*, 18, 446, 1972.
60. **Heller, J., Baker, R. W., Gale, R. M., and Rodin, J. O.**, Controlled release by polymer dissolution. I. Partial esters of maleic anhydride copolymers, properties and theory, *J. Appl. Polym. Sci.*, 22, 1991, 1978.
61. **Haward, R. N.**, The nature of glassy state in polymers, in *Molecular Behaviour and the Development of Polymeric Materials*, Ledwith, A. and North, A. M., Eds., Chapman and Hall, London, 1975, 404.
62. **Flory, P. J.**, *Principles of Polymer Chemistry*, Cornell University Press, Ithaca, N.Y., 1953.
63. **Dusek, K. and Prins, W.**, Structure and elasticity of non-crystalline polymer networks, *Adv. Polym. Sci.*, 6, 1, 1969.
64. **Kincl, F. A., Benagiano, G., and Angee, I.**, Sustained release hormonal prepartions. I. Diffusion of various steroids through polymer membranes, *Steroids*, 11, 673, 1968.
65. **Zentner, G. M., Cardinal, J. R., and Kim, S. W.**, Progestin permeation through polymer membranes. I. Diffusion studies in plasma-soaked membranes, *J. Pharm. Sci.*, 69, 1349, 1978.
66. **Chien, Y. W., Jefferson, D. M., Cooney, J. G., and Lambert, H. J.**, Controlled drug release from polymeric delivery devices. V. Hydroxyl group effect on drug release kinetics and thermodynamics, *J. Pharm. Sci.*, 68, 689, 1979.
67. **Roseman, T. J.**, Release of steroid from a silicone polymer, *J. Pharm. Sci.*, 61, 46, 1972.
68. **Zentner, G. M., Cardinal, J. R., Feijen, J., and Song, S.**, Progestin permeation through polymer membranes. IV. Mechanism of steroid permeation and functional group contribution to diffusion through hydrogel films, *J. Pharm. Sci.*, 68, 970, 1979.
69. **Most, C. F.**, Some filler effects on diffusion in silicone rubber, *J. Appl. Polym. Sci.*, 14, 1019, 1970.
70. **Most, C. F.**, Co-permeant enhancement of drug transmission rates through silicone rubber, *J. Biomed. Mater. Res.*, 6, 3, 1972.
71. **Warrick, E. L., Pierce, O. R., Polmanteer, K. E., and Saam, J. C.**, Silicone elastomer developments 1967—1977, *Rubber Chem. Technol.*, 52, 437, 1979.
72. **Folkman, J. and Long, D. M.**, The use of silicone rubber as a carrier for prolonged drug therapy, *J. Surg. Res.*, 4, 139, 1964.
73. **Long, D. M. and Folkman, J.**, Polysiloxane Carriers for Controlled Release of Drugs and Other Agents, U.S. Patent 3,279,996, 1966.
74. **Kratochvil, P., Benagiano, G., and Kincl, F. A.**, Sustained release hormonal preparations. VI. Permeability constants of various steroids, *Steroids*, 15, 505, 1970.
75. **Fu, J. C., Kale, A. K., and Moyer, D. L.**, Drug-incorporated silicone discs as sustained release capsules. I. Chloroquine diphosphate, *J. Biomed. Mater. Res.*, 7, 71, 1973.
76. **Gaginella, T. S. and Vallner, J. J.**, Indomethacine release from silicone rubber: a suggested method for continuous inhibition of prostaglandin synthesis, *Res. Commun. Chem. Pathol. Pharmacol.*, 11, 323, 1975.
77. **Bass, P., Purdon, R. A., and Wiley, J. N.**, Prolonged administration of atropine or histamine in a silicone rubber implant, *Nature (London)*, 208, 591, 1965.
78. **Chien, Y. W. and Lambert, H. J.**, Microsealed Pharmaceutical Delivery System, U.S. Patent 3,946,106, 1976.
79. **Chien, Y. W., Rozek, L. F., and Lambert, H. J.**, Microsealed drug delivery systems. I. *In vitro-in vivo* correlation of subcutaneous release of desoxycorticosterone acetate and prolonged hypertension animal model for cardiovascular studies, *J. Pharm. Sci.*, 67, 214, 1978.

80. **Riffee, W. H., Wilcox, R. E., Anderson, J. A., and McGinity, J. W.**, Polydimethylsiloxane pellets for sustained delivery of morphine in mice, *J. Pharm. Sci.*, 69, 980, 1980.
81. **Kalkwarf, D. R., Sikov, M. R., Smith, L., and Gordon, R.**, Release of progesterone from polyethylene devices *in vitro* and in experimental animals, *Contraception*, 6, 423, 1972.
82. **Pharriss, B. B., Erickson, R., Bashaw, J., Hoff, S., Place, V. A., and Zaffaroni, A.**, Progestasert: a uterine therapeutic system for long-term contraception. I. Philosophy and clinical efficacy, *Fertil. Steril.*, 25, 915, 1974.
83. **Martinez-Manautou, J.**, Contraception by intrauterine release of progesterone. Clinical results, *J. Steroid Biochem.*, 6, 889, 1975.
84. **Fu, J. C., Moyer, D. L., and Hagemeier, C.**, Effect of comonomer ratio on hydrocortisone diffusion from sustained-release composite capsules, *J. Biomed. Mater. Res.*, 12, 249, 1978.
85. **Folkman, M. J. and Long, R. S.**, Systems for Controlled Release of Macromolecules, U.S. Patent 4,164,560, 1979.
86. **Farhadieh, B., Borodkin, S., and Buddnehagen, J. D.**, Drug release from methyl acrylate-methyl methacrylate copolymer matrix. I. Kinetics of release, *J. Pharm. Sci.*, 60, 209, 1971.
87. **Farhadieh, B., Borodkin, S., and Buddnehagen, J. D.**, Drug release from methyl acrylate-methyl methacrylate copolymer matrix.II. Control of release rate by exposure to acetone vapor, *J. Pharm. Sci.*, 60, 212, 1971.
88. **Olanoff, L., Koinis, T., and Anderson, J. M.**, Controlled release of tetracycline. I. *In vitro* studies with a trilaminate 2-hydroxyethyl methacrylate-methyl methacrylate system, *J. Pharm. Sci.*, 68, 1147, 1979.
89. **Olanoff, L. and Anderson, J. M.**, Controlled release of tetracycline. II. Development of an *in vivo* flow limited pharmacokinetic model, *J. Pharm. Sci.*, 68, 1151, 1979.
90. **Welch, A. B.**, Antibiotics in acrylic bone cement, *In vitro* studies, *J. Biomed. Mater. Res.*, 12, 679, 1978.
91. **Welch, A. B.**, Antibiotics in acrylic bone cement, *In vivo* studies, *J. Biomed. Mater. Res.*, 12, 843, 1978.
92. **Lautenschlager, E. P., Marshall, G. W., Marks, K. E., Schwartz, J., and Nelson, C. L.**, Mechanical strength of acrylic bone cements impregnated with antibiotics, *J. Biomed. Mater. Res.*, 10, 837, 1976.
93. **Nelson, R. C., Hoffman, R. O., and Burton, T. A.**, The effect of antibiotic additions on the mechanical properties of acrylic bone cement, *J. Biomed. Mater. Res.*, 12, 473, 1978.
94. **Wahlig, H., Dingeldein, E., Bergman, R., and Reuss, K.**, The release of gentamycin from polymethylmethacrylate beads. An experimental and pharmokinetic study, *J. Bone Jt. Surg. Br. Vol.*, 60b, 270, 1978.
95. **Yoshida, M., Kumakura, M., and Kaetsu, I.**, Controlled release of biofunctional substances by radiation-induced polymerization. I. Release of potassium chloride by polymerization of various vinyl monomers, *Polymer*, 19, 1375, 1978.
96. **Yoshida, M., Kumakura, M., and Kaetsu, I.**, Drug entrapment for controlled release in radiation-polymerized beads, *J. Pharm. Sci.*, 68, 628, 1979.
97. **Johansson, J. A. O. and Ulmsten, U.**, U.K. Patent Application GB 2,041,220A, 1980.
98. **Ceskoslovenska Akademie Ved.**, Improvements in or Relating to Sustained-Release Medicaments, British Patent 1,135,966, 1968.
99. **Akkapeddi, M. K., Halpern, B. D., Davis, R. H., and Balin, H.**, Cervical hydrogel dilator: a new delivery system for prostaglandins, in *Controlled Release of Biologically Active Agents*, Tanquary, A. C. and Lacey, R. E., Eds., Plenum Press, London, 1974, 165.
100. **Hudgin, D. E. and Blair, E. A.**, Hydrophilic or Hydrogel Carrier Systems such as Coatings, Body Implants and Other Articles, U.S. Patent 3,975,350, 1976.
101. **Ciba-Geigy, A. G.**, Water-Insoluble Hydrophilic Copolymers, British Patent 1,511,563, 1978.
102. **King, P. A.**, Sterile, Transparent, Moisture-Retaining Dressings Based on Poly(alkylene oxides), U.S. Patent 3,419,006, 1968.
103. **Graham, N. B., McNeill, M. E., Zulfiqar, M., and Embrey, M. P.**, Hydrogels for the controlled release of prostaglandin $E_2$, *Am. Chem. Soc. Polym. Prepr.*, 21(1), 104, 1980.
104. Imperial Chemical Industries Limited, Delivery Means for Biologically Active Agents, British Patent 1,551,620, 1979.
105. **Embrey, M. P., Graham, N. B., and McNeill, M. E.**, Induction of labour with a sustained-release prostaglandin $E_2$ pessary, *Br. Med. J.*, 281, 901, 1980.
106. **Harris, A. S., Kirstein-Pedersen, A., Stenberg, P., Ulmsten, U., and Wingerup, L.**, Preparation, characterisation, and stability of new prostaglandin $E_2$ gel for local administration, *J. Pharm. Sci.*, 69, 1271, 1980.
107. **Lee, E. S., Kim, S. W., Kim, S. H., Cardinal, J. R., and Jacobs, H.**, Drug release from hydrogel devices with rate-controlling barriers, *J. Membr. Sci.*, 7, 293, 1980.
108. **Cowsar, D. R., Tarwater, O. R., and Tanquary, A. C.**, Controlled release of fluoride from hydrogels for dental applications, in *Hydrogels for Medical and Related Applications*, Andrade, J. D., Ed., Am. Chem. Soc. Symp. Ser. 33, American Chemical Society, Washington, D.C., 1976, 180.
109. **Fildes, F. J. T.**, Biologically Active Compositions Suitable for Feeding to Ruminant Animals, British Patent 1,440,217, 1976.

110. **Gould, F. E.,** Polyurethane Polymers Characterized by Lactone Groups and Hydroxyl Groups in the Polymer Backbone, U.S. Patent, 4,156,067, 1979.
111. **Theeuwes, F.,** Elementary osmotic pump, *J. Pharm. Sci.*, 64, 1987, 1975.
112. **Theeuwes, F.,** Osmotic System with Volume Amplifier for Increasing Amount of Agent Delivered Therefrom, U.S. Patent 4,203,439, 1980.
113. **Theeuwes, F.,** Osmotically Triggered Device with Gas Generating Means, U.S. Patent, 4,203,441, 1980.
114. **Capozza, R. C.,** An implantable, self-powered pump for drug delivery use in laboratory animals, in *Polymeric Drug Delivery Systems,* Kostelnik, J. R., Ed., Gordon and Breach, London, 1978, 261.
115. **Will, P. C., Cortright, R. N., and Hopper, U.,** Poly (ethylene glycols) as solvents in implantable osmotic pumps, *J. Pharm. Sci.*, 69, 747, 1980.
116. **Wood, D. A.,** Biodegradable drug delivery systems—a review article, *Int. J. Pharm.*, 7, 1, 1980.
117. **Taylor, L.,** Degradable plastics: solution or illusion?, *Chemtech*, 9, 542, 1979.
118. **Stenlake, J. B., Urwin, J., Waigh, R. D., and Hughes, R.,** Biodegradable neuromuscular blocking agents, *Eur. J. Med. Chem.*, 14, 77, 1979.
119. **Blok, T.,** Methods for measurement of biodegradability of chemical compounds, *Int. Biodeterior. Bull.*, 11, 78, 1975.
120. **Dagley, S.,** Determinants of biodegradability, *Q. Rev. Biophys.*, 11, 577, 1978.
121. **Huang, S. J., Bansleben, D. A., and Knox, J. R.,** Biodegradable polymers: chymotryptic degradation of low molecular weight poly(ester-urea) containing phenylalanine, *J. Appl. Polym. Sci.*, 23, 429, 1979.
122. **Tokowa, Y., Suzuki, T., and Ando, T.,** Synthesis of copolyamide-esters and some aspects involved in their hydrolysis by lipase, *J. Appl. Polym. Sci.*, 24, 1701, 1979.
123. **Wade, C. W. R., Hegyeli, A. F., and Kulkarni, R. K.,** Standards for *in vitro* and *in vivo* comparison and qualification of bioabsorbable polymers, *J. Test. Eval.*, 5, 397, 1977.
124. **Heller, J. and Trescony, P. V.,** Controlled drug release by polymer dissolution. II. Enzyme-mediated delivery device, *J. Pharm. Sci.*, 68, 919, 1979.
125. **Ramwell, P. W.,** Intrauterine Device that Bioerodes in Response to the Environment of the Uterus and Methods of Using, U.S. Patent 3,993,057, 1976.
126. **Schmitt, E. E.,** Poly(orthoester) Co- and Homopolymers and Poly(orthocarbonate) Co- and Homopolymers Having Carbonyloxy Functionality, U.S. Patent, 4,070,347, 1978.
127. **Choi, N. S. and Heller, J.,** Poly(carbonates), U.S. Patent 4,079,038, 1978.
128. **Choi, N. S. and Heller, J.,** Drug Delivery Devices Manufactured from Poly(orthoesters) and Poly(orthocarbonates), U.S. Patent 4,093,709, 1978.
129. **Heller, J.,** Controlled release of biologically active compounds from bioerodible polymers, *Biomaterials*, 1, 51, 1980.
130. **Choi, N. S.,** Drug delivery by bioerodible polymer systems, *Pollimo*, 4, 35, 1980.
131. **Capozza, R. C., Sendelbeck, L., and Balkenhol, W. J.,** Preparation and evaluation of a bioerodible naltrexone delivery system, in *Polymeric Delivery Systems*, Kostelnik, R. J., Ed., Gordon and Breach, London, 1978, 59.
132. **Benagiano, G., Schmitt, E., Wise, D., and Goodman, M.,** Sustained release hormonal preparations for the delivery of fertility-regulating agents, *J. Polym. Sci. Polym. Symp.*, 66, 129, 1979.
133. **Brady, J. M., Cutright, D. E., Miller, R. A., Battistone, G. C., and Hunsuck, E. E.,** Resorption rate, route of elimination and ultrastructure of implant site of polylactic acid in the abdominal wall of the rat, *J. Biomed. Mater. Res.*, 7, 155, 1973.
134. **Frazza, E. J. and Schmitt, E. E.,** A new absorbable suture, *J. Biomed. Mater. Res. Symp.*, 1, 43, 1971.
135. **Wise, D. L., Fellmann, T. D., Sanderson, J. E., and Wentworth, R. L.,** Lactic/glycolic acid polymers, in *Drug Carriers in Medicine and Biology*, Gregoriadis, G., Ed., Academic Press, London, 1979, 237.
136. **Wise, D. L., Rosenkrantz, H., Gregory, J. B., and Esber, H. J.,** Long-term controlled delivery of levonorgestrel in rats by means of a small biodegradable cylinder, *J. Pharm. Pharmacol.*, 32, 399, 1980.
137. **Beck, L. R., Cowsar, D. R., Lewis, D. H., Cosgrove, R. J., Riddle, C. T., Lowry, S. L., and Epperly, T.,** A new long-acting injectable microcapsule system for the administration of progesterone, *Fertil. Steril.*, 31, 545, 1979.
138. **Woodland, J. H. R., Yolles, S., Blake, D. A., Helrich, M., and Meyer, F. J.,** Long-acting delivery systems for narcotic antagonists, *J. Med. Chem.*, 16, 897, 1973.
139. **Yolles, S., Elridge, J., Leafe, T., Woodland, J. H. R., Blake, D. R., and Meyer, F. J.,** Long-acting delivery systems for narcotic antagonists, in *Controlled Release of Biologically Active Agents*, Tanquary, A. C. and Lacey, R. E., Eds., Plenum Press, London, 1974, 177.
140. **Wise, D. L., Schwope, A. D., Harrigan, S. E., McCarthy, D. A., and Howes, J. F.,** Sustained delivery of a narcotic antagonist from lactic/glycolic acid copolymer implants, in *Polymeric Delivery Systems*, Kostelnik, R. J., Ed., Gordon and Breach, London, 1978, 75.
141. **Wise, D. L., McCormick, G. J., Willet, G. P., Anderson, L. C., and Howes, J. F.,** Sustained release of sulphadiazine, *J. Pharm. Pharmacol.*, 30, 686, 1978.

142. **Wise, D. L., Gresser, J. D., and McCormick, G. J.,** Sustained release of a dual antimalarial system, *J. Pharm. Pharmacol.*, 31, 201, 1979.
143. **Yolles, S., Leafe, T. D., and Meyer, F. J.,** Timed-release depot for anticancer agents, *J. Pharm. Sci.*, 64, 115, 1975.
144. **Yolles, S.,** Time-release depot for anticancer drugs: release of drugs covalently bonded to polymers, *J. Parenteral Drug Assoc.*, 32, 188, 1978.
145. **Gilding, D. K. and Reed, A. M.,** Biodegradable polymers for use in surgery, polyglycolic/poly(lactic acid) homo- and copolymers, *Polymer*, 20, 1459, 1979.
146. **Moiseev, Y. V., Daurova, T. T., Voronkova, O. S., Gumargalieva, K. Z., and Privalova, L. G.,** The specificity of polymer degradation in the living body, *J. Polym. Sci. Polym. Symp.*, 66, 269, 1979.
147. **Yolles, S., Leafe, T., Ward, L., and Boettner, F.,** Controlled release of biologically active drugs, *Bull. Parenteral Drug Assoc.*, 30, 306, 1976.
148. **Doddi, N., Versfelt, C. C., and Wasserman, D.,** Synthetic Absorbable Surgical Devices of Poly(dioxanone), U.S. Patent 4,052,988, 1977.
149. **Graham, N. B., et al.,** to be published.
150. **Marck, K. W., Wildevuur, C. R. H., Sederel, W. L., Bantjes, A., and Feijen, J.,** Biodegradability and tissue reaction of random copolymers of L-leucine, L-aspartic acid, and L-aspartic acid esters, *J. Biomed. Mater. Res.*, 11, 405, 1977.
151. **Miyamae, T., Mori, S., and Takeda, Y.,** Filaments and Surgical Sutures of Poly(L-glutamic acid) Partly Esterified with Lower Alkanols and Processes Therefor, U.S. Patent 3,371,069, 1968.
152. **Anderson, J. M., Gibbons, D. F., Martin, R. L., Hiltner, A., and Woods, R.,** The potential for poly α-amino acids as biomaterials, *J. Biomed. Mater. Res. Symp.*, 5, 197, 1974.
153. **Rowland, G. F., O'Neill, G. J., and Davies, D. A. L.,** Suppression of tumour growth in mice by a drug-antibody conjugate using a novel approach to linkage, *Nature (London)*, 255, 487, 1975.
154. **Hirano, T., Klesse, W., and Ringsdorf, H.,** Polymeric derivatives of activated cyclophosphamide as drug delivery systems in antitumor chemotherapy, *Makromol. Chem.*, 180, 1125, 1979.
155. **Leonard, F.,** Hemostatic applications of alpha cyanoacrylates: bonding mechanism and physiological degradation of bonds, in *Adhesion in Biological Systems*, Manly, R. S., Ed., Academic Press, London, 1970, 185.
156. **Leonard, F., Collins, J. A., and Porter, H. J.,** Interfacial polymerization of *n*-alkyl α-cyanoacrylate homologs, *J. Appl. Polym. Sci.*, 10, 1617, 1966.
157. **Mungiu, C., Gogalniceanu, D., Leibovici, M., and Negulescu, I.,** On the medical use of cyanoacrylic esters: toxicity of pure *n*-butyl α-cyanoacrylate, *J. Polym. Sci. Polym. Symp.*, 66, 189, 1979.
158. **Florence, A. T., Whateley, T. L., and Wood, D. A.,** Potentially biodegradable microcapsules with poly(alkyl 2-cyanoacrylate) membranes, *J. Pharm. Pharmacol.*, 31, 422, 1979.
159. **Chang, T. M. S.,** Encapsulation of enzymes, cell contents, cells, vaccines, antigens, antiserum, cofactors, hormones and proteins, in *Biomedical Applications of Immobilized Enzymes and Proteins*, Vol. 1, Chang, T. M. S., Ed., Plenum Press, New York, 1977, 69.
160. **Wood, D. A., Whateley, T. L., and Florence, A. T.,** Microencapsulation of histidase for enzyme replacement therapy, *J. Pharm. Pharmacol.*, 31, 79p, 1979.
161. **Couvreur, P., Kante, B., Roland, M., Guiot, P., Bauduin, P., and Speiser, P.,** Polycyanoacrylate nanocapsules as potential lysosomotropic carriers: preparation, morphology and sorptive properties, *J. Pharm. Pharmacol.*, 31, 331, 1979.
162. **Couvreur, P., Kante, B., Roland, M., and Speiser, P.,** Adsorption of antineoplastic drugs to polyalkylcyanoacrylate nanoparticles and their release in calf serum, *J. Pharm. Sci.*, 68, 1521, 1979.
163. **Couvreur, P., Kante, B., Lenaerts, V., Scialteur, V., Roland, M., and Speiser, P.,** Tissue distribution of antitumour drugs associated with polyalkylcyanoacrylate nanoparticles, *J. Pharm. Sci.*, 69, 199, 1980.
164. **Zaffaroni, A.,** Selective Administration of Drug with Ocular Therapeutic System, U.S. Patent, 4,186,184, 1980.
165. **Hayton, W. L.,** Rate-limiting barriers to intestinal drug absorption: a review, *J. Pharmacokinet. Biopharm.*, 8, 321, 1980.
166. **Park, J. Y., Langworthy, M., Behl, C. R., Higuchi, W. I., Flynn, G. L., and Ho, N. F. H.,** Vaginal drug absorption in the rhesus monkey: bioavailability method and assessment of estrogens, *Int. J. Pharm.*, 2, 215, 1979.
167. **DeBlaey, C. and Polderman, J.,** Rationales in the design of rectal and vaginal delivery forms of drugs, in *Drug Design*, Vol. 9, Ariens, E. J., Ed., Academic Press, London, 1980, 237.
168. **Robb, E. C., Fitz, D. G., and Nathan, P.,** Delivery of the topical antimicrobial agents silver sulfadiazine, gentamycin and nystatin to infected burn wounds in rats from preloaded synthetic dressing, *Trans. Am. Soc. Artif. Intern. Organs*, 26, 533, 1980.
169. **Gregoriadis, G., Ed.,** *Drug Carriers in Biology and Medicine*, Academic Press, London, 1979.
170. **Papahadjopoulos, D.,** Liposomes as drug carriers, *Annu. Rep. Med. Chem.*, 14, 250, 1979.

171. **Kimelberg, H. K. and Mayhew, E. G.,** Properties and biological effects of liposomes and their uses in pharmacology and toxicology, *Crit. Rev. Toxicol.*, 6, 25, 1978.
172. **Vernon-Roberts, B.,** The macrophage, in *The Cell in Medical Science*, Vol. 4, Cellular Control Mechanisms Cellular Responses to the Environment, Beck, F. and Lloyd, J. B., Eds., Academic Press, London, 1976, 223.
173. **Walters, M. N. and Papadimitriou, J. M.,** Phagocytosis: a review, *Crit. Rev. Toxicol.*, 5, 377, 1978.
174. **Ryser, H. J. P.,** Uptake of protein by mammalian cells: an underdeveloped area, *Science*, 159, 390, 1968.
175. **Sbarra, A. J., Shirley, W., and Bardawil, W. A.,** Piggyback phagocytosis, *Nature (London)*, 194, 255, 1962.
176. **Widder, K., Flouret, G., and Senyei, A.,** Magnetic microspheres: synthesis of a novel parenteral drug carrier, *J. Pharm. Sci.*, 68, 79, 1979.
177. **Widder, K. J., Senyei, A. E., and Ranney, D. F.,** Magnetically responsive microspheres and other carriers for the biophysical targeting of antitumor agents, *Adv. Pharmacol. Chemother.*, 16, 213, 1979.
178. **Folkman, J., Reiling, W., and Williams, G.,** Chronic analgesia by silicone rubber diffusion, *Surgery*, 66, 194, 1969.
179. **Brand, K. G., Johnson, K. H., and Buoen, L. C.,** Foreign body tumorigenesis, *Crit. Rev. Toxicol.*, 4, 353, 1976.
180. **Malaga, B. F., Yasenchak, L. P., and Salthouse, T. N.,** Tissue response to implanted polymers: the significance of sample shape, *J. Biomed. Mater. Res.*, 10, 391, 1976.
181. **Sprincl, L., Kopecek, J., and Lim, D.,** Effect of porosity of heterogeneous poly(glycol monomethacrylate) gels on the healing-in of test implants, *J. Biomed. Mater. Res.*, 4, 447, 1971.
182. **Sprincl, L., Kopecek, J., and Lim, D.,** Effect of the structure of poly(glycol monomethacrylate) gel on the calcification of implants, *Calcif. Tissue Res.*, 13, 63, 1973.
183. **Bagnall, R. D.,** Implant biocompatibility, *Biomaterials*, 1, 97, 1980.
184. **Beugeling, P. E., Bantjes, A., and Kolar, Z.,** Adsorption of a synthetic heparinoid polyelectrolyte on anion-exchanging surface, *J. Colloid Interface Sci.*, 62, 35, 1977.
185. **Regelson, W.,** The biologic activity of polyanions: past history and new perspectives, *J. Polym. Sci. Polym. Symp.*, 66, 483, 1979.
186. **Breslow, D. S.,** Divinyl ether-maleic anhydride copolymer: its structure and biological properties, *Am. Chem. Soc. Polym. Prepr.*, 21, 1, 1980.
187. **Muck, K. F., Rolly, H., and Burg, K.,** Herstellung und antivirale Wirksamkeit von Polyacrylsaure und Polymethacrylsaure, *Makromol. Chem.*, 178, 2773, 1977.
188. **Vasko, K. A., Riley, A. M., and DeWall, R. A.,** Poloxalkol (Pluronic F-68): a priming solution for cardiopulmonary bypass, *Trans. Am. Soc. Artif. Intern. Organs*, 18, 526, 1972.
189. **Maggio, B., Ahkong, Q. F., and Lucy, J. A.,** Poly(ethylene glycol), surface potential and cell fusion, *Biochem. J.*, 158, 647, 1976.
190. **Abuchowski, A., McCoy, J. R., Palczuk, N. C., Van Es, T., and Davis, F. F.,** Effect of covalent attachment of poly(ethylene glycol) on immunogenicity and circulating life of bovine liver catalase, *J. Biol. Chem.*, 252, 3582, 1977.
191. **Sundqvist, T., Magnusson, K. E., Sjodahl, R., Stjernstrom, I., and Tagesson, C.,** Passage of molecules through the wall of the gastrointestinal tract. II. Application of low molecular weight poly(ethylene glycol) and a deterministic mathematical model for determining intestinal permeability in man, *Gut*, 21, 208, 1980.

Chapter 9

# MODIFICATIONS TO POLYMER SURFACES TO IMPROVE BLOOD COMPATIBILITY

**Christine Fougnot, Denis Labarre, Jacqueline Jozefonvicz, and Marcel Jozefowicz**

## TABLE OF CONTENTS

I. Introduction ..... 216

II. Blood Coagulation Induced at the Interface with Materials ..... 217
 A. Platelet Activation ..... 218
 B. Intrinsic Pathway ..... 218
 C. Extrinsic Pathway ..... 220
 D. Clot Formation ..... 220
 E. Control Systems ..... 221
  1. Plasmin Formation ..... 221
  2. Action of Enzymatic Inhibitors ..... 221
  3. Antithrombin III and Heparin ..... 222
 F. Conclusion ..... 223

III. Biomaterials with Improved Blood Compatibility ..... 224
 A. Protein Coatings ..... 224
 B. Bioprostheses ..... 225
 C. Hydrogels ..... 225
 D. Discussion ..... 226

IV. Antithrombogenic Coatings ..... 226
 A. Surfaces Containing Platelet Aggregation Inhibitors ..... 227
 B. Surface Coating with Plasminogen Activators ..... 227
 C. Coating with Heparin ..... 227
 D. Discussion ..... 229

V. Heparin-Like Materials ..... 229

VI. General Conclusion ..... 233

References ..... 233

## I. INTRODUCTION

The way of life in highly industrialized countries generates stresses to which people are submitted. As a result some diseases, for instance atherosclerosis, have developed and have become today a major cause of mortality. The healing of those diseases implies the development of implantable devices in order to replace the wounded organs totally or partially. The research in this field started about 30 years ago, leading to the common practice of implantation of prosthetic devices like artificial heart valves and aortic prostheses.

Cardiovascular biomaterials, i.e., materials used to make any device intended for contact with blood, have to function under biological constraints. These include heart valves, arteries and veins, and generally speaking, all blood vessels for long- and short-term use either as implants or as circulatory assistance devices. This means that they have to possess particular properties which are gathered under the generic term of biocompatibility.

Biocompatible biomaterials have to be first nontoxic and noncarcinogenic. They should not induce any cellular or tissue inflammatory response after implantation. They should neither be allergenic nor induce immunological response. Generally speaking, they should not induce any rejection process or diseases in the host after implantation. Moreover, the biocompatibility of a given biomaterial implies that the degradation or released products that might appear after implantation should also be biocompatible.

In the case of biomaterials in contact with blood, the major tissue response induced at the interface is the blood clotting and the resulting thrombosis of the artificial cardiovascular devices. Therefore, during the past decades, research efforts were essentially devoted to the control of thrombus formation at the interface between materials' surfaces and blood.

In the first step, biomaterials used for artificial heart valves and aorta segments were essentially metals and classical technical polymers, such as polyester (Mylar®, Dacron®), polytetrafluoroethylene (Teflon®), silicone rubber (Silastic®), polyamide (Nylon®)[1]. In these cases, the biocompatibility of the devices is due to the high flow rates of the circulating blood which decreases clot formation rate and the adjuvant long-term anticoagulant therapy delivered to the host patients. However, this therapy is the source of long-term complications for the patient.

Moreover, in case of small artery and vein replacement, i.e., those in which internal diameter is about 2 mm, artificial devices have failed even when the patients received adjuvant anticoagulant. Thus, in a second step, research was devoted to a better understanding of the events occurring at the interface between blood and materials when they are placed in contact. In turn, the better knowledge of the induced reactions has allowed one to achieve surface treatments on materials which improve their blood compatibility, i.e., their ability to prevent clot formation at the interface. Essentially two possibilities have been carefully studied. The first approach has attempted to reduce the thrombogenicity of the surface by coating the materials' surfaces with proteins or hydrogels. In the second, surface treatments were achieved in order to obtain anticoagulant surfaces. This was mainly done by:

1. Ionic binding of heparin onto the surfaces
2. Covalent binding of heparin onto polymers
3. Chemical modification of polymer surfaces in order to confer on them heparin-like properties

In the following, after a schematic analysis of blood clotting processes induced at the interface with materials (including induction and control of thrombus formation), we shall analyze the different ways proposed in order to design and tailor blood compatible biomaterials.

**Table 1**
**COMPOSITION OF NORMAL HUMAN BLOOD**

| Cell | Number/mℓ |
|---|---|
| Erythrocytes | |
| Men | $(5.4 \pm 0.8)\ 10^6$ |
| Women | $(4.8 \pm 0.6)\ 10^6$ |
| Leucocytes | 7,500 ± 2,500 |
| Polynuclear neutrophile | 60 ± 8% |
| Polynuclear eosinophile | 2 ± 1% |
| Polynuclear basophile | 0.5 ± 0.25% |
| Lymphocyte | 32 ± 6% |
| Monocyte | 6 ± 2% |
| Platelets | 300,000 ± 50,000 |

**Table 2**
**IONIC COMPOSITION OF NORMAL HUMAN PLASMA**

| Cations | meq/ℓ | Anions | meq/ℓ |
|---|---|---|---|
| $Na^+$ | 142 | $Cl^-$ | 103 |
| $K^+$ | 5 | $HCO_3^-$ | 27 |
| $Ca^{++}$ | 5 | $HPO_4^{2-}$ | 2 |
| $Mg^{++}$ | 3 | $SO_4^{2-}$ | 1 |
| | | Organic acids | 6 |
| | | Proteins | 16 |
| Total | 155 | Total | 155 |

**Table 3**
**PROTEIN CONCENTRATIONS IN NORMAL HUMAN PLASMA**

| Protein | g/ℓ |
|---|---|
| Albumin | 35—55 |
| Fibrinogen | 2—4 |
| Immunoglobulins | |
| IgG | 9.2—14.8 |
| IgA | 1.42—2.62 |
| IgM | 0.88—1.84 |
| Coagulation factors | |
| XII | 0.030 |
| XI | 0.004 |
| X | 0.005 |
| V | 0.020 |
| II | 0.150 |
| Prekallikrein | 0.050 |
| High molecular weight kininogen | 0.070 |

## II. BLOOD COAGULATION INDUCED AT THE INTERFACE WITH MATERIALS

Blood is a suspension of cells in an aqueous solution called plasma (Table 1, 2, and 3). Among the cells, as far as coagulation is concerned, only platelets are of major importance. Blood plasma is an aqueous solution with pH about 7.3. Ions present in solution are mainly chloride, sodium, carbonate, potassium, calcium, and magnesium ions (compare Table 2). Besides the salts, lipids and phospholipids, polysaccharides and proteins or glycoproteins are present at low or very low concentrations.

In normal native blood, there are no or very low enzymatic activities available. With regard to their biological activity, the proteins and glycoproteins are different. Some of them, zymogens, are able to undergo an activation process which results in enzyme formation; this is, for instance, the case of the so-called coagulation factors, except fibrinogen.

Others, like fibrinogen, for instance, play the role of enzyme substrates, i.e., are able to undergo an hydrolysis reaction catalyzed enzymatically. A third category consists of enzymes inhibitors (Table 4) which are able to form specifically inactive complexes with given enzymes.

Finally, plasma contains proteins and glycoproteins which are neither zymogens nor substrates or inhibitors. They are typically represented by albumin and immunoglobulins.

**Table 4**
**CONCENTRATION OF SOME COAGULATION INHIBITORS IN NORMAL HUMAN PLASMA**

| Inhibitor | g/ℓ |
|---|---|
| $\alpha_1$-Antitrypsin | 2—4 |
| $\alpha_2$-Antiplasmin | 0.085 |
| $C_1$-Inactivator | 0.250 |
| $\alpha_1$-Antichymotrypsin | 0.480 |
| Inter-α-trypsin inhibitor | 0.500 |
| $\alpha_2$-Macroglobulin | 3 |
| Antithrombin III | 0.150—0.250 |

Their biological activities are essential, but they seem to be of less importance as far as the coagulation process is concerned.

When blood is placed in contact with any foreign, natural, or artificial surface, excluding the normal natural endothelium of a healthy natural blood vessel, the following sequence of events occurs. First, in a short time, 1 min or less, a rapid competitive adsorption of proteins or glycoproteins onto the surface occurs. Depending on the nature of the proteins and of the surface, this competition results in a complex protein coating on the surface. Some of these adsorption processes appear to be partially or completely reversible.[2-4] The composition of the protein coating is a function of a large number of parameters:

1. Protein-surface affinities
2. Adsorption kinetics
3. Flow rate of the blood through the vessel
4. Reversibility of the protein adsorption
5. Relative abundance of the proteins in the blood plasma

Furthermore, the nature of the protein coating seems to influence the blood coagulation process. This particular point will be discussed later on. Depending on the nature of the materials, some of the deposited proteins may initiate different so-called coagulation pathways (Figure 1).

### A. Platelet Activation

First, a protein called von Willebrand factor (a polymer of factor VIII) mediates platelet attachment onto the surface. Then, platelets undergo aggregation and release, i.e., some internal platelet constituents are released into the plasma or made available. These transformations are under control of prostaglandins, calcium ions, cyclic adenyl monophosphate (cyclic AMP), and adenyl diphosphate (ADP) concentration levels in the activated platelets[5] (Figure 2). These transformations are essential in that they make available both specific proteins and platelet membrane phospholipids (platelet factor 3) which are essential for acceleration and control of the coagulation process.[6]

It is noteworthy that platelet activation is also initiated by thrombin, ADP, and other reactants, as, for instance, it can also be produced by turbulences of the blood flow and is a function of the flow rate. On the reverse, the platelet activation sequence is under the control of inhibitors (acetyl salicylic acid, for instance), the most potent of them being prostacyclin ($PGI_2$).[5]

### B. Intrinsic Pathway

When artificial surfaces, at least negatively charged ones, are placed in contact with blood,

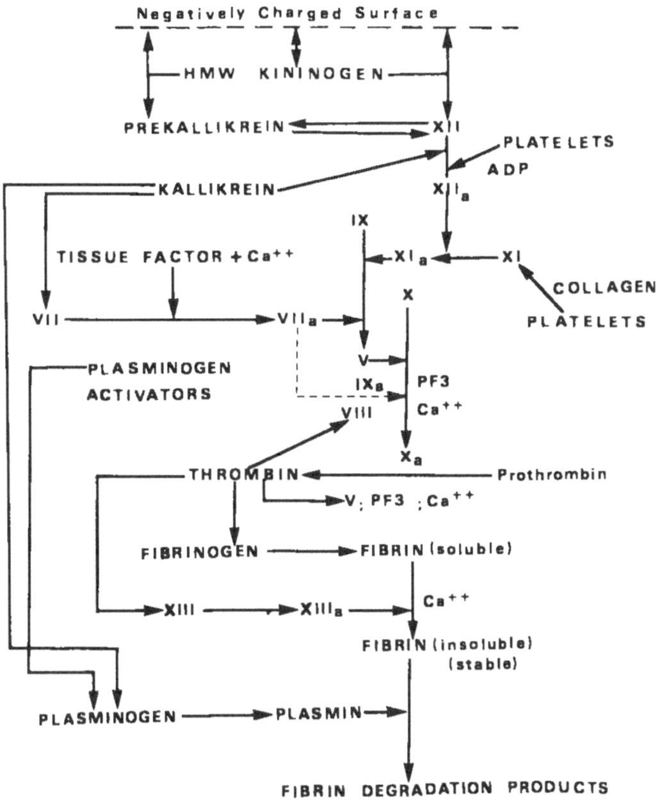


FIGURE 1. The coagulation cascade and its interrelationships. (From Fareed, J., Messmore, H. L., Fenton, J. W., II, and Brinkhovs, K. M., *Perspectives in Hemostasis*, Pergamon Press, Elmsford, N.Y., 1981, 6. With permission.)

the initial protein coating contains more or less two proteins, high molecular weight kininogen (HMWK) and factor XII. Their adsorption induces a sequence of enzymatic reactions[7] (Figure 3), so-called contact activation phase. First prekallikrein forms a complex with bound HMWK. This complex has a low specific enzymatic activity, i.e., is responsible for the cleavage of factor XII into factor $XII_a$. In turn factor $XII_a$ is a potent enzyme able to catalyze the hydrolysis of prekallikrein engaged in the HMWK complex into a potent protease kallikrein. This latter enzyme as shown on Figure 3 activates factor XII. As a result, a cycle of enzymatic reactions occurs and a short time after blood exposure to the artificial surface, a large amount of factor $XII_a$ is available on the surface or in its vicinity.

It is noteworthy that kallikrein has other biological activities than the one described above. Its action is essential in the activation of the kinin system, principally through the formation of bradykinin which controls the circulatory blood pressure.

HMWK is also able to form a complex with factor XI. An enzymatic reaction involving factor $XII_a$ is able to cleave this complex leading to an enzymatic species called factor $XI_a$[7] (Figure 4).

The latter enzyme is able to activate factor IX which is its preferred substrate. The product factor $IX_a$ is able to catalyze enzymatically the transformation of factor X in its enzymatic active form (factor $X_a$). The rate of the latter reaction is slow. It is calcium dependent, i.e., the rate is slowed down when plasma is calcium depleted. The rate is increased when platelet factor 3 and factor $VIII_a$ (a very potent catalyst) are present.

In turn factor $X_a$, by a similar mechanism in presence of platelet factor 3, calcium ions,

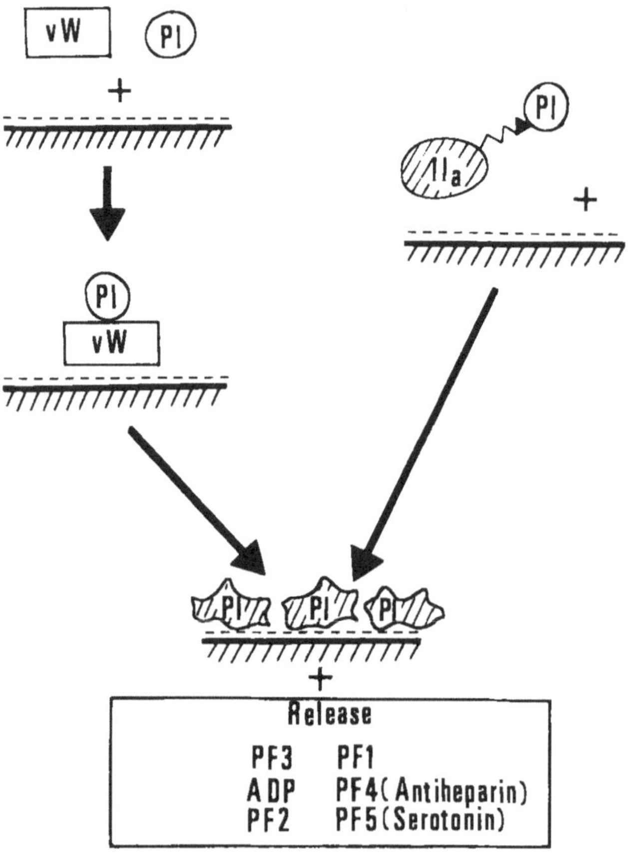


FIGURE 2. Activation of platelets (Pl). (From Griffin, J. H., Scripps Clinic, La Jolla, Calif. With permission.)

and factor $V_a$ (the cocatalyst), catalyzes the enzymatic transformation of factor II (prothrombin) into thrombin (factor $II_a$).

### C. Extrinsic Pathway

There is an alternative pathway for the activation of blood coagulation called "the extrinsic pathway". This one is involved when blood contacts natural surfaces excluding the normal vessel endothelium. In presence of the so-called tissue factor (or factor III) or tissue thromboplastin and calcium ions, factor VII is activated and the resulting enzymatic activity causes factors IX and X activation. Then, thrombin formation occurs as in the intrinsic pathway (Figure 4).

### D. Clot Formation

Thus, it appears that when materials are in contact with blood, a cascade of enzymatic reactions leads, whatever the pathways, to the formation of thrombin which might be the central figure in the coagulation of blood and plays numerous roles. First of all, it is able to induce platelet aggregation and release. This enzyme has the possibility to catalyze its own formation. When small amounts of thrombin become available, increasing quantities of it will be readily formed. Moreover, thrombin is mainly responsible for activation of factor XIII.

It is also necessary to activate fibrinogen in an hydrolysis reaction which leads to fibrinopeptides A and B and the soluble fibrin polymer. The cross-linking of soluble fibrin which

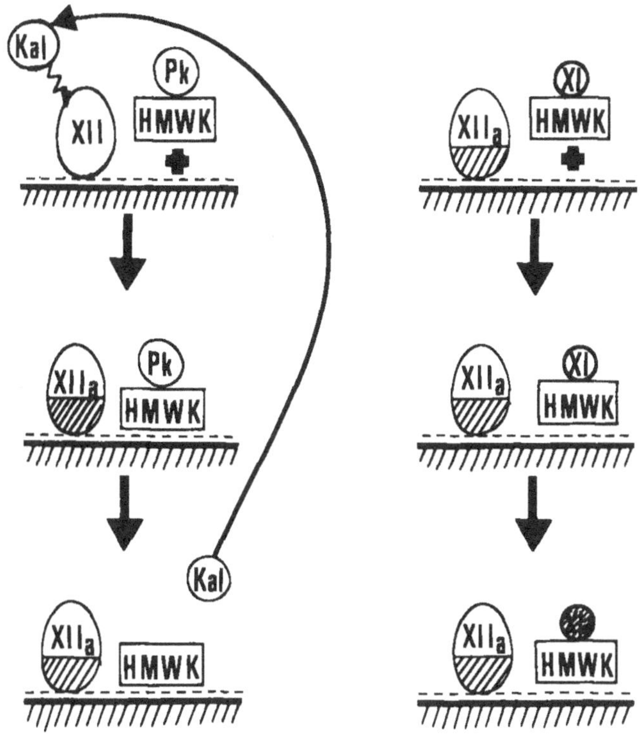


FIGURE 3. Phase contact. (From Griffin, J. H., Scripps Clinic, La Jolla, Calif. With permission.)

occurs is catalyzed by factor XIII forming the insoluble fibrin clot. It is noteworthy that aggregated platelets, other blood cells, and plasma proteins are entrapped in the cross-linked fibrin network.

## E. Control Systems

It has to be emphasized that the clot formation is self-controlled by the blood. If not, any contact between blood and any natural or artificial foreign surfaces would result in general thrombosis of the blood circulation system. As it is not the case, it is obvious that blood contains some control systems. Moreover, the normal healthy vessel endothelium should be able to prevent the coagulation system activation.

This is really the case as shown by Weksler et al.,[8] who proved that endothelium cells are able to release prostacyclin ($PGI_2$), the most potent known inhibitor of platelet aggregation. The question whether endothelium cells may act to prevent coagulation by other ways remains today completely open. On the other hand, blood appears to be able to control the clot formation by different numerous and sophisticated ways.

### *1. Plasmin Formation*

First activation of plasminogen occurs under the action of factor $XII_a$, leading to plasmin formation. This latter protein has an enzymatic activity and catalyzes a sequence of reactions which result in fibrinolysis and subsequent destruction of the fibrin clots. It is of interest to note that some foreign proteins, as, for instance, urokinase, kinase, and streptokinase, have an enzymatic ability to activate plasminogen and to promote plasmin formation. It is interesting that with enzymes being entrapped in the fibrin cross-linked network, the fibrin clot has been considered as one of the enzymatic inhibiting coagulation system.

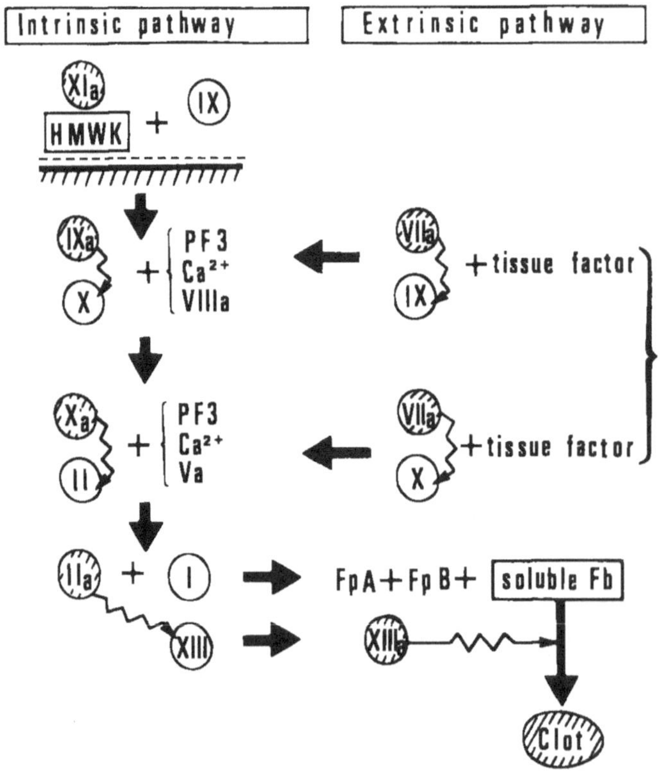


FIGURE 4. Coagulation cascade. (From Griffin, J. H., Scripps Clinic, La Jolla, Calif. With permission.)

### 2. *Action of Enzymatic Inhibitors*

Second, enzyme inhibitors are always present in blood plasma in relatively large amounts. These inhibitors are proteins (or glycoproteins) and are able to react specifically with given enzymes, thus forming inactive stable complexes. This is the case, for instance, of $\alpha_1$-antitrypsin, $\alpha_2$-macroglobulin, $C_1$-inactivator, $\alpha_2$-antiplasmin, and antithrombin III. The action of the plasmatic enzymatic inhibitors is generally slow when compared to the blood clotting time.

### 3. *Antithrombin III and Heparin*

The most potent enzyme inhibitor, as far as the coagulation of blood is concerned, appears to be antithrombin III ($AT_{III}$) so called because of its ability to form an inactive stable complex with thrombin through the irreversible reaction:

$$AT_{III} + T \rightarrow T - AT_{III}$$

The exact nature of the bonds involved in the complex structure is not known. However, it has been proved that the serine residue of the thrombin active site and an arginine residue of $AT_{III}$ are involved in the complex formation.[9-11]

The reaction is normally slow. However, heparin, a natural mucopolysaccharide, is able to catalyze this reaction, enhancing tremendously the reaction rate. Heparin (Figure 5) is a sequenced polymer in which monomeric units are essentially iduronic and glucuronic acids, *N*-acetyl and *N*-sulfate glucosamines.

Numerous studies have been devoted to the mechanism of the catalytic action of hepar-

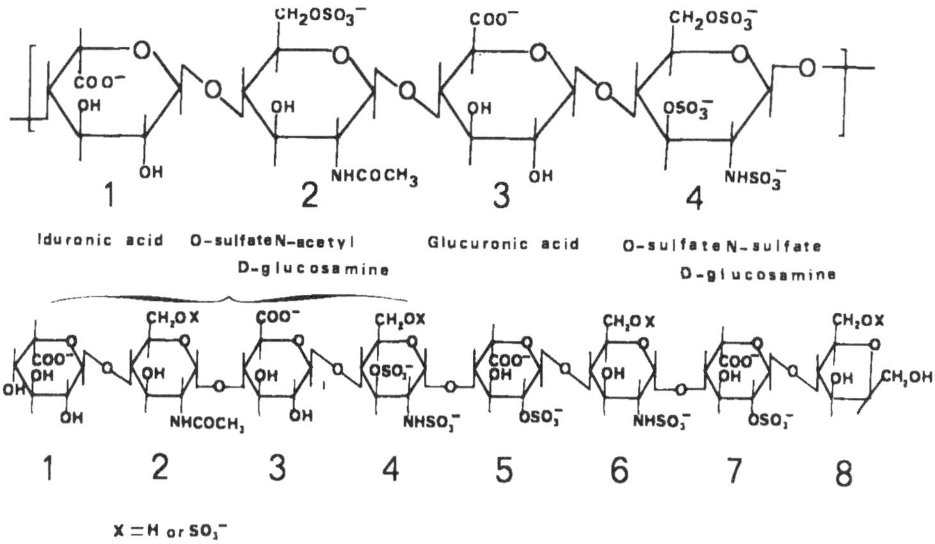


FIGURE 5. Tentative structures for the antithrombin-binding tetrasaccharide or octasaccharide derived from heparin.[15,17,18]

in.[12-18] According to Rosenberg, et al.,[12,14] the catalysis mainly consists in the reversible formation of a complex between heparin and $AT_{III}$ (Figure 6). It has been shown that the complex formation implies a site to site interaction between the protein and the polysaccharides. It is believed that this interaction involves the carboxylic functional groups of the heparin active site[19,20] and a lysin residue of the $AT_{III}$ macromolecule. According to this proposed mechanism, the heparin-antithrombin complex undergoes a rapid irreversible reaction with thrombin.

$$H \ldots AT + T \rightarrow AT\text{-}T + H$$

It should be emphasized that $AT_{III}$ is able to inhibit not only thrombin, but also most of the serine proteases involved in the blood clotting process, namely, factors $X_a$, $IX_a$, $XI_a$, $XII_a$, and plasmin. Moreover, it appears that these reactions are catalyzed to some extent by heparin and that the same model accounts for the mechanism of the observed reactions.[21-23]

This model fits well with the observed reaction kinetics and with the results of equilibrium studies performed on purified systems. Due to the fact that multiple interactions may compete, this model is too simple when control of whole blood or plasma coagulation is under consideration. Nevertheless, it has been used as a useful tool by biomaterials scientists involved in preventing thrombosis formation at the interface between blood and materials.

## F. Conclusion

Clot formation appears to be the normal consequence of the contact between blood and any foreign surface except that of the normal healthy vessel endothelium. It is initiated by the contact itself and depending on the nature of the surface, it may develop through different pathways, but the final result is thrombin formation which might well be the central event in coagulation.

Blood itself (and the natural endothelium of the vessel wall) is able to control the coagulation process. One of the potent control systems is the heparin catalysis of the antithrombin III-coagulation serin proteases complex formation reaction.

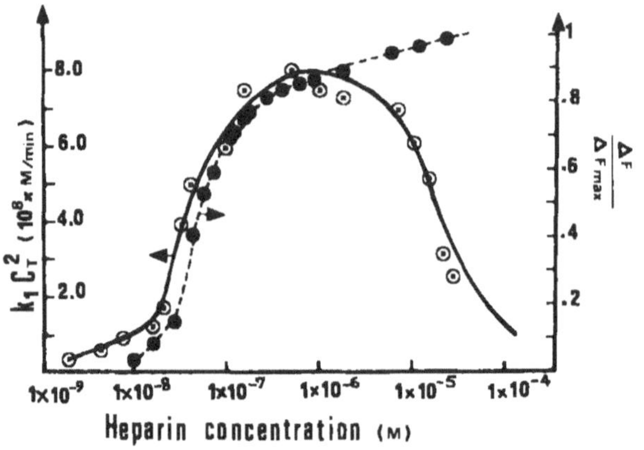


$$\frac{dT}{dt} = k[HAT][T]$$

$$k^{HAT}_{Diss.} = \frac{[H][AT]}{[HAT]} \qquad k^{HT}_{Diss.} = \frac{[T][H]}{[TH]}$$

$$C_{AT} = [AT]+[HAT]+[TAT]$$

$$C_T = [T]+[HT]+[TAT]$$

$$C_H = [H]+[HT]+[HAT]$$

$$k^{H\text{-}TAT}_{Diss} = 0$$

$$-\frac{dT}{dt} = k(1-c)[T]^2$$

$$\frac{1}{[T]} = -k_1 t - \frac{1}{C_T} \qquad v = k_1 C_T^2 \text{ (initial velocity)}$$

FIGURE 6. The extent of binding of heparin species to antithrombin. (From Jordon, R., Beeler, D., and Rosenberg, R., *J. Biol. Chem.*, 254(8), 2902, 1979. With permission.)

It has to be emphasized that coagulation of blood is a complicated process. Its kinetics cannot completely be described by the models schematically mentioned above, the kinetics of the clot formation being a complex combination of each of the involved reactions. Moreover, these kinetics are strongly dependent on the flow rate of the circulating blood. This adds to the multiparameter system which controls the blood coagulation, one more parameter which is able to modify to some extent the whole feature of the clot formation process.

## III. BIOMATERIALS WITH IMPROVED BLOOD COMPATIBILITY

### A. Protein Coatings

At the beginning of the two past decades, with the progress in cardiovascular surgery, many prosthetic devices were proposed and developed. It was generally assumed that the nature of the protein coatings which occurs when blood contacts a material surface determines its blood compatibility. Thus, there was considerable interest in developing the identification of the protein deposits and determining their relative amounts and rates of adsorption.[2-4] It should be pointed out that a great majority of studies were made on the most abundant plasma proteins, namely, albumin, fibrinogen, and γ-globulin. From consideration of these proteins, the studies led to the conclusion that when hydrophobic materials are concerned, the coating is essentially made of native proteins. Their relative amounts were slightly dependent on the chemical nature of the materials.[3] Looking for correlations between blood compatibility and the nature of the protein deposit at the blood-material interface, some authors postulated that the amount of native albumin deposited on the material surface is essential to improve its blood compatibility.[24-28] Based on this hypothesis, a great number of surface treatments were developed in order to improve the blood compatibility of cardiovascular prosthetic devices. In such treatments, the surface is activated generally by using glutaraldehyde[25,27] and treated with albumin, whose amine functional groups are involved in the linkage formation. In some cases, cross-linking of the protein coating is achieved by using glutaraldehyde in order to improve the protein coating stability.[27]

Different materials have been treated in such ways, including glass and polystyrene,[26] silicone rubber,[27] anionic polyether urethane lattices,[28] cuprophane,[29] and polyester (Dacron®).[30]

Those surfaces with albumin coating reduce platelet adhesion onto the materials,[25,26] but still have a significant tendency to initiate thrombosis in vivo.[24,25,28,30] However, today, no prosthetic devices of practical uses based on this principle have been proved to be suitable to reach the market.

### B. Bioprostheses

Other polymers have been developed as prostheses for a number of cardiovascular devices and have demonstrated significant thromboresistance. Thus, Carpentier et al.[31] have prepared cardiac valves with tissue collagen and glycoprotein extracts cross-linked with glutaraldehyde. The authors have reported a high degree of blood biocompatibility after the implantation of these valves. Such valves avoid the need for an adjuvant permanent anticoagulant therapy for the patient.

Unfortunately, the mechanical properties of the prosthesis change with time so that their average lifetime is only about 5 years after implantation. Moreover, biopolymers have failed when used for the construction of small internal diameter tubing for vascular prostheses.[32] Indeed, in vivo tests performed on animals showed that thrombosis occurs with a significant probability in the case of small vessels prepared with human umbilical collagen cross-linked with glutaraldehyde.

### C. Hydrogels

One approach to the preparation of blood-compatible surfaces involves hydrogels. However, it appears that the thromboresistance of such materials is rather poor. Usually, it was suggested that the performance of hydrogels depends on hydrophilicity and the existence of an organized water structure. Therefore a large number of studies were performed where a grafting technique was developed to improve the hydrophilicity of hydrophobic materials such as polyurethanes. The technique permits treatment of polymer surfaces by grafting hydrogels onto the surface.

Biomaterials of these types have been synthesized by grafting of 2-hydroxymethylmethacrylate (HEMA)[33,34] or HEMA and ethylmethacrylate copolymers[35,36] onto silicone rubber or polystyrene by using radiochemical or chemical methods.

In other cases, grafting of 2-hydroxyethylmethacrylate onto silicone rubber or polyurethane tubing surfaces was induced by cerium (IV) peroxidation.[37] A method of coating with an acrylonitrile and dimethylaminoethyl methacrylate hydrogel was also performed by other authors.[38]

Later on, ionic 2-hydroxyethylmethacrylate-methacrylic acid copolymers[39] and stereoregular poly(hydroxyethylmethacrylate) hydrogels[40] were synthesized. The properties of these biomaterials have been extensively studied. The authors suggest that the amount of blood cell adhesion onto the hydrophilic surfaces is fairly low.[41] The thrombogenicity is decreased[42,43] when compared to that of the nontreated hydrophobic material surfaces. Such results are observed whatever the starting material, silicone rubber[43] or polypropylene,[44] for instance. Moreover, it was claimed that the balance between hydrophilicity and hydrophobicity of microphase structures on the surface controls the inhibition of platelet adhesion and aggregation onto the surface. Platelet adhesion and aggregation especially should be suppressed when films made of block PHEMA-polystyrene or polydimethylsiloxane copolymers presented microphase separated structures of alternate lamellae about 300 to 500 Å width.[36] Similar results are claimed by the authors in the case of polystyrene-polyamine comb-type copolymers.[45]

The hydrogels have been proved to induce a tissue response varying with the acidic, basic, or neutral nature. It appears that some weeks after implantation, hydrogels are progressively encapsulated into fibrous collagenic structures.[46,47] Due to their properties, hydrogels have seemed to be suitable for the construction of implantable prosthetic cardiovascular devices. These possibilities have been reviewed extensively.[48-52]

### D. Discussion

As mentioned above, in the first minutes after the blood contact with any polymeric surface, protein adsorption occurs. The composition of the protein coating is essentially variable, but it was claimed that it is essential to determine the thrombogenicity of the biomaterial's surface. Unfortunately, for sake of simplicity, only the most abundant plasma protein adsorption, i.e., albumin, fibrinogen, and γ-globulins, was extensively studied.

Despite the numerous studies devoted to the subject, no correlation could be firmly established between the ability of a surface to bind one of these proteins and its thrombogenicity, and no general theory was found to account for it. This is probably due to the fact that, as seen above, the thrombus formation at the blood-material interface is not directly induced either by albumin or by the other two proteins, but by other plasmatic factors, as, for instance, HMWK or von Willebrandt factor.

However, the adsorption of the major plasmatic components onto the surfaces competes clearly with the latter ones and influences their adsorption and as a consequence, the activation of the blood coagulation initiation at the interface. This influence is not a simple direct correlation, but rather a multiparameter system in which the activation of the blood coagulation is a complex function of the relative affinities of the proteins towards the surface, the kinetic of their deposition, the plasma and surface composition, the flow rate of the blood in the vessel, the tissue, etc. Therefore, it is not surprising that satisfactory theories to explain the thrombogenic properties of biomaterials have not yet been developed.

On the other hand, despite this failure, the carefully performed experiments have empirically shown that it is possible for some at present unknown reasons to improve the blood compatibility of materials by use of surface treatments such as bonding of albumin onto the surface or making them more hydrophilic. Further studies performed in the future in order to make clear the exact mechanisms of the competitions in protein adsorption at the interface, including those of the coagulation factors, will probably allow development of biomaterials with improved blood compatibility. However, this general aim may be also reached by other ways, i.e., the search for antithrombogenic materials. In contrast to the approach described above, this latter one consists of synthesizing materials which will inhibit the coagulation process. This can be achieved either by covering a given material with an antithrombogenic coating or by synthesizing materials which intrinsically possess anticoagulant properties.

## IV. ANTITHROMBOGENIC COATINGS

A large amount of work has been devoted during the past 20 years to the surface modification performed to deposit antithrombogenic coatings onto appropriate polymeric substrates. The general idea involved was that binding on a surface of substances known to inhibit in solution the development of a thrombus, or to lyse it, could be the best way to obtain antithrombogenic surfaces. So, the most popular strategy has been coating of surfaces with heparin.[53-56]

### A. Surfaces Containing Platelet Aggregation Inhibitors

Another approach to preparation of antithrombogenic surfaces has been immobilizing platelet-aggregation inhibitors. The works of Marconi et al.,[57] MacRea and Kim,[58] and Grode et al.[59] illustrate this point of view.

The former immobilized ditazole and dipyridamole which are known as antiaggregating platelet drugs on cellulose or nylon. The author observed that after implantation in the canine femoral vein, these surfaces were free from thrombus. The latter incorporated and immobilized prostaglandins on surfaces of polymers such as polyvinylchloride and polyetherurethane and used those surfaces as drug delivery systems in extracorporeal circuits.

### B. Surface Coating with Plasminogen Activators

As seen above, activation of plasminogen results in plasmin, an enzyme able to lyse the clots. Therefore a thrombus lysing surface can be achieved by fixing plasminogen activators such as urokinase or streptokinase onto surfaces. This was achieved by Sugitachi,[60] who fixed these enzymes onto polyvinylchloride and silicone rubber tubes. These surface treatments improve significantly the antithrombogenic surface properties as reported from a large number of tests.

### C. Coating With Heparin

As seen above (Section II), heparin is an anionic polyelectrolyte (more precisely a sulfated mucopolysaccharide) which acts as a catalyst, accelerating the inhibition of thrombin and similar proteases by antithrombin III.[61,62] It is noticeable, but not yet certain, that heparin is a platelet aggregating drug, this activity (as the catalytic one) being a function of its macromolecular size and ability to bind $AT_{III}$.[63]

The strong anionic character of heparin can be used to bind it ionically to cationic surfaces as reviewed by several authors. The first report of surface-attached heparin was by Gott and co-workers in 1963;[64] a graphite surface was created on a rigid polymer and a quaternary ammonium salt, i.e., benzalkonium chloride, was adsorbed on the graphite. After complexing by heparin, a thromboresistant surface, namely, graphite-benzalkonium-heparin (GBH), was obtained. Using chloromethylated polystyrene, Leininger et al.[65] also prepared heparinized surfaces after treatment of the modified polystyrene by *N,N*-dimethyl aniline, quaternization and soaking in a heparin solution. By treatment of cellulose membranes with ethylene imine and subsequent treatment by heparin, Britton[66] prepared antithrombogenic cellulose films and tubings. However, none of these coatings could be applied to flexible materials without further modification. Imai and Masuhara[67] and Uy and Kammermeyer[68] adapted the GBH treatment to flexible materials, and Grode[69] developed modified polystyrene analogs (by radiation grafting of 4-vinylpyridine or dimethylaminoethylmethacrylate to the surface) and tridodecylmethylammonium chloride treatment (TDMAC) on several polymers. This latter method was successfully used for preparing a shunt for aneurysmectomy. Another route to heparinization of silicone rubber was the quaternization of the silica filler by reacting with δ-aminopropyltriethoxysilane (APTES) followed by treatment with heparin, as described by Merker et al.[70]

Heparin was also incorporated in epoxy resins and polyurethanes by Salyer et al.[71] or complexed as GBH and associated with urokinase by Kusserow et al.[72] or complexed with methyltricaprylammonium chloride (MTCAC) by Rea et al.[73] Other authors[74] simultaneously incorporated heparin and prostaglandins in order to combine the respective thrombin and platelet aggregation inhibiting effects of the release drugs. Heparin was also incorporated either in cellulose membranes[75] or elastomers[76] by use of ammonium binding sites. These materials show slow heparin releases and have proved to be efficient for short-term uses (days or weeks). Finally, heparin was incorporated in cross-linked collagen tubings of small diameter. This incorporation was claimed to improve the blood compatibility of the devices.[32]

An alternate route was to prepare copolymers containing aminated units capable of being quaternized. This was described by several authors: Yen and Rembaum[77] and Pusineri[78] with polyurethanes, Idezuki et al.[79] with a graft copolymer (HRSD), Courtney et al.[80] with an acrylonitrile-dimethylaminoethyl methacrylate copolymer, Ferruti and Provenzale [81] with polyamidoamine-styrene, and Paik Sung et al.[82] with polymethylacrylate-*N*-acrylhexamethylene-diamine.

The common feature for all of these ionic coatings is a leaching of heparin when placed in contact with blood, the wall acting as a heparin reservoir. So, they are suitable only for short-term uses.

In order to stabilize this heparin layer ionically bonded on a surface, Lagergren and Eriksson[83] proposed an additional treatment with glutardialdehyde. Further experiments with such a surface using $^{35}S$-labeled heparin demonstrated a satisfactory stability of the heparin layer[84] and a still good antithrombogenicity.[85]

While these works on ionically bonded heparin were developed, a few attempts to obtain covalently bonded surfaces were made; Halpern and Shibakawa[86] modified polystyrene, introducing isocyanate groups able to react with amino or hydroxyl groups of heparin. Merrill et al.[87] prepared a hydrogel in which heparin was bonded to poly(vinyl alcohol) through acetal bridges. Iverius,[88] Schmer,[89] and Danishefsky and Tzeng[90] developed methods to obtain gels in which heparin was bonded to agarose. They activated or modified agarose either by cyanogen bromide or by thiophosgene or carbodiimide. Hoffman et al.[91] used a similar technique to bind heparin on the surface of a hydrogel radiation-grafted on a silicone elastomer. Chawla and Chang[92] prepared hemodialysis cellulose membranes radiation-grafted with heparin. Falb et al.[93] also prepared aminated silicone rubber that they reacted with a cyanuric chloride-heparin adduct.

While the gels were very promising, with regards to their biological activities, they could not be used for purposes of implantation because of their lack of mechanical resistance. However, concerning the biological activities of covalently bonded heparin solid surfaces, the results were disappointing. The disappearance of the activity of heparin bonded on such surfaces was generally attributed either to a lack of mobility of the heparin chain or to the masking of the active site dramatically reducing its ability to interact with antithrombin III.

Nevertheless, an increasing number of papers were published describing new methods either to bind heparin in order to preserve anticoagulant activity or to reinforce the mechanical properties of the gels. Labarre et al.[94] described the preparation of heparin-poly(methylmethacrylate) copolymers. The heparin molecule, used as the initiator of polymerization after reaction with cerium IV salts, was located at one end of the resulting copolymer. Baquey et al.[95] described a similar method, using the macroradicals resulting from the action of $\gamma$-rays on heparin as initiator of polymerization of acrylic acid. Mester et al.[96] and Plate et al.[97] proposed methods to acylate heparin, the former using methacryloyl chloride and the latter using acrylic chloranhydride. The resulting chains of heparin, bearing in average one double bond, were able to copolymerize with acrylic monomers, the copolymers possessing a high anticoagulant activity. A Japanese patent by Nakashima and Takakura[98] proposed the attachment of heparin on poly(hydroxyethylmethacrylate-glycidylmethacrylate) copolymer. By modifying the surface of an elastomer, i.e., poly(methylacrylate), Dincer[99] created carboxylated or aminated surfaces able to be coupled to different types of heparin by use of either a hydrosoluble carbodiimide or glutardialdehyde. After reinforcing a poly(vinyl alcohol) gel by $\gamma$-ray treatment, Peppas and Merrill[100] bonded heparin on its surface through glutardialdehyde treatment. Goosen and Sefton[101] used the same cross-linking agent, but started from an elastomer, i.e., styrene-butadiene-styrene. After hydroxylation of the surface through peracetic acid treatment, poly(vinyl alcohol) and heparin were bonded.

The properties and activity of covalently bonded heparin surfaces were carefully studied. Unlike the ionically bonded heparin surfaces, they will not release heparin. Nevertheless, Labarre et al.[102] and Boffa et al.[103] demonstrated that when active the mode of action of heparin-poly(methylmethacrylate) insoluble copolymers is similar to that of soluble heparin.

These copolymers, as all the others above described in which heparin is covalently bonded, are able to form a complex with antithrombin III. However, as shown by Salzman et al.,[104] though having this ability, some of them are unable to catalyze the thrombin inhibiting reaction. Thus, it appears that, depending on the reaction involved in binding covalently heparin to a polymer, the resulting material may or not have an antithrombogenic activity.

On the other hand, Lindon et al.[105,106] examined the interactions between platelets and heparin-coated surfaces. The authors concluded that platelet adhesion and aggregation occur on each of these surfaces, but that those which have a high ability to bind antithrombin III and to catalyze the inhibition reaction of thrombin with antithrombin III are passivated towards platelets when treated by a preincubation with plasma or with antithrombin III aqueous solutions. Similar observations were made later on by Olsson et al.[85] in the case of a glutardialdehyde-stabilized heparin-coated surface.

On this basis, Salzmann et al.[104] and Larsson et al.[107] suggested that artificial materials with improved thromboresistant properties may be prepared by heparin coating of a material in such a way that the ability of heparin to bind and activate antithrombin III should be preserved. According to the same authors,[63,107] achievement of the goal should be facilitated by the use of low molecular weight heparin with high affinity for antithrombin III which minimizes the unfavorable activation of platelets induced by the commercially available heparin.

### D. Discussion

Antithrombogenic coatings, whatever the involved drug, are generally efficient, i.e., allow one to generate surfaces with activity which prevents thrombus formation. However, their use is somewhat limited. First, the drugs used are generally expensive, this being a limitation for large-sale manufacture of cardiovascular artificial devices.

Moreover, when released, the drugs used may be hazardous. This is particularly the case for the plasminogen activators, like streptokinase and urokinase, and makes the use of surface treatment employing these drugs difficult.

In some cases, the drugs used to make the coating are unstable either in the short or the long term. This is the case for prostacyclin whose biological half-time when soluble is less than 1 min. Even in the case of covalently bonded heparin surfaces, the anticoagulant activity decreases slowly after several months, probably because of the action of either platelets, factor 4 (antiheparin), or plasmatic hypothetic enzymes such as heparinase. For these reasons, prostacyclin-treated surfaces have not yet been employed for practical purposes. Heparin-coated surfaces, ionically or covalently bonded, might be used for short-term devices such as catheters. In that case, the former act as heparin reservoirs and the latter act as anticoagulant surfaces. They are unsuitable for long-term practical uses.

This review of the work done to achieve efficient antithrombogenic coatings shows indeed that it remains a promising field, but even in the case of heparin coatings which appears to be the most exciting, difficulties remain. This is probably the reason why people started early to synthesize anticoagulant and heparin-like materials.

## V. HEPARIN-LIKE MATERIALS

Natural heparin and heparin analogs of lower anticoagulant activity such as heparan or chondroitin-6 sulfate are polyanionic derivatives of polysaccharides. Thus, it is not surprising that a number of efforts were devoted to the preparation of synthetic polyanions having antithrombogenic or heparin-like properties.

As early as 1951, Lovelock and Porterfield[108] described the anticoagulant properties of polystyrene sulfonates, reacting a polystyrene tubing with sulfuric acid. Then they observed that the blood clotting time was longer (41 min) in the treated tubing than in the starting one (29 min) or in a glass tubing (9 min). Since that time, the anticoagulant properties of sulfonated soluble polymers, specially polystyrene and polyethylene sulfonates, have been described by Gregor.[109]

Furthermore Machovich[110] has shown that polymethacrylic acid has an anticoagulant activity in aqueous solutions. He demonstrated that a sample of this polymer at the concentration of 2 μg/mℓ increases by a factor of five the rate of the thrombin-antithrombin III reaction.

Later on, anticoagulant materials were obtained by reaction of *N*-chlorosulfonyl isocyanate with an unsaturated polymer followed by a suitable basic hydrolysis.[111-113] For instance, a soluble polyelectrolyte derived from 1,4-*cis* polyisoprene:

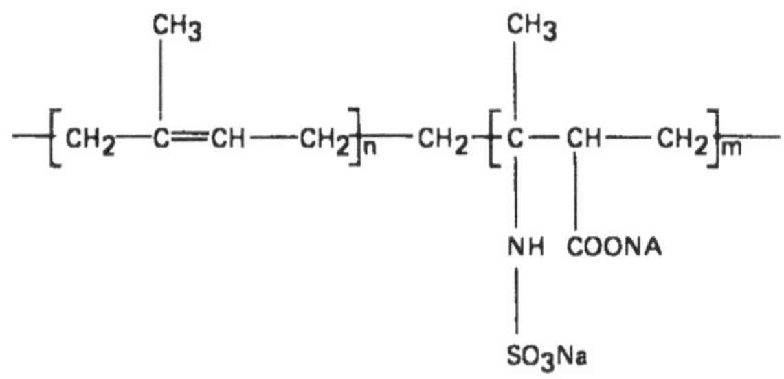


It was shown to possess an anticoagulant activity which was 10% of that of heparin.

Such compounds were used to make ionic coatings covering thrombogenic hydrophobic surfaces of polyvinylchloride, for instance, by use of a coupling agent, the tridodecylmethyl ammonium chloride (TDMAC).[114] These coatings resulted in a drastic decrease of the platelet activation (adhesion and/or aggregation) on the surfaces. It appears to be stable since the polyelectrolyte seems to be not or only slightly eluted in presence of plasma or of an aqueous 0.15 *M* sodium chloride solution.

Other methods were employed by the same authors in order to coat insoluble substrates with the same anticoagulant materials.[112] In a first approach, a block styrene-isoprene copolymer is reacted with *N*-chlorosulfonyl isocyanate. In the second approach, the soluble polyelectrolyte is partially reacted with a diol, the resulting cross-linked network being insoluble.

Whatever the approach, these insoluble materials either in film or powder form show a rather high anticoagulant activity. However, tests performed on plasma suspensions indicate that some release of anticoagulant soluble products may occur.[112]

Other anticoagulant materials were developed by copolymerization of acrylic acid with ethylene[115,116] or ethylene with vinyl sulfonates[115] or again vinyl acetate and crotonic acid.[115] In vivo testing by means of the Gott test showed that the blood compatibility depends on the net negative charge of the macromolecular chains and the hydrophilicity of materials.[115] In these series, the best materials appear to be the polyvinylacetate — 2% crotonic acid — 60% sodium ionomer. The nature of the counter ions influences also the antithrombogenicity of the materials, calcium or magnesium ions being less favorable than sodium ions.[116]

Several works[93,117-123] were devoted to polyelectrolyte and polyelectrolyte complexes, i.e., insoluble materials formed from complexation of positively and negatively charged polymers. Generally, polystyrene derivatives were under consideration, as, for instance, the

Ioplex complexes of sodium poly(styrene sulfonate) and poly(vinylbenzyltrimethyl ammonium) chloride,[117,118] in variable proportions with regard to their net charge. These complexes can be either positive, negative, or neutral.

With regard to their thrombogenicity, the best materials seem to be those bearing a low negative net charge, the neutral or highly positively or negatively charged polymers being highly thrombogenic. Besides the net charge, the charge distribution on the surface seems to determine the thrombogenicity, some of them being favorable, others not.[124]

Other polyelectrolyte complexes have been prepared by mixing chitosan either with carboxymethyldextran[120,121] or with a mixture of carboxylmethyldextran and sodium dextran sulfate.[122] All these insoluble complexes whose composition depends on the preparation conditions have anticoagulant properties when prepared at pH between 3 and 6.5.

Some polyelectrolytes complexes have been prepared by mixing polyanions and polycations derived from acrylonitrile-sodium methallyl sulfonate copolymer and acrylonitrile-2-methyl 5-vinyl pyridine copolymer:

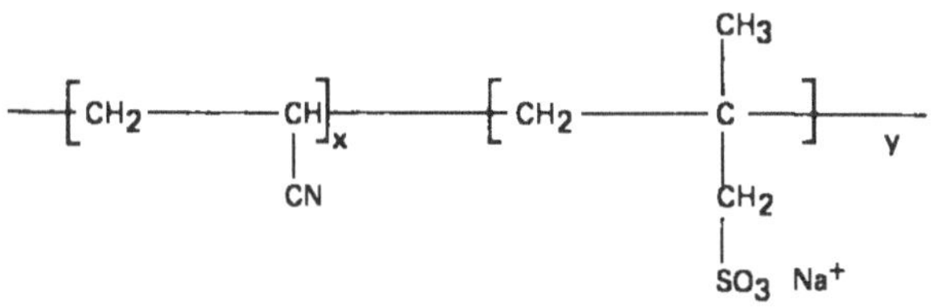


Polyanion

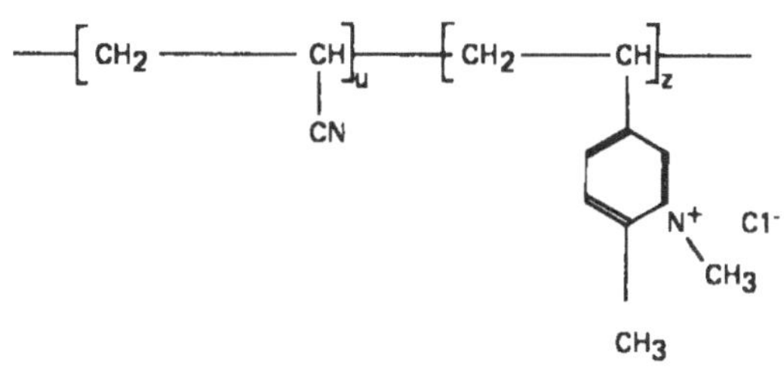


Polycation

By varying the polyanion/polycation ratio, a large number of polyelectrolyte complexes can be prepared with an excess of anionic or cationic sites. Platelet adherence and protein adsorption in such materials were measured using the blood flow system in vitro described by Baumgartner.[129] No correlation between these two parameters could be established.[119] The authors concluded that cell and protein attachment could not be used as an index of thrombogenicity. Negatively charged polymers activate the contact phase and positively charged polymers develop an anticoagulant surface-dependent effect observed in a global cephalin clotting time and prothrombin time.[125]

The mechanism of the action of the above described polyelectrolytes was generally not elucidated. Despite this fact, except the case of polyelectrolytes composites whose mode of action might be different, these materials can be assumed to act as heparin-like materials, i.e., to catalyze the $AT_{III}$ thrombin inhibition reaction. This assumption is based on the results of the detailed studies performed by Jozefowicz and al.[19,20,126-128]

The authors have obtained resins with significant antithrombogenic activity by binding sulfonate or amino acid sulfamide groups onto cross-linked polystyrene, cross-linked polysaccharides, or polystyrene-polyethylene graft copolymers. They have proved this activity to be antithrombic and to involve a plasmatic component by use of tests performed on platelet-poor plasma suspensions of the polymers.

**Table 5**
**ACTIVITY COEFFICIENT OF SOME SUBSTITUENTS LINKED TO POLYSTYRENE RESINS**

| Substituent | Activity coefficient ($meq^{-1}$) for 1 NIH thrombin unit |
|---|---|
| $-SO_3^-$ | 50—70 |
| $-SO_2$NH butyl | 0 |
| $-SO_2$ 11-amino undecanoic acid | 40—80 |
| $-SO_2$ alanine | 60—80 |
| $-SO_2$ glycine | 100—120 |
| $-SO_2$ hydroxyproline | 120—150 |
| $-SO_2$ proline | 120—150 |
| $-SO_2$ methionine | 120—150 |
| $-SO_2$ threonine | 120—150 |
| $-SO_2$ *N*-benzyloxycarbonyl lysine | 120—150 |
| $-SO_2$ β-alanine | 140—200 |
| $-SO_2$ ε-amino caproic acid | 300—350 |
| $-SO_2$ glutamic acid | 350—400 |
| $-SO_2$ δ-amino valeric acid | 400—450 |
| $-SO_2$ γ-amino butyric acid | 500—600 |
| $-SO_2$ aspartic acid | 500—600 |

This activity has also been proved to be located on the surface of the polymers. Moreover, by direct kinetic studies performed on suspensions of the polymers in aqueous solutions of $AT_{III}$ and thrombin, they were able to show that the antithrombic activity is heparin-like.

The anticoagulant activities of the materials are strongly dependent upon the contents of amino acid sulfamide groups and sulfonate groups and the nature of the involved amino acid. The authors were able to show that the substituting groups have either a cooperative or an additive effect. Given these observations, they were able to assess activity parameters for each of the substituents (Table 5), the higher activity coefficients having been observed for aspartic acid sulfamide substituting groups.

Based on these results, it was suggested that the isolated functional groups borne by polysaccharide chain of heparin, rather than the secondary or tertiary structure of this mucopolysaccharide, are responsible for its interaction with $AT_{III}$ or thrombin. Moreover, the comparison and discussion of the estimates of activity parameters substantiate the hypothesis that carboxylic functions are essential for heparin-like activity.

Ex vivo animal experiments were performed on surface-treated small-diameter tubing made of polystyrene-polyethylene copolymers. They showed that no significant platelet adhesion and aggregation could be observed on the wall of the tubing despite their poor mechanical properties. The studies described above have shown that biomaterials with high heparin-like activities can be synthesized.

They have proved to be antithrombic when suspended in plasma, i.e., they act as heterogeneous catalysts with regard to the $AT_{III}$-thrombin inhibiting reaction. These materials are stable and it can be assumed, despite the fact that only short-term experiments were performed, that their antithrombogenic properties will remain unchanged in the long term.

Moreover, surface treatments can be easily derived from the synthesis described in the above mentioned studies. Ex vivo preliminary studies have shown that they might be used for the making of all kinds of cardiovascular devices including small-diameter vessels for long-term uses. Thus, it appears that synthesizing heparin-like materials might well be in the future one of the ways to obtain suitable materials for the making of cardiovascular prosthetic devices.

## VI. GENERAL CONCLUSION

The biomaterials used until now are only suitable for the making of some cardiovascular devices, for instance, aorta or cardiac valves. Their use implies that a permanent adjuvant anticoagulant therapy is delivered to the host patient.

Improvements to the blood compatibility of the devices can be achieved if the mechanical compliance of the biomaterial used for the making of the device is comparable to that of the natural, normal healthy vessel as shown in elsewhere in this series. This can also be a result of surface treatments.

As seen above, improvements to the ability to bind albumin or of the hydrophilicity of the biomaterials surfaces have been proposed in order to improve their blood compatibility. Discussions of the published results suggest that fundamental studies of the competitive protein adsorption that occurs at the blood materials interface are necessary in order to make clear the materials structure, coagulation factors, and adsorption relationships responsible for the observed overall thrombogenicities.

Other surface treatments based on the binding of natural coagulation inhibitors such as platelet aggregation inhibitors, plasminogen activators, or heparin were also proposed. Among them, the heparin releasing biomaterials appear to be the most promising for the making of cardiovascular devices for short-term uses, for instance, as catheters.

However, it appears that synthesis of heparin-like materials in the near future will allow one to achieve surface treatments on biomaterials of suitable mechanical compliance. In turn, these materials will probably be suitable for the making of cardiovascular devices suitable for short- and long-term uses either as implants or not.

Moreover, this review of the published data suggests that in order to tailor blood compatible biomaterials, it is necessary to take into account the reactions that occur at the blood material interface. Then it is possible to design materials which are able to control the coagulation process initiated at the interface.

## REFERENCES


1. **Leininger, R. I.,** Changes in properties of plastics during implantation, in Plastics in Surgical Implants, Am. Soc. Testing Mater., Spec. Tech. Publ. No. 386, 1964, 71.
2. **Brash, J. L. and Lyman, D. J.,** Adsorption of plasma proteins in solution to uncharged, hydrophobic polymer surfaces, *J. Biomed. Mater. Res.*, 3, 175, 1969.
3. **Lyman, D. J., Brash, J. L., Chaikin, S. W., Klein, K. G., and Carini, M.,** The effect of chemical structure and surface properties of synthetic polymers on the coagulation of blood. II. Protein and platelet interaction with polymer surfaces, *Trans. Am. Soc. Artif. Intern. Organs,* 14, 250, 1968.
4. **Boffa, G. A., Lucien, N., Faure, A., Boffa, M. C., Jozefonvicz, J., Szubarga, A., Larrieu, M. J., and Mandon, P.,** Polytetrafluoroethylene — N-vinyl-pyrrolidone graft copolymers: affinity with plasma proteins, *J. Biomed. Mater. Res.*, 11, 317, 1977.
5. **Marcus, A. J.,** The role of lipids in platelet function: with particular reference to the arachidonic acid pathway, *J. Lipid Res.*, 19, 793, 1978.
6. **Schmid-Schonbein, H.,** Der Gegenwaertige Stand der Haemorheologie: Methoden, Befunde und Bedeutung fuer die Physiologie und Pathologie des Blutkreislaufs, *Wien. Klin. Wochenschr.*, 90(8), 245, 1978.
7. **Griffin, J. H. and Cochrane, C. G.,** Recent advances in the understanding of contact activation reactions, *Semin. Thromb. Hemostas.*, 5(4), 254, 1979.
8. **Weksler, B. B., Marcus, A. J., and Jaffe, E. A.,** Synthesis of prostaglandin $I_2$ (prostacyclin) by cultured human and bovine endothelial cells, *Proc. Natl. Acad. Sci. U.S.A.*, 74, 3922, 1977.
9. **Elion, J., Downing, M. R., Butkowski, R. J., and Mann, K. G.,** Structure of human thrombin: comparison with other serine proteases, in *Chemistry and Biology of Thrombin,* Lundblad, R., Fenton, J. W., and Mann, K. G., Eds., Ann Arbor, Mich., 1977, 97.

10. **Rosenberg, R. E. and Damus, P. S.,** The purification and mechanism of action of antithrombin-heparin cofactor, *J. Biol. Chem.*, 248, 6490, 1973.
11. **Petersen, T. E., Dudek-Wojciechowska, G., Sottrup-Jensen, L., and Magnusson, S.,** Primary structure of antithrombin III (heparin cofactor). Partial homology between $\alpha_1$ antitrypsin and antithrombin III, in *The Physiological Inhibitors of Coagulation and Fibrinolysis*, Collen, D., Wiman, B., and Verstraete, M., Eds., Elsevier/North-Holland, Amsterdam, 1979, 43.
12. **Rosenberg, R. D., Oosta, G. M., Jordan, R. E., and Gardner, W. T.,** The interaction of heparin with thrombin and antithrombin, *Biochem. Biophys. Res. Commun.*, 96, 1200, 1980.
13. **Rosenberg, R. D., Jordan, R. E., Favreau, L. V., and Lam, L. H.,** Highly active heparin species with multiple binding sites for antithrombin, *Biochem. Biophys. Res. Commun.*, 86, 1319, 1979.
14. **Jordan, R., Beeler, D., and Rosenberg, R.,** Fractionation of low molecular weight heparin species and their interaction with antithrombin, *J. Biol. Chem.*, 254(8), 2902, 1979.
15. **Lindahl, U., Bäckström, G., Hook, M., Thunberg, L., Fransson, L. A., and Linker, A.,** Structure of the antithrombin binding site in heparin, *Proc. Natl. Acad. Sci. U.S.A.*, 76(7), 3198, 1979.
16. **Holmer, E.,** Anticoagulant properties of heparin and heparin fractions, *Scand. J. Haematol. Suppl.*, 25(36), 25, 1980.
17. **Thunberg, L., Bäckström, G., Grundberg, H., Riesenfeld, J., and Lindahl, U.,** The molecular size of the antithrombin-binding sequence in heparin, *FEBS Lett.*, 117(1), 203, 1980.
18. **Rosenberg, R. D. and Lam, L.,** Correlation between structure and function of heparin, *Proc. Natl. Acad. Sci. U.S.A.*, 76(3), 1218, 1979.
19. **Fougnot, C., Jozefonvicz, J., Samama, M., and Bara, L.,** New heparin-like insoluble materials. I, *Ann. Biomed. Eng.*, 7, 429, 1979.
20. **Fougnot, C., Jozefowicz, M., Samama, M., and Bara, L.,** New heparin-like insoluble materials. II, *Ann. Biomed. Eng.*, 7, 441, 1979.
21. **Jordan, R. W., Oosta, G. M., Gardner, W. T., and Rosenberg, R. D.,** The binding of low molecular weight heparin to hemostatic enzymes, *J. Biol. Chem.*, 255(21), 10,073, 1980.
22. **Jordan, R. W., Oosta, G. M., Gardner, W. T., and Rosenberg, R. D.,** The kinetics of hemostatic enzyme-antithrombin interactions in the presence of low molecular weight heparin, *J. Biol. Chem.*, 255(21), 10,081, 1980.
23. **Lau, H. K. and Rosenberg, R. D.,** The isolation and characterization of a specific antibody population directed against the thrombin-antithrombin complex, *J. Biol. Chem.*, 255(12), 5885, 1980.
24. **Murty, C. V. S. N. and Sastri, N. V. S.,** Protein bonded polymers: an approach to anti-thrombogenic surfaces, *Curr. Sci.*, 45(10), 364, 1976.
25. **Domurado, D., Thomas, D., and Broun, G.,** New method for producing protein coatings, *J. Biomed. Mater. Res.*, 9(1), 109, 1975.
26. **Coleman, D. L., Atwood, A. I., and Andrade, J. D.,** Platelet retention by albuminated glass and polystyrene beads, *J. Bioeng.*, 1(1), 33, 1976.
27. **Guidoin, R. G., Awad, J., Brassard, A., Domurado, D., Lawny, F., Wetzer, J., Barbotin, J. N., Calvot, C., and Broun, G.,** Blood compatibility of silicone rubber chemically coated with cross-linked albumin, *Biomater. Med. Devices Artif. Organs*, 4(2), 205, 1976.
28. **Kambic, H. E., Barenburg, S., Harasaki, H., Gibbons, D., Kiraly, R. J., and Nose, Y.,** Glutaraldehyde — protein complexes as blood compatible coatings, *Trans. Am. Soc. Artif. Intern. Organs*, 24, 426, 1978.
29. **Limber, G. K. and Mason, R. G.,** Proteins elutable from cuprophane exposed to human plasma, *Thromb. Res.*, 6(5), 421, 1975.
30. **Domurado, D., Guidoin, R. G., Marois, M., Martin, L., Gosselin, C., and Awad, J.,** Albuminated Dacron protheses as improved blood vessel substitutes, *J. Bioeng.*, 2(1-2), 79, 1978.
31. **Carpentier, A., Deloche, A., Prigent, C., Beuzelin, J. P., Lessana, A., Salamagne, J. C., Guerin, L., Danan, M., Bedrane, Z., Carpentier, S., Baron, M., Macher, J. P., and Dubost, C.,** Utilisation de biopolymères dans la construction de prothèses valvulaires cardiaques, *C.R. Acad. Sci. Paris*, 272, 178, 1971.
32. **Loisance, D., Moczar, M., and David, Ph.,** A new microarterial biograft, 3rd Annu. Meet. International Society for Artificial Organs, Paris, July 1981.
33. **Ratner, B. R. and Hoffman, A. S.,** Radiation grafted hydrogels on silicone rubber as new biomaterials, *Polym. Sci. Technol.*, 7, 159, 1975.
34. **Ratner, B. R. and Hoffman, A. S.,** Radiation grafting of Hydrogels onto Organic Polymer Substrates, German Offen., 2,515,671, 1975.
35. **Sasaki, T., Ratner, B. R., and Hoffman, A. S.,** Radiation-induced co-graft polymerization of 2-hydroxyethyl methacrylate and ethyl methacrylate onto silicone rubber films, *Am. Chem. Soc. Symp. Ser.*, 31, 283, 1976.
36. **Okano, T., Kataoka, K., Sakurai, Y., Shimada, M., Miyahara, M., Akaide, T., and Shinohara, I.,** Molecular design of block and graft copolymers having the ability to suppress platelet adhesion, *Artif. Organs*, 5(Suppl.), 468, 1981.

37. **Ratner, B. D., Balisky, T., and Hoffman, A. S.,** A flow grafting apparatus for chemically modifying the luminal surface of polymeric tubing, *J. Bioeng.*, 1(2), 115, 1977.
38. **Hood, R. G., Courtney, J. M., Gaylor, J. D. S., and Gilchrist, T.,** A Method of Coating Granules, British Patent 1,496,330, 1974.
39. **Ilavsky, M., Dusek, K., Vacik, J., and Kopecek, J.,** Deformational, swelling, and potentiometric behavior of ionized gels of 2-hydroxyethylmethacrylate methacrylic acid copolymers, *J. Appl. Polym. Sci.*, 23(7), 2073, 1979.
40. **Gregonis, D. E., Russel, G. A., Andrade, J. D., and De Visser, A. C.,** Preparation and properties of stereoregular poly(hydroxyethylmethacrylate) polymers and hydrogels, *Polymer*, 19(11), 1279, 1978.
41. **Ratner, B. D., Horbett, T., Hoffman, A. S., and Hauschka, S. D.,** Cell adhesion to polymeric materials. Implications with respect to biocompatibility, *J. Biomed. Mater. Res.*, 9(5), 407, 1975.
42. **Donetski, I. A., Belomestnaya, Z. M., Chepurov, A. K., and Shumakov, V. I.,** Use of hydrophilic gels as thromboresistant coatings, *Aktual. Probl. Transplantol. Iskusstv. Organov*, 246, 1975.
43. **Ratner, B. D., Hoffman, A. S., and Whiffen, J. D.,** Blood compatibility of radiation-grafted hydrogels, *Biomater. Med. Devices Artif. Organs*, 3(1), 115, 1975.
44. **Abrahams, R. A. and Ronel, S. H.,** Blood compatible coatings of Hydron hydrophilic polymers. Techniques for application and testing, *Soc. Plast. Eng. Tech. Pap.*, 21, 570, 1975.
45. **Kataoka, K., Okano, T., Sakurai, Y., Nishimura, T., Akaide, T., and Tsuruta, T.,** Development of new biomaterials for blood cell separator, *Artif. Organs*, 5(Suppl.), 532, 1981.
46. **Rosen, J. J., Gibbons, D. F., and Culp, L. A.,** Fibroblast interactions with hydrogels surfaces, *Polym. Prepr. Am. Chem. Soc. Div. Polym. Chem.*, 16(2), 555, 1975.
47. **Rosen, J. J., Gibbons, D. F., and Culp, L. A.,** Fibrous capsule formation and fibroblast interactions at charged hydrogels interfaces, *Am. Chem. Soc. Symp. Ser.*, 31, 329, 1976.
48. **Hoffman, A. S. and Ratner, B. D.,** Synthetic hydrogels for biomedical applications, a review, *Polym. Prepr. Am. Chem. Soc. Div. Polym. Chem.*, 16(2), 272, 1975.
49. **Hoffman, A. S.,** Hydrogels — a broad class of biomaterials, *Polym. Sci. Technol.*, 8, 33, 1975.
50. **Hoffman, A. S.,** Surface modifications of polymers for biomedical applications, *Sci. Technol. Polym. Process. Proc. Int. Conf. 1977*, Suh, N. P. and Sung, N.-H., Eds., MIT Press, Cambridge, Mass., 1979, 200.
51. **Ott, Z.,** Biological testing of surface-modified acrylic polymers, *Polym. Med.*, 5(2), 167, 1975.
52. **Abrahams, R. A. and Ronel, S. H.,** Blood compatible coatings of Hydron hydrophilic polymers, *Polym. News*, 3(1), 11, 1976.
53. **Leininger, R. I.,** Polymeric materials that don't clot blood, *Chem. Technol.*, 5(3), 172, 1975.
54. **Ehrlich, J.,** Long term thromboresistance of heparinized surfaces, *Polym. Eng. Sci.*, 15(4), 281, 1975.
55. **Guidoin, R. G., Bergeron, P., and Awad, J. A.,** Heparinization of non-biologic surfaces, *Union Med. Can.*, 103(1), 71, 1974.
56. **Gott, V. L.,** Wall-bonded heparin — historical background and current clinical applications, *Adv. Exp. Med. Biol.*, 52, 351, 1975.
57. **Marconi, W., Bartoli, F., Mantovani, E., Pittalis, F., Settembri, L., Cordova, C., Musca, A., and Alessandri, C.,** Development of new antithrombogenic surfaces by employing platelet antiaggregating agents: preparation and characterization, *Trans. Am. Soc. Artif. Intern. Organs*, 25, 280, 1979.
58. **MacRea, J. C. and Kim, S. W.,** Characterization of controlled release of prostaglandin from polymer matrices for thrombus prevention, *Trans. Am. Soc. Artif. Intern. Organs*, 24, 746, 1978.
59. **Grode, G. A., Pitman, J., Crowley, J. P., Leininger, R. I., and Falb, R. D.,** Surface-immobilized prostaglandin as a platelet protective agent, *Trans. Am. Soc. Artif. Intern. Organs*, 20, 38, 1974.
60. **Sugitachi, A., Tanaka, M., Kawahara, T., Kitamura, N., and Takagi, K.,** A new type of drain tube, 3rd Annu. Meet. of the International Society for Artificial Organs, Paris, July 1981.
61. **Rosenberg, R. D.,** Biologic actions of heparin, *Sem. Hematol.*, 14(4), 427, 1977.
62. **Lindahl, U., Bäckström, G., Thunberg, L., and Leder, I. G.,** Evidence for a 3-O-sulfated D-glucosamine residue in the antithrombin-binding sequence of heparin, *Proc. Natl. Acad. Sci. U.S.A.*, 77(11), 6551, 1980.
63. **Salzman, E. W., Rosenberg, R. D., Smith, M. H., Lindon, J. N., and Favreau, L.,** Effect of heparin and heparin fractions on platelet aggregation, *J. Clin. Invest.*, 65(1), 64, 1980.
64. **Gott, V. L., Whiffen, J. D., and Dutton, R. C.,** Heparin bonding on colloidal graphite surfaces, *Science*, 142, 1297, 1963.
65. **Leininger, R. I., Falb, R. D., and Grode, G. A.,** Blood compatible plastics, *Ann. N.Y. Acad. Sci.*, 146(1), 11, 1968.
66. **Britton, R. A., Merrill, E. W., Gilliand, E. R., Salzman, E. W., Austen, W. G., and Kemp, D. S.,** Antithrombogenic cellulose film, *J. Biomed. Mater. Res.*, 2(4), 429, 1968.
67. **Imai, Y. and Masuhara, E.,** Preparation of nonthrombogenic plastic materials, *Rep. Inst. Med. Dent. Eng.*, 3, 72, 1969.

68. **Uy, S. and Kammermeyer, K.**, Nonthrombogenic surface preparation for silicone rubber, *J. Biomed. Mater. Res.*, 3, 587, 1969.
69. **Grode, G. A., Anderson, S. J., Grotta, H. M., and Falb, R. D.**, Biocompatible materials for use in the vascular system, *Trans. Am. Soc. Artif. Intern. Organs*, 15, 1, 1969.
70. **Merker, R. L., Elyash, L. J., Mayhew, S. H., and Wang, J. Y. C.**, in *Artificial Heart Program Conference Proceedings*, Hegyeli, R. J., Ed., National Heart Institute Artificial Heart Program, Washington, D.C., 1969, 29.
71. **Salyer, I. O., Blardinelli, A. J., Ball, G. L., Weesner, W. E., Gott, V. L., Ramos, M. D., and Furuse, A.**, New blood-compatible polymers for artificial heart applications, *J. Biomed. Mater. Res., Symp.*, 1, 105, 1971.
72. **Kusserow, B. K., Larrow, R., and Nichols, J.**, Urokinase — heparin bonded synthetic surface. Creation of a prosthetic surface, *Trans. Am. Soc. Artif. Intern. Organs*, 17, 1, 1971.
73. **Rea, W. J., Eberle, J. W., Watson, J. T., Ecker, R. R., and Sugg, W. L.**, Gas transfer in a heparinized membrane oxygenator, *Surg. Forum*, 22, 188, 1971.
74. **Kim, S. W., Ebert, C. D., and MacRea, J. C.**, Pharmacological modification of polymers for the enhancement of blood compatibility, 3rd Annu. Meet. of the International Society for Artificial Organs, Paris, July 1981.
75. **Hasenfratz, H. and Knaup, G.**, Improvement of the blood compatibility of cellulosic membranes through the immobilization of heparin and measurement of biological heparin activity, 3rd Annu. Meet. of the International Society for Artificial Organs, Paris, July 1981.
76. **Pusineri, C., Eloy, R., Baquet, J., Belleville, J., and Farges, J. B.**, In vivo evaluation of heparinized catheters in dogs, 3rd Annu. Meet. of the International Society for Artificial Organs, Paris, July 1981.
77. **Yen, S. P. S. and Rembaum, P.**, Complexes of heparin with elastomeric positive polyelectrolytes, *J. Biomed. Mater. Res. Symp.*, 1, 83, 1971.
78. **Pusineri, C.**, Polymers for Use in Medical Apparatus, French Demande 2,264,939, 1978.
79. **Idezuki, Y., Watanabe, H., Hagiwara, M., Kanasugi, K., Mori, Y., Nagaoka, S., Hagio, M., Yamamoto, K., and Tanzawa, H.**, Mechanism of antithrombogenicity of a new heparinized hydrophilic polymer. Chronic in vivo studies and clinical application, *Trans. Am. Soc. Artif. Intern. Organs*, 21, 436, 1975.
80. **Courtney, J. M., Park, G. B., Fairweather, I. A., and Lindsay, R. M.**, Polymer structure and blood compatibility evaluation-application of an acrylonitrile copolymer, *Biomater. Med. Devices Artif. Organs*, 4(3-4), 263, 1976.
81. **Ferruti, P. and Provenzale, L.**, A new nonthrombogenic polymeric material, *Transplant. Proc.*, 8(1), 103, 1976.
82. **Paik Sung, C. S., Bush, J., MacKie, D. B., and Merril, E. W.**, Copolymers containing aminohexyl residues in side chains for attaching heparin, *J. Appl. Polym. Sci.*, 20, 2603, 1976.
83. **Lagergren, H. R. and Eriksson, J. C.**, Plastics with a stable surface monolayer of crosslinked heparin: preparation and evaluation, *Trans. Am. Soc. Artif. Intern. Organs*, 17, 10, 1971.
84. **Larsson, R. L., Hjelte, M. B., Eriksson, J. C., Lagergren, H. R., and Olsson, P.**, The stability of glutardialdehyde-stabilized $^{35}S$ heparinized surfaces in contact with blood, *Thromb. Haemostas.*, 37, 262, 1977.
85. **Olsson, P., Lagergren, H., Larsson, R., and Radegran, K.**, Prevention of platelet adhesion and aggregation by a glutardialdehyde-stabilized heparin surface, *Thromb. Haemostas.*, 37, 274, 1977.
86. **Halpern, B. D. and Shibakawa, P.**, Heparin covalently bonded to polymer surface, *Adv. Chem. Serv.* , 87, 197, 1968.
87. **Merril, E. W., Salzman, E. W., Wong, P. S. L., Ashford, T. P., Brown, A. H., and Austen, W. G.**, Polyvinyl alcohol — heparin hydrogel "G", *J. Appl. Physiol.*, 29(5), 723, 1970.
88. **Iverius, P. H.**, Coupling of glycosaminoglycans to agarose beads, *Biochem. J.*, 124, 677, 1971.
89. **Schmer, G.**, The biological activity of covalently immobilized heparin, *Trans. Am. Soc. Artif. Intern. Organs*, 18, 321, 1972.
90. **Danishefsky, I. and Tzeng, F.** Preparation of heparin-linked agarose and its interaction with plasma, *Thromb. Res.*, 4, 237, 1974.
91. **Hoffman, A. S., Schmer, G., Harris, C., and Kraft, W. G.**, Covalent binding of biomolecules to radiation-grafted hydrogels on inert polymer surfaces, *Trans. Am. Soc. Artif. Intern. Organs*, 18, 10, 1972.
92. **Chawla, A. S. and Chang, T. M. S.**, Nonthrombogenic surface by radiation grafting of heparin: preparation, in vitro and in vivo studies, *Biomater. Med. Devices Artif. Organs*, 2(2), 157, 1974.
93. **Falb, R. D., Leininger, R. I., and Crowley, J. P.**, Materials with chemically active substituents, *Ann. N.Y. Acad. Sci.*, 283, 396, 1977.
94. **Labarre, D., Boffa, M. C., and Jozefowicz, M.**, Preparation and properties of heparin-poly(methylmethacrylate) copolymers, *J. Polym. Sci. Symp.*, 47, 131, 1974.
95. **Baquey, C., Blanquet, P., and Ducassou, D.**, Reactivity of paramagnetic species created in heparin under γ rays action, *Ann. Phys. Biol. Med.*, 9(2), 131, 1975.

96. **Mester, L., Amit-Amaya, A., and Mester, M.,** Heparin derivatives of high molecular weight. Carbohydrate sulfates, *Am. Chem. Soc. Symp. Ser.*, 77, 113, 1978.
97. **Plate, N. A., Valuev, L. I., and Gumirova, F. K.,** Covalent immobilization of modified heparin on synthetic polymers, *Dokl. Akad. Nauk. SSSR*, 244(6), 1505, 1979.
98. **Nakashima, T. and Takakura, K.,** Antiblood clotting medical polymers, Japanese Kokai, 75 03 494, 1975.
99. **Dincer, A.,** Covalent Coupling of Heparin to Synthetic Polymer Surfaces, Thesis, Massachusetts Institute of Technology, Cambridge, 1977.
100. **Peppas, N. A. and Merrill, E. W.,** Development of semicristalline P.V.A. hydrogels for biomedical applications, *J. Biomed. Mater. Res.*, 11, 423, 1977.
101. **Goosen, M. F. A. and Sefton, M. V.,** Heparinized styrene-butadiene-styrene elastomers, *J. Biomed. Mater. Res.*, 13, 347, 1979.
102. **Labarre, D., Jozefowicz, M., and Boffa, M. C.,** Properties of heparin-poly(methyl methacrylate) copolymers. II, *J. Biomed. Mater. Res.*, 11, 283, 1977.
103. **Boffa, M. C., Labarre, D., Jozefowicz, M., and Boffa, G. A.,** Interactions between human plasma proteins and heparin-poly (methyl methacrylate) copolymer, *Thromb. Haemostas. (Stuttgart)*, 41(2), 346, 1979.
104. **Salzman, E., Silane, M., Lindon, J., Brier-Russel, K., Dincer, A., Rosenberg, R., Labarre, D., and Merrill, E.,** Thromboresistance of heparin coated surfaces, in *The Chemistry and Biology of Heparin*, Brown, W. V., Mann, K. G., Roberts, H. R., and Lundblad, R. L., Eds., Elsevier/North-Holland, New York, 1980, 435.
105. **Lindon, J., Rodvien, R., Brier, D., Greenberg, R., Merrill, E., and Salzman, E.,** In vitro assessment of interaction of blood with model surfaces, *J. Lab. Clin. Med.*, 92(6), 904, 1978.
106. **Lindon, J., Rosenberg, R., Merrill, E., and Salzman, E.,** Interaction of human platelets with heparinized agarose gels, *J. Lab. Clin. Med.*, 91(1), 47, 1978.
107. **Larsson, R., Olsson, P., and Lindahl, U.,** Inhibition of thrombin on surfaces coated with immobilized and heparin-like polysaccharides: a crucial nonthrombogenic principle, *Thromb. Res.*, 19, 43, 1980.
108. **Lovelock, J. E. and Porterfield, J. S.,** Blood coagulation: its prolongation in vessels with negatively charged surfaces, *Nature (London)*, 167, 39, 1951.
109. **Gregor, H. P.,** Anticoagulant activity of sulfonate polymers and copolymers, *Polym. Sci. Technol. U.S.A.*, 7, 51, 1975.
110. **Machovich, R. and Horvath, I.,** Heparin-like effect of polymethacrylic acid on the reaction between thrombin and antithrombin III, *Thromb. Res.*, 11, 765, 1977.
111. **van der Does, L., Hofman, J., van Utteren, T. E. C.,** Reaction of N-chlorosulfonylisocyanate with unsaturated polymers — route to a synthesis of polyampholytes, *Polym. Lett.*, 11, 169, 1973.
112. **Beugeling, T., van der Does, L., Bantjes, A., and Sederel, W. L.,** Antithrombin activity of a polyelectrolyte synthesized from cis-1,4-polyisoprene, *J. Biomed. Mater. Res.*, 8, 375, 1974.
113. **Beugeling, T., van der Does, L., Rejda, B. V., and Bantjes, A.,** Antithrombogenic polymers synthesized from polyisoprenes, in *Biocompatibility of Implant Materials*, Williams, D., Ed., Sector, London, 1976, 187.
114. **Beugeling, T.,** The interaction of polymer surfaces with blood, *J. Polym. Sci. Polym. Symp.*, 66, 419, 1979.
115. **Byck, J. S., Kwiatkowski, G. T., Gonsior, L. S., Creasy, W. S., Stewart, D. D., and Lundell, E. O.,** Factors affecting the blood compatibility of polyelectrolytes, *Soc. Plast. Eng. Tech. Pap.*, 21, 563, 1975.
116. **Costello, M., Stanczewski, B., Vriesman, P., Lucas, T., Srinivasan, S., and Sawyer, P. N.,** Correlations between electrochemical and antithrombogenic characteristics of polyelectrolyte materials, *Trans. Am. Soc. Artif. Intern. Organs*, 16, 1, 1970.
117. **Marshall, D., Cross, R., and Bixler, H.,** An evaluation of polyelectrolyte complexes as biomedical materials, *J. Biomed. Mater. Res.*, 4, 357, 1970.
118. **Nelsen, L., Cross, R. A., Vogel, M. A., Gott, V. L., and Fadali, A. M.,** Synthetic thromboresistant surfaces from sulfonated polyelectrolyte complexes, *Surgery*, 67(5), 826, 1970.
119. **Sultan, Y., Maisonneuve, P., and Pusineri, C.,** Characterization of artificial surfaces by protein adsorption and platelet adherence, *1st World Biomater. Congr.*, Baden near Vienna, April 1980, Abstr. No. 2.5.4, European Society for Biomaterials, Vienna, 1980.
120. **Fukuda, H. and Kikuchi, Y.,** In vitro clot formation on the polyelectrolyte complexes of sodium dextran sulfate with chitosan, *J. Biomed. Mater. Res.*, 12, 531, 1978.
121. **Fukuda, H. and Kikuchi, Y.,** Polyelectrolyte complexes of chitosan with sodium carboxymethyldextran, *Bull. Chem. Soc. Jpn.*, 51(4), 1142, 1978.
122. **Kikuchi, Y. and Onishi, Y.,** Polyelectrolyte complex consisting of carboxymethyldextran, sodium dextran sulfate, and chitosan, *Nippon Kagaku Kaishi*, 1, 127, 1979.

123. **Larsson, R., Eriksson, J. C., and Olsson, P.,** The effects of surface localized sulfate groups on the clotting time in plasma, *Thromb. Res.*, 14(6), 941, 1979.
124. **Leonard, F.,** Rationale for the preparation of non-thrombogenic materials, *Trans. Am. Soc. Artif. Intern. Organs*, 15, 15, 1969.
125. **Boffa, M. C., Farges, J. P., Dreyer, B., Conche, B., Pusineri, C., and Vantard, G.,** Striking differences observed in the effect of amphipathic polyelectrolytes on contact and other clotting factors, in *Advances in Biomaterials*, Vol. 3, Winter, G. D., Gibbons, D. F., and Plenk, H., Eds., John Wiley & Sons, 1982, 399.
126. **Mauzac, M., Fougnot, C., and Jozefonvicz, J.,** New antithrombogenic polysaccharides, *Artif. Organs*, 5(Suppl.), 504, 1981.
127. **Fougnot, C., Baquey, C., Serne, H., Labarre, D., Basse-Cathalinat, B., Ducassou, D., and Jozefowicz, M.,** Heparin-like polymer surfaces used in extracorporeal shunts, *Artif. Organs*, 5 (Suppl.), 524, 1981.
128. **Fougnot, C., Jozefonvicz, J., and Jozefowicz, M.,** New heparin-like insoluble materials, in *Advances in Biomaterials*, Vol. 3, Winter, G. D., Gibbons, D. F., and Plenk, H., Ed., John Wiley & Sons, 1982, 465.
129. **Baumgartner, H. R.,** Morphometric quantitation of adherence of platelets to an artificial surface and components of connective tissue, *Thromb. Diath. Haemorrh.*, 60, 39, 1974.

Chapter 10

# TISSUE REACTIONS WITH SILICONE RUBBER IMPLANTS (MORPHOLOGICAL, MICROCHEMICAL, AND CLINICAL INVESTIGATIONS IN HUMANS AND LABORATORY ANIMALS)

**G. Mikuz, G. Hoinkes, A. Propst, and P. Wilflingseder**

## TABLE OF CONTENTS

I. Introduction .......... 240

II. Material and Methods .......... 240

III. Results .......... 241
A. Human Material — Light Microscopy .......... 241
B. Observations in Guinea Pigs (Light Microscopy) .......... 241

IV. Electron Microscopy Findings .......... 243

V. Microchemical Findings .......... 245

VI. Statistical Evaluation .......... 245

VII. Discussion .......... 247

References .......... 248

# I. INTRODUCTION

In plastic reconstructive surgery, silicone rubber implants are used to maintain space for a secondary autologous graft or to augment soft tissue defects. This material, a polydimethylsiloxane, is perhaps one of the most widely used synthetic polymers for surgical devices.[1] There is a fibrous tissue formation around such not truly "inert" implants which is a characteristic basic response of the organism — in this way a large "indigestible" foreign body is apparently extraterritorialized.

In particular cases, as, for example, in tendon repair, a silicone spacer is removed after 6 to 8 weeks or within a year and the fibrous sheet is filled with the appropriate autograft.[2-6] According to Smahel,[7] the fibrous sheet ("capsule") resolves after removal of the implant.

In augmentation-mammaplasties, when a permanent retromammary or subpectoral silicone polymer implant is used, this encapsulation, however, impairs the primary good results of the operation to a various degree, because the shrinkage of the "capsule" changes the shape, position, and consistency of the augmented breast, less or more, in all cases. This tissue reaction has been widely studied in humans as well as in laboratory animals during the last decade by various investigators.[5,6,8-10]

# II. MATERIAL AND METHODS

We have studied 158 capsulotomy and capsulectomy specimens. The biopsies from the capsule surrounding the mammary prostheses were taken at various intervals after implantation. The breast prostheses in the investigated cases were of two types: (1) inflatable silicone rubber bags filled with saline ($n = 93$) and (2) silicone rubber bags filled with short chain silastomers ($n = 65$).

For histology, these specimens were fixed in 4% formaldehyde, embedded in paraffin, and stained with hematoxylin-eosin, PAS, and van Gieson.

For electron microscopy, 33 specimens were fixed with phosphate-buffered (0.15 *M*, pH 7) 6.5% glutaraldehyde, followed by postfixation with Dalton's chrome-osmium tetroxide. The specimens were dehydrated with acetone, embedded in Durcopan®, and sectioned on a Reichert® ultramicrotome OMU 2 and Ultracut®. Thin sections were stained with uranyl acetate, lead citrate, and examined with Zeiss® electron microscopes EM 9 A and 9 S.

Microchemical investigations were carried out in 21 cases — 6 specimens were embedded in paraffin, 5 were embedded in Durcopan®, and 10 were freeze-dried. The specimens were carbon-coated for conduction. The electron probe was an ARL-SEMQ (Scanning Electron Microscope Quantometer). The operating conditions of the electron microprobe were 15 KeV accelerating voltage and 2 nA sample current. The element Si was qualitatively identified by energy dispersive X-ray analysis, using an Si (Li) detector in a Kevex® system. X-ray scans of specimens were recorded by means of wavelength-dispersive spectrometers using RbAP crystals for Si (K).

For statistical investigation, the time interval between implantation and capsulotomy, i.e., the formation of a clinically marked contracting capsule, was correlated with the frequency of capsulotomies at those intervals. For comparison, the incapsulations of silicone polymer were also studied in guinea pigs. Patches of the envelope and gel (short chain silastomers) separately from a mammary prostheses (Dow® Corning Corporation, Catalog Number 971, Lot Number HH 3718, Volume 165 $cm^3$) were implanted subcutaneously, in i.m. pockets and in peritoneal folds of 15 animals. The animals were sacrificed 15, 45, and 120 days after implantation. The tissue surrounding the implants was investigated by means of light and electron microscopy (methods described above).

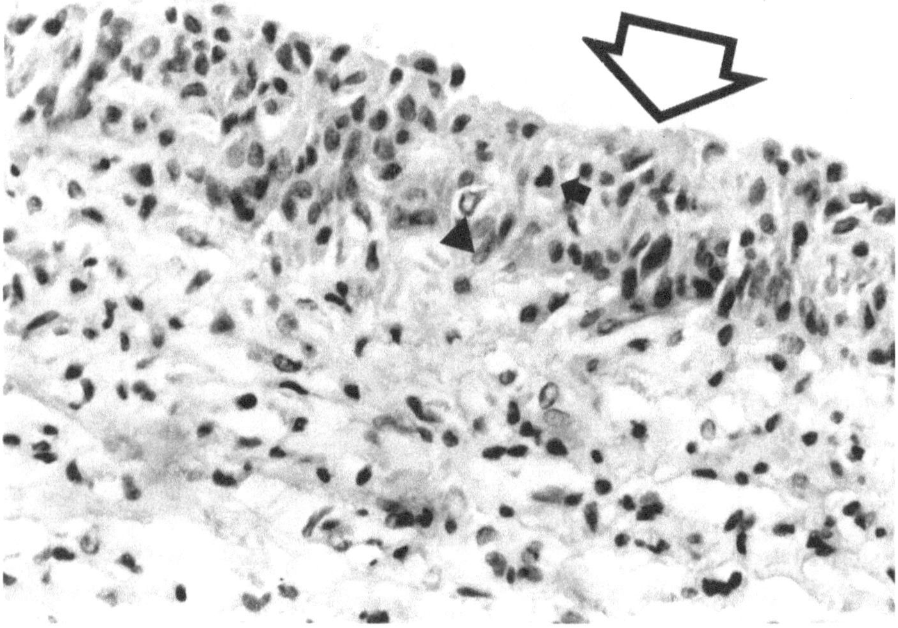

FIGURE 1. Fibrous capsule surrounding a silicone rubber mammary prosthesis. The innermost part is lined by histiocytes (bordered arrow). The black arrow indicates a mitosis and the arrow head indicates a small particle deriving from the prosthesis. (Human material; capsulotomy 6 months after mammaugmentation). (HE; magnification × 400.)

## III. RESULTS

### A. Human Material — Light Microscopy

All capsules showed the same basic histological structure. The significant morphological differences were related to the clinical features.

In 29 (18.4%) incidences, the operation had been indicated because of leakage of the inflatable prostheses. In these cases the fibrous capsule was thin (1 mm). The innermost stratus, contacting the prostheses, was lined with fibrocytes and scattered small histiocytes. The outer stratus consisted of parallel collagenous fibers.

The other 129 specimens derived from capsulotomies which have been carried out because of deformation, dislocation, an increasing consistency of the augmented breast, and in some cases because of severe pain. In 109 (65.8%) cases, the fibrous capsules were 2 to 3 mm thick. The collagenous fibers were not parallel, but interlaced. Foci of young granulation tissue with fibroblasts and inflammatory cells were also scattered among these fibers. The innermost stratus was lined with histiocytes arranged like epithelium, which formed one or more layers, and sometimes the cytoplasm contained small, PAS positive granules (Figure 1). Single foreign body giant cells were also found in these specimens. In 25 (15.8%) cases, which demonstrated clinically a severe disfigurement of the breast, the capsules were thicker than in the previous group (about 5 mm). The innermost stratus was lined with histiocytes or it was necrotic and covered with fibrin. In the outer part, small foreign body granulomas and small neuromas were seen.

### B. Observations in Guinea Pigs (Light Microscopy)

In guinea pigs two different reactions were seen. The implanted silicone gel caused a classic foreign body reaction. The drops of liquid material were surrounded by foamy cells and giant cells (Figure 2). These "siliconomas" did not change the morphology during the observation time. The implanted silicone envelope caused a reaction very similar to that

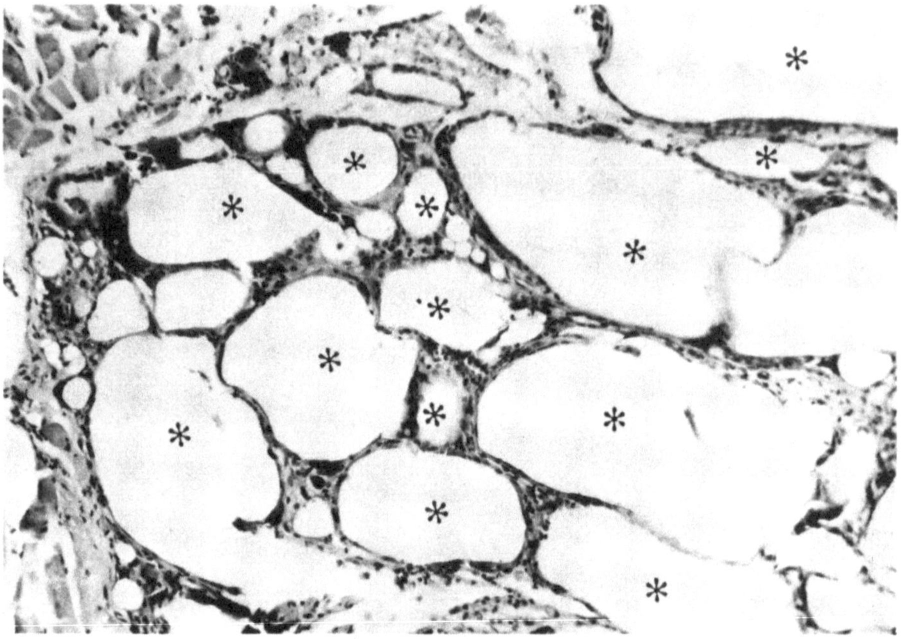

FIGURE 2. Typical siliconoma after implantation of silicone gel into subcutis of guinea pigs. Gel droplets (*) are surrounded by histiocytes and foreign body giant cells. (HE; magnification × 40.)

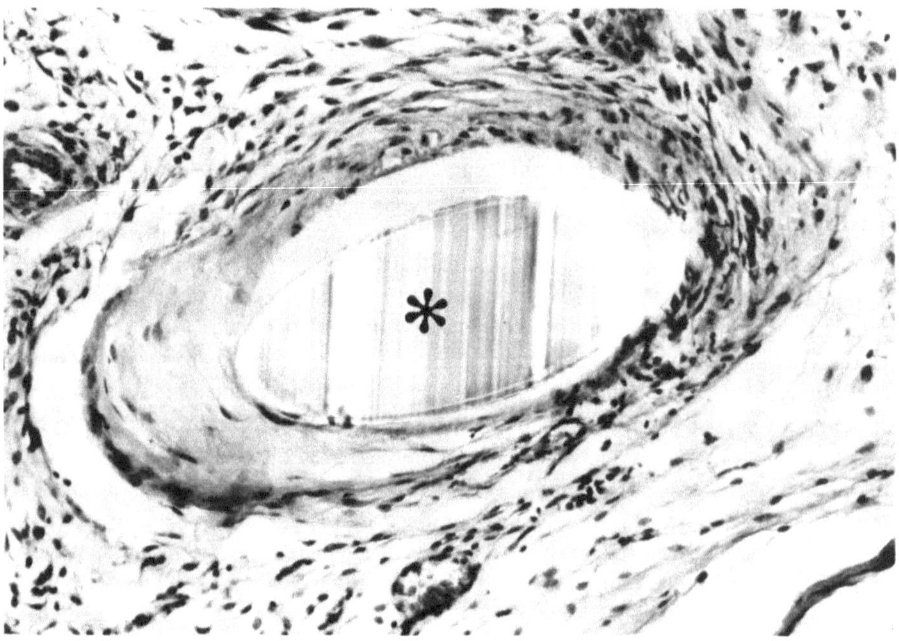

FIGURE 3. Fibrous capsule surrounding a silicone rubber patch (*) 1 month after implantation into the subcutis of a guinea pig. (HE; magnification × 100.)

seen in the human material. In 14 days after implantation, a capsule arose around the solid implants. It consisted of one to two layers of fibrocytes and some collagenous fibers. The innermost stratus, contacting the silicone rubber, was lined with histiocytes (Figure 3). In

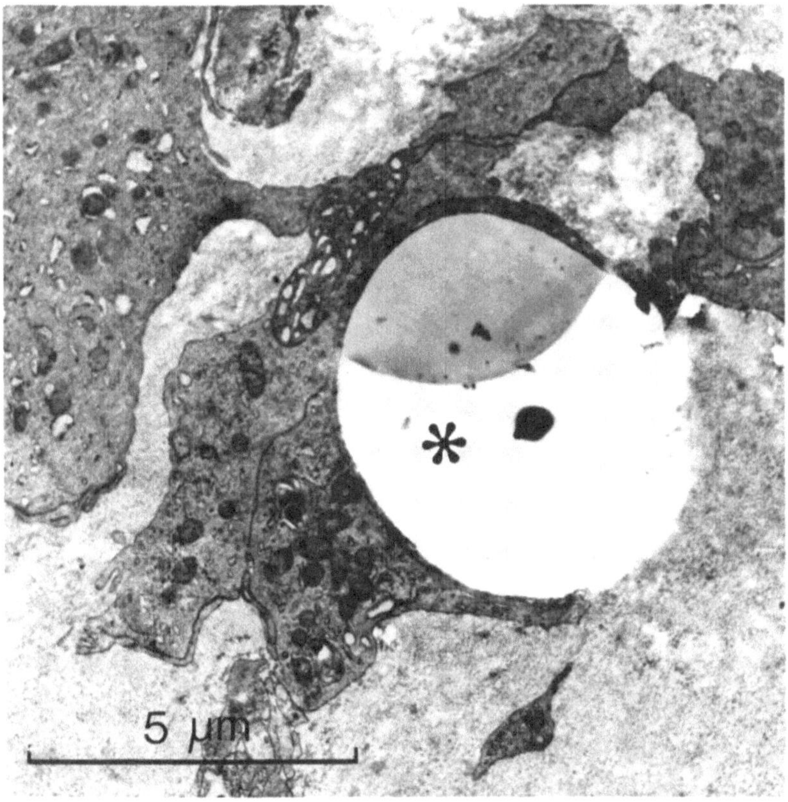


FIGURE 4. Phagocytosis of silicone rubber (*) by histiocytes (in guinea pig). (Magnification × 4600.)

1 to 1.5 months following the implantation, the capsule became thicker and showed more collagenous fibers. The contacting stratus, however, was the same as 14 days after implantation. There were no gross morphological differences in the reactions between the different recipient sites.

## IV. ELECTRON MICROSCOPY FINDINGS

In human as well as in guinea pig specimens, the capsules showed the same ultrastructural features. The innermost stratus cells were macrophages with an oval indented nucleus and dense peripheral chromatin. The cell surface had numerous microvilli and pseudopods characteristic of phagocytic activity. The microvilli sometimes extended towards extracellular round-shaped foreign 2 to 25-μm diameter size particles (Figure 4). In the cytoplasm of macrophages, various quantities of round-shaped, membrane-bound, partly electron dense, and partly translucent particles were found. These particles had the same appearance as the extracellular foreign material. The electron beam does burn out such intra- and extracellular particles, leaving round-shaped transparent cavities (Figure 5). The number of macrophages loaded with phagocytized material was constant in laboratory animals. In the human specimens, however, the number of macrophages loaded with the foreign material seemed to be correlated to the intensity of the reaction. Occasionally foreign body giant cells appeared to be loaded with that material (Figure 6). In guinea pigs, translucent particles were also found in the pericytes. Towards the outside of the capsule, the cellularity decreased and the bundles of collagenous fibers are predominant. Between these fibers, macrophages (with foreign

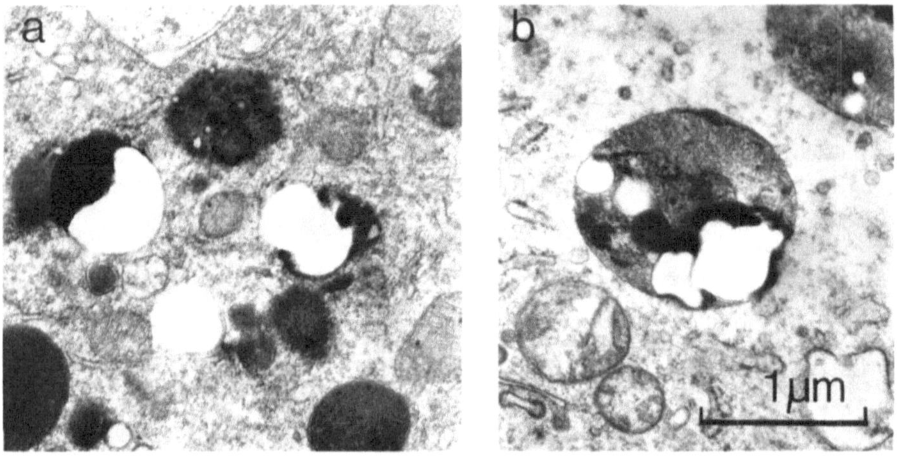


FIGURE 5. Phagosomes with phagocytized silicone rubber. The different size and electron density of the particles suggest an intracellular decomposition of the rubber. (a) Human material; (b) guinea pigs. (Magnification × 22,000.)

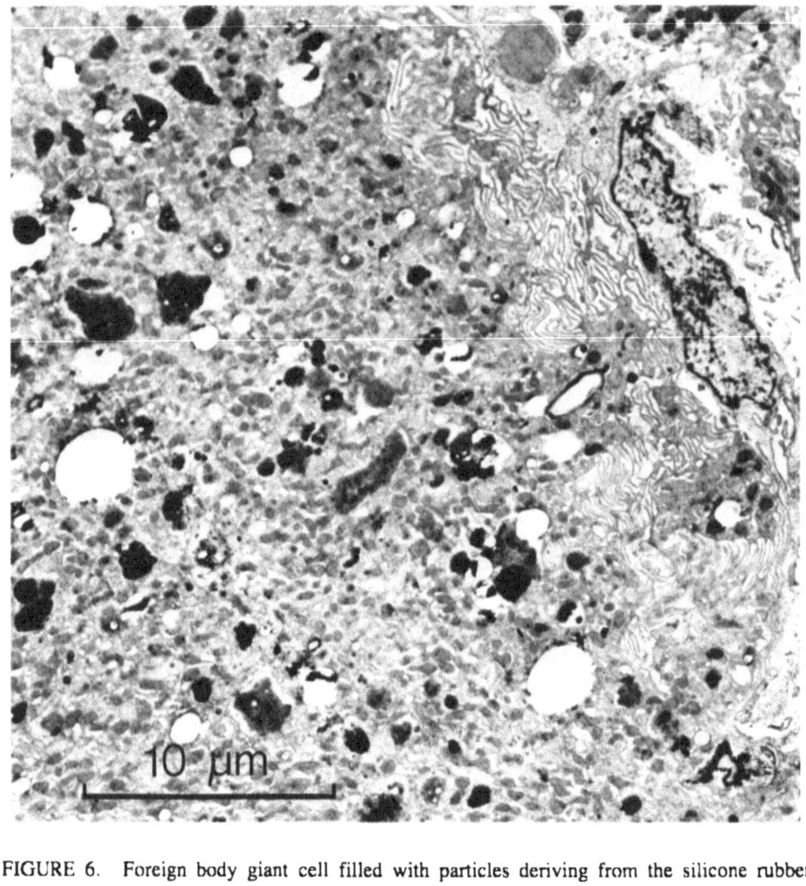


FIGURE 6. Foreign body giant cell filled with particles deriving from the silicone rubber implant (human material). (Magnification × 1700.)

materials in the cytoplasm), fibroblasts, and fibrocytes, as well as other inflammatory cells, were seen (Figure 7). Single myofibroblasts were also constantly present.

The vascularization of the capsules in human material was very scanty. In guinea pigs,

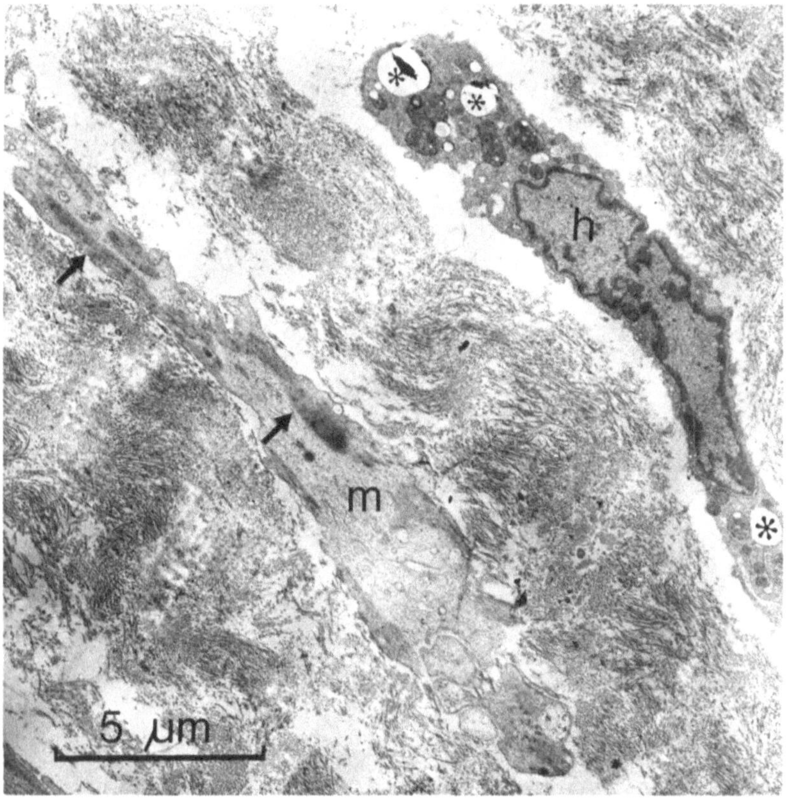


FIGURE 7. The outer part of the capsule surrounding a mammary prosthesis consisting of collagenous fibers, histiocytes (h) with phagocytized silicone rubber (*), and myofibroblasts (m) responsible for capsule contraction (arrows indicate myofibrils). (Magnification × 2000).

on the other hand, a lot of vessels appeared 2 weeks after implantation, but they disappeared after 1 month.

## V. MICROCHEMICAL FINDINGS

The capsule specimens showed SiK-α-radiation emitted from intra- or extracellular granules containing Si. The smallest radiating granules were 10 nm in size. There was no significant difference in radiation quantity between the different histological types of capsules. In some specimens, the radiation rate was highest in the innermost part of the capsule (adjacent to the implant) and decreased gradually in the outer parts. The semiquantitative count rates showed a Si concentration of 5 to 30% in the capsule specimens. As a comparison, this concentration was 47% in Silicon ($SiO_2$) and 30 to 50% in material from silicone prostheses (Figure 8).

## VI. STATISTICAL EVALUATION

About 50% of the capsulotomies were carried out within the 1st year after implantation and 75% at the end of the 2nd year. There was a significant negative correlation ($r = -0.8390$) between the frequency of capsulotomies and the time interval at which the re-operations were carried out. Between saline- and gel-filled prostheses, no significant difference in correlation existed (saline filled, $r = -0.9706$; gel-filled, $r = -0.8671$) (Figure 9).

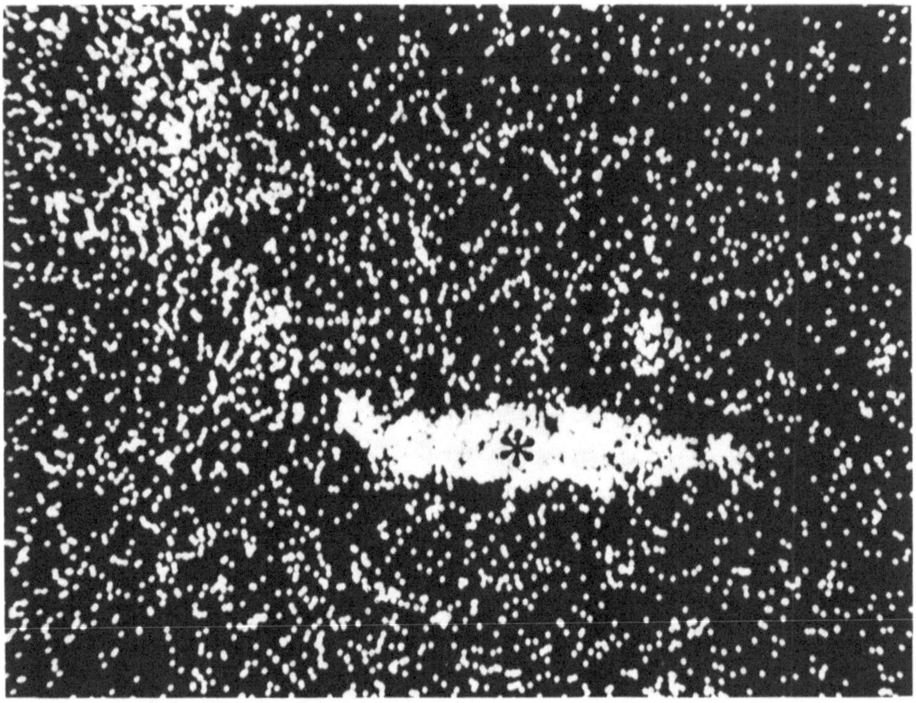

FIGURE 8. Distribution of SiK α-radiation in fibrous tissue adjacent to the prosthesis. The cell-shaped domain is an extracellular silastic rubber particle. (Magnification × 1000.)

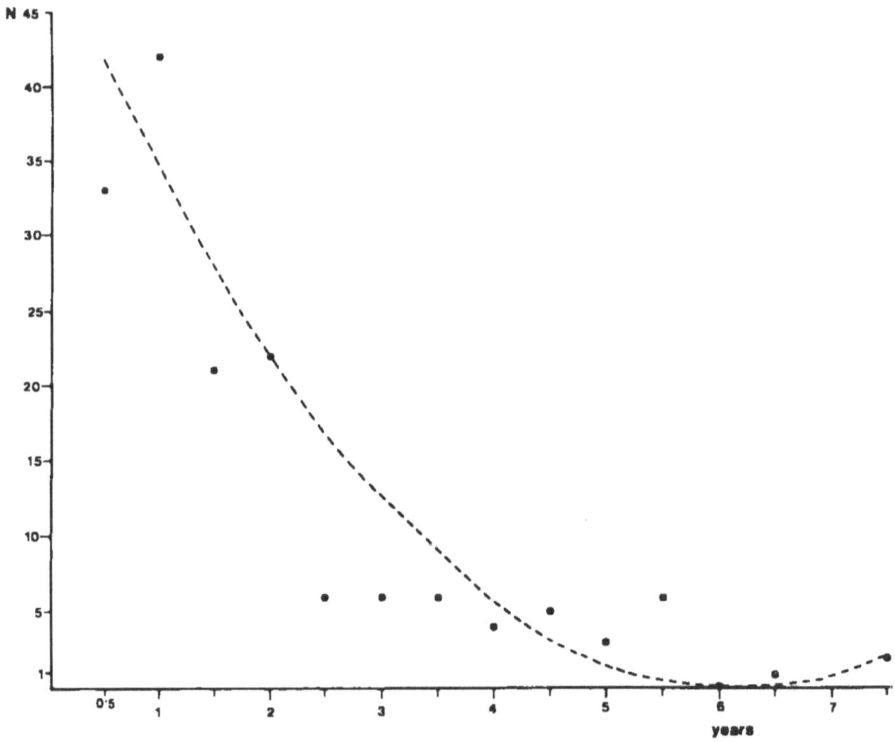


FIGURE 9. Negative correlation (r = −0.8390) between the frequency of capsulotomies (y axis) and the time interval (x axis) at which the reoperations had to be carried out.

## VII. DISCUSSION

In the early days, soft tissue augmentation with liquid silicone (silicone oil) injections were tried. Tissue reaction toward this material is a foreign body granuloma called "siliconoma",[11-14] which consists of liquid silicone droplets surrounded by foamy macrophages, foreign body giant cells, and other inflammatory cells. The introduction of the softer, natural touch mammary prostheses, polydimethylsiloxane bags filled with saline or silicone gel, did not prevent but as it seems increased the specific, sometimes severe tissue reaction and subsequent complications.[8,9,15-19] The fibrous capsule surrounding the implants contracts and deforms the prostheses into an unnatural globe. The morphological substrate for active capsule shrinkage consists of myofibroblasts as demonstrated by several investigators.[20-22] Several terms for these complications, as "capsule contraction"[9] or "fibrous contraction",[23] have been frequently used. The term "constrictive fibrosis" was introduced by Wilflingseder et al.[8] and reflects exactly the clinical and morphological conditions which can be graded by four types.

Type I has no clinical signs. The specimens were gained only accidently when a retromammary prosthesis had to be replaced because of rupture, or a silicone spacer from the hands or the facial region was removed. This very thin and uncontracted fibrous capsule represents the "ideal reaction". Types II and III represent the true "constrictive fibrosis" with the characteristic morphologically and clinical signs: disfiguration, dislocation, and increasing firmness of the breast. The only morphological difference between Type II and III is the absence or presence of giant cells in the innermost stratus. Clinically these two morphologic types are not different.

Type IV shows in addition inflammatory cells, vascularization, and foreign body granulomas in the capsule. The neuromas that can occasionally be found are probably responsible for the pain which occurs sometimes in these patients. In several cases, apathogenic bacilli could be isolated in the fibrous tissue.

The electron microscope investigation of human and guinea pig material suggests that the particles found extracellulary in the capsule and intracellulary within the macrophages derive from the silicone prostheses. In view of the different size, shape, and electron density of such particles, we assumed that intracellular silicone rubber can be decomposed. The anhydridic Si-O-Si bindings are relatively unprotected and therefore exposed to the action of hydrolytic enzymes. The microchemical analyses are related to the distribution of silicone in the tissue. This finding previously reported in 1975[5] was confirmed by Wickham et al.[24]

We have found silicone-containing deposits in all four histological types of capsules, and there is no doubt that silicone rubber is decomposed by macrophages to molecules beyond the resolving power of the electron microscope. Experiments with rabbits (unpublished data) showed a time-related increase of silicone concentration, which means that the decomposition of silicone does not stop. According to Smahel,[19] the thickness and structure of the capsule seems to correlate with its Si contamination. It is obvious that the Si containing material must have been grazed off from the surface of the saline-filled prosthesis and is probably also diffused through the semipermeable envelope of the gel-filled prosthesis.

In our material there are no morphological differences between the tissue reaction towards the saline- or gel-filled bags. Therefore the so-called "bleeding" of the gel-filled envelopes[9,25-27] seems not to be a decisive factor in the morphogenesis of the tissue reaction. In rabbit experiments, Lilla and Vistnes[28] came to the same conclusion: the type of filling in the prosthesis had no demonstrable effect on the encapsulation. As we could demonstrate in guinea pigs, silicon gel does produce a typical siliconoma, whereas the higher molecular envelope produces the typical encapsulation as previously described. There is also the possibility of physical and chemical deterioration of silicone polymers due to absorption of fatty acids esters or contact with trialkyl phosphates.[29,30]

According to our statistical investigation, a significant negative correlation exists between the capsulotomy frequency and the time of the clinical marked capsule contraction and subsequent reoperation. This means that friction contrary to the assertion of Smahel[7] is only a negligible factor. If friction alone were responsible for the development of the capsule, a positive correlation would be expected. We presume that Si-loaded macrophages are responsible for the generation of the encapsulation and the self-preservation of the pathogenetic mechanism. Heppleston and Styles[31] showed in vitro experiments that silica damages the macrophages, which subsequently produces a factor which stimulates the fibroblasts to produce collagen.

The transportation of silicone oil to distant organs[32,33] and of silicone gel to the regionary lymph nodes[34,35,35a] have been observed. Uretsky et al.[36] also reported the occurrence of a severe illness, with fever, joint swelling, lung infiltrates, and decreased renal function, 1 day after augmentation mammaplasty and quick recovery after the prosthesis had been removed. Klinge[37] observed a case of therapy resistent angiitis after augmentation mammaplasty. These findings suggest that — contrary to Stern et al.,[38] who found no immunogenicity with silicone polymer implants in animals — this material can have immunogenic site effects. It is known that Si-loaded macrophages can create inflammatory fibrotic and general immunologic reactions.[39] Therefore the possibility of immunologic reaction in the encapsulation process of the silicone polymer implants must be kept in mind.

## REFERENCES


1. **Braley, S.,** The chemistry and properties of the medical-grade silicones, *J. Macromol. Sci. Chem.*, A4, 529, 1970.
2. **Calnan, I.,** The use of inert plastic material in reconstructive surgery, *Br. J. Plast. Surg.*, 16, 1, 1963.
3. **Thomson, H. G.,** The fate of the pseudosheath pocket around silicone implants, *Plast. Reconstr. Surg.*, 51, 667, 1973.
4. **Anderl, H., Bauer, M., and Mikuz, G.,** The take of bone, cartilage and dermis within a capsule preformed by a silastic implant, *Chir. Plast. (Berl.)*, 4, 73, 1977.
5. **Wilflingseder, P., Propst, A., Mikuz, G., and Hoinkes, G.,** Constrictive fibrosis post augmentation mammaplasty, in *Trans. 6th Int. Congr. of Plastic and Reconstructive Surgery*, Masson, Paris, 1976, 535.
6. **Wilflingseder, P., Bauer, M., and Miller, B.,** The use of plastics in the primary repair of facial injuries, *Riv. Ital. Chir. Plast.*, 8, 243, 1976.
7. **Smahel, J.,** Structure of the capsules around silicone elastomer implants, *Chir. Plast. (Berl.)*, 5, 119, 1980.
8. **Wilflingseder, P., Propst, A., and Mikuz, G.,** Constrictive fibrosis following silicone implants in mammary augmentation, *Chir. Plast. (Berl.)*, 2, 215, 1974.
9. **Gayou, R. and Rudolph, R.,** Capsular contraction around silicone mammary prosthesis, *Ann. Plast. Surg.*, 2, 62, 1979.
10. **Mikuz, G., Propst, A., Hoinkes, G., and Wilflingseder, P.,** Granulomatose Gewebsreaktion bei intrazellularem Kunststoff (Silikongummi-) Abbau, *Verh. Dtsch. Ges. Pathol.*, 64, 279, 1980.
11. **Ben-Hur, N. and Neuman, A.,** Siliconoma — another cutaneous response to dimethylpolysiloxane, *Plast. Reconstr. Surg.*, 36, 629, 1965.
12. **Rees, T. D., Platt, J., and Ballantyne, D. L.,** An investigation of cutaneous response to dimethylpolysiloxame (silicone liquid) in animals and humans. A preliminary report, *Plast. Reconstr. Surg.*, 35, 131, 1965.
13. **Nosanchuk, J. S.,** Silicone granuloma in the breast — case report, *Arch. Surg.*, 97, 583, 1968.
14. **Pearl, R. M., Laub, D. R., and Kaplan, E. N.,** Complications following silicone injections for augmentation of the contours of the face, *Plast. Reconstr. Surg.*, 61, 888, 1978.
15. **Gurdin, M. and Carlin, G. A.,** Complications of breast implantation, *Plast. Reconstr. Surg.*, 40, 530, 1967.
16. **De Cholnoky, T.,** Augmentation mammaplasty — survey of complications in 10,941 patients by 265 surgeons, *Plast. Reconstr. Surg.*, 45, 573, 1970.

17. **Silver, H. L.**,Treating the complications of augmentation mammaplasty, *Plast. Reconstr. Surg.*, 49, 637, 1972.
18. **Vinnik, C. A.**, Spherical contracture of fibrous capsules around breast implants, *Plast. Reconstr. Surg.*, 58, 555, 1976.
19. **Smahel, J.**, Histology of the capsules causing constrictive fibrosis around breast implants, *Br. J. Plast. Surg.*, 30, 324, 1977.
20. **Montandon, D., Gabbiani, G., Ryan, G., and Majno, G.**, The contractile fibroblasts, *Plast. Reconstr. Surg.*, 52, 286, 1973.
21. **Imber, G., Schwager, R. G., Guthrie, R. H., Jr., and Gray, G. F.**, Fibrous capsule formation after subcutaneous implantation of synthetic materials in experimental animals, *Plast. Reconstr. Surg.*, 54, 183, 1974.
22. **Rudolph, R., Abraham, J., and Vecchione, T. R.**, Myofibroblasts and free silicon around breast implants, *Plast. Reconstr. Surg.*, 62, 185, 1978.
23. **Freeman, B. S.**, Successful treatment of some fibrous envelope contractures around breast implants, *Plast. Reconstr. Surg.*, 50, 107, 1972.
24. **Wickham, M. G., Ross, R., and Abraham, J. L.**, Silicon identification in prosthesis-associated fibrous capsule, *Science*, 199, 437, 1978.
25. **Domanskis, E. J. and Owsley, J. Q., Jr.**, Histological investigation of the etiology of capsule contracture following augmentation mammaplasty, *Plast. Reconstr. Surg.*, 58, 689, 1976.
26. **Barker, D. E., Retsky, M. I., and Schultz, S.**, "Bleeding" of silicone from bag. Gel breast implants and its clinical relation to fibrous capsule reaction, *Plast. Reconstr. Surg.*, 61, 836, 1978.
27. **Smahel, J.**, Foreign material in the capsules around breast prosthesis and the cellular reaction to it, *Br. J. Plast. Surg.*, 32, 35, 1979.
28. **Lilla, J. A. and Vistnes, L. M.**, Long term study of reactions to various silicone breast implants in rabbits, *Plast. Reconstr. Surg.*, 57, 637, 1976.
29. **Braley, S. and Boone, J. L.**, Resistance of silicone rubber to body fluids, *Rubber Chem. Technol.*, 39, 1293, 1966.
30. **Chan, S., Salzbrenner, R. and Predecki, P.**, Attack of silicone rubber by trialkyl phosphates, *J. Biomed. Mater. Res.*, 7, 485, 1973.
31. **Heppleston, A. G. and Styles, J. A.**, cited by **Carr, I.**, *The Macrophage*, Academic Press, London, 1973, 84.
32. **Solomons, E. T. and Jones, J. K.**, The determination of polydimethylsiloxane (silicon oil) in biological materials; a case report, *J. Forensic Sci.*, 20, 191, 1975.
33. **Ellenbogen, R., Ellenbogen, R., and Rubin, L.**, Injectable fluid silicone therapy. Human morbidity and mortality, *JAMA*, 234, 308, 1975.
34. **Wintsch, W., Smahel, J., and Clodius, L.**, Local and regional lymph node response to ruptured gel-filled mammary prostheses, *Br. J. Plast. Surg.*, 31, 349, 1978.
35. **Hausner, R. J., Schoen, F. J., and Pierson, K. K.**, Foreign body reaction to silicone gel in axillary lymph nodes after an augmentation mammaplasty, *Plast. Reconstr. Surg.*, 62, 381, 1978.
35a. **Mikuz, G.**, Unpublished observation.
36. **Uretsky, B. F., O'Brien, J. J., Courtiss, E. H., and Baker, M. D.**, Augmentation mammaplasty associated with a severe systemic illness, *Ann. Plast. Surg.*, 3, 445, 1979.
37. **Klinge, O.**, personal communication.
38. **Stern, I. J., Kapsalis, A. A., DeLuca, B. L., and Pieczynski, W.**, Immunogenic effects of foreign materials on plasma proteins, *Nature (London)*, 238, 151, 1972.
39. **Pernis, B., Silicosis**, in *Textbook of Immunopathology*, Vol. 1, 2nd ed., Miescher, P. E. and Muller-Eberhard, H. J., Eds., Grune & Stratton, New York, 1976, 443.
40. **Carr, I.**, *The Macrophage*, Academic Press, London, 1973, 84.

# *Index*

# INDEX

## A


Abrasion (wear) resistance, 11
Abrasive particles, 4
Absorbable pins, 122
Accessible body cavities, inserts into, 183
Acetal bridges, 228
Acetic acid, 36
Acetylcellulose (cellophane), 193
Acid-base titrations, of organic bioresorbable internal fixation devices, 121
Acrylic acid, 228, 230
Acrylic cement, 4, 10—12
failure, 11
standards, 13
Acrylic chloranhydride, 228
Acrylonitrile, 225
Acrylonitrile-dimethylaminoethyl methacrylate, 163, 164, 228
Acrylonitrile-sodium methallyl sulfonate (AN69), 162, 163
RP-Acrylonitrile-sodium methallyl sulfonate (AN 69), 162
Addition polymerization, 62
Adhesion, surface chemistry for, 8
Adsorption, 9, 226
B-Alanine, 161
$-SO_2$ Alanine, 232
$-SO_2$ $\beta$ Alanine, 232
L-Alanine, 161
Albumin, 149, 217, 224—226
bonding of, 226
coating, 225
Alkaline phosphatase, 196
Alza Corporation, 195
Amine, 200
Amine-modified polyacrylonitrile, 163
Amino acid sulfamide groups, 231
$-SO_2$ $\gamma$-Amino butyric acid, 232
$-SO_2$ $\epsilon$-Amino caproic acid, 232
δ-Aminopropyltriethoxysilane (APTES), 227
$-SO_2$ 11-Amino undecanoil acid, 232
$-SO_2$ δ-Amino valeric acid, 232
Amorphous polymers, 10
AN69, see Acrylonitrile-sodium methallyl sulfate
Angitis, 248
Animal testing, on bioresorbable internal fixation devices, 120—123, 127—134
Anionic polyelectrolytes, 207
Anionic polyether urethane lattices, 225
Annealing, 126
Antiarthritic activity, 207
Antibiotics, 196
Anticancer drugs, 196, 203
Anticoagulant activity, 230
Anticoagulant material, 230
Antimalarials, 203
Antimicrobial therapy, 206
Antithrombin III (ATIII), 222—223, 227, 228, 232
affinity of low molecular weight heparin for, 229
binding and activation, 229
Antithrombogenic activity, 231
Antithrombogenic coatings, 226—229
Antithrombogenicity, 228, 230
Antithrombogenic materials, 226
Antithrombogenic properties, 227
Apathogenic bacilli, 247
Approaches to membrane design, 152—153
Arginine, 161
Aromatic polyamide, 166
"Artificial cells", 205
Artificial kidney, 145
Ascorbic acid, 161
$-SO_2$ Aspartic acid, 232
Asymmetric membranes, 153
cellulose acetate, 159
Atropine, 195
Augmentation, 247
Augmentation-mammaplasties, 240, 248


## B


Bacilli, apathogenic, 247
Bacitracine, 161
Bad solvents, 24
Barium sulfate, 13
Basophile, polynuclear, 217
Batch to batch variations, 15
Beads, 184
Benzene, 36
$-SO_2$ *N*-Benzyloxycarbonyl lysine, 232
Biconvex tablets, 188
Bioadsorbable, 202
Biocompatibility
problems, 4
stereopolymers, 133
Biocompatible biomaterials, 216
Biodegradable, 202
Biodegradable drug delivery system, 203
Biodegradable matrix, 205
Biodegradable polymers, 202, 205
Biodegradable systems, 202—205
Biodegradation, of stereocopolymers, 133
Bioerodible, 202
Bioerodible polymers, 203
Bioerodible systems, 202—205
Biologically active polymers, 207—208
Biological performance
effect of surface properties on, 8—10
mechanical properties and, 10—11
Biomaterials, 216, 22
biocompatible, 216
improved blood compatibility and, 224—226
Bioplast, 97
clinical applications, 99—105

preparation, 97, 98
properties, 98—99
Biopolymers, see also Polyelectrolytes, 23
Bioprosthesis, 225
Bioresorbability, 120
Bioresorbable internal fixation devices, 120
organic, history, 120—127
poly(α-hydroxy-acids), 124—127
Bioresorbable plastic materials, biofunctional investigation, 134—140
Blank solutions, 51
Blank solvents, 51
Block copolymers, see also Copolymers, light scattering; Copolymers, viscometry, 50
of poly(oxyethylene)-poly(oxypropylene), 207
Blood
flow rate, 224, 226
human, composition, 217
Blood biocompatibility, 225
Blood coagulation
induced at interface with materials, 217—224
inhibitors, concentrations in normal human plasma, 218
Blood compatibility(ies), 9, 12, 216, 233
improved, biomaterials with, 224—226
membrane, 152
molecular properties of macromolecules and, see also Molecular properties, 23
Blood-material interface, 226
Blow out fractures, of orbital floor, 121
Bond(s), hydrogen, 58
Bond dissociation energies, of polymers, 58, 60
Bonded surfaces, covalently, 288
Bonding, 60
of polymers, 58—60
Bone cavities, 206
osteomyelitic, 114
Bone cysts, filling of, 114
Bone fracture(s), 120
blow out, of orbital floor, 121
mandibular, 121, 137
skull, 137, 140
Bone growth, sequential labeling of, 127
Bone infection, chronic, 196, 206
Bone plate(s), 122
composite, 137
PLA 100, 133
Bone regeneration, and collagen fleece, 114
Boundary-layer, 184, 190
Boundary-layer controlled release, 189—190
Bovalbumin, 162
Bovine albumin, 149
serum, 189
Breast, augmented, 241
Brittleness, 23
"Burst", effects, 186
$-SO_2NH$ Butyl, 232
Butyl 2-cyanoacrylate, 205
Butyl methacrylate, 13
Butyl methacrylate-acrylic acid membranes, 166

## C

Cadophane, 156
Calcification, 207
Calcium, 160
Cancer chemotherapy, 205, 206
Capillary(ies), 188—189
Capillary viscometers, 30, 31
Capsule(s), 240, 241, 243, 245
contraction, 248
electron microscopy findings, 243
fibrous, 247
granulomas in, 247
histological types of, 247
microchemical findings, 245
particles in, 247
vascularization of, 244, 245
Capsulectomy, 240
Capsulotomy, 240, 245, 246, 248
Carbodiimide, 228
Carbon fibers, coated with poly(L-lactic acid), 123
Carboxylic functions, 232
Carboxymethyldextran, 231
Cardiac valves, 225
Cardiovascular prosthetic devices, 232
Casting, 184
Catalase, 196, 207
Catheters, 229
Cationic surfaces, ionic binding of heparin to, 227
Cell membrane, 8
Cellulose, structure of, 153, 154
Cellulose acetate, deacetylation of, 156
Cellulose acetate membranes, 159
asymmetrical, 159
controlling ultrafiltration with, 159
reinforcement of, 159
Cellulose film, identification of, 154
Cellulose membrane(s), 153—159
alternatives to, 158—159
manufacture of, 154, 156
regenerated, 150, 155
solute transport characteristics, 154
strengthening of, 156
ultrathin, 157, 158
Cellulose nitrate membranes, 160
Ceramic fibers, 123
Chain conformation, 7
Characterization, 22, 27
"Chemical carriers", 184
Chemical degradation of macromolecules, 126
Chemical nature of materials, 225
Chitosan, 231
Chloroquine, 195
Citric acid, 161
Clot
formation, 220, 221
kinetics of, 224
lysing of, 227
Coagulation factors, 217
Coatings, antithrombogenic, 227—229

Coil dialyzer, 145, 147
Cold flow, 23
  in polyethylene, 4
Collagen, 112
  conversion to soluble form, 115
  cross-linked, 197
  microcrystalline, 113, 114
  preparation of, 113—114
  purification of, 113
  purified, 114
  structure of, 112, 113
Collagen articles, preparation of, 114
Collagen fibril(s), 112—114
Collagen filament(s), 114
Collagen film(s), 114
Collagen fleece, 114
Collagen membranes, 160—161
Collagenous fiber, 241, 245
Collagen products
  fleece-like, 114
  reconstituted fibrous, 112
Collagen sutures, 112, 113
Colligative properties, 31—36
  of copolymers, 49
Compatibility, 226
Complex association, 53
Composite(s), 5, 11
  bone plates, clinical applications, 137
  fatigue in, 11
Composite material, 4, 123, 137, 140
  resorbable, 135
Compression tests, 83
Concentrated solutions, 31
Concentration boundary layers, 150
Configurational structure, 126
  poly(α-hydroxy acids), 124
Conformation, 5
"Constrictive fibrosis", 247
Contamination, Si, 247
Contraceptives, steroidal, 203
Control processes, 8
Copolyether-ester membranes, half-time escape rate ratios, 161
Copolymer(s), see also Inhomogeneity, 47
  colligative properties, 49
  degradation rates, 122
  formation, 62
  of glutamic acid, 205
  of leucine, 205
  molecular weight (distribution), 48—51
  polydiastereoisomer, 124
  sequence distribution, see also Inhomogeneity, 47
  size exclusion chromatography, 50
  ultracentrifugation, 50
  viscometry, 50
Copolymerization, 62
Copolymer membranes, polyether block, 161
Cortical bone, 127
Cortical repair, 131
Corticosterone, 194
Cortisol, 194
Counterions, 51
  binding of, 53
Covalently bonded surfaces, 228
Crack behavior, 13
Cracking, of polymer surface, 10—11
Crazing, of polymers, 83—84
Creatinine, 149, 160—162
Creep, 10
  in polyethylene, 4
Creep modulus, 83
Creep ratio, of polymers, 81—82
Creep rupture data, for polymer, 84, 86
Creep stress, 83
Critical stress intensity factor, 87
Cross-fractionation, 51
Cross-link(s), 68
  density, 79
Cross-linked collagen, 197
Cross-linked matrices, 184
Cross-linking
  density of, 192
  polymers, 65, 66
Crotonic acid, 230
Cryometry, 35—36
Cryoscopic constant, 36
Cryoscopy, 31
Crystal, 62
  amorphous ratio, 8
  melting point, 66
  unit cell, 73
Crystalline, 192
  polymers, 10
Crystalline state, 67, 191
  of polymer, 74
Crystallinity, see also Semicrystalline state, 7, 22, 126, 129, 134
  influence on crystalline polymers, 23
  of organic bioresorbable internal fixation devices, 121
  of polymers, 75
Crystallite(s), 198
Crystallization, see also Fractionation, 47
  PET, 77
  polymer, 73, 75
Cuenophane, 156
Cup(s), 116
Cuprammonium process, 154, 156
Cuprophan, 158, 225
Cuprophan membrane(s), 155—157
  sorbent, 170
Cyanoacrylates, 5
Cyanogen bromide, 228
Cyclazocine, 204
Cyclophosphamide, 205
Cylinders, 188
  diffusional release of solutes from, 187
  implantable, 203
Cytotoxicity, 202

## D

Dacron, see Polyester
Daunomycin, 206
Deacetylation, of cellulose acetate, 156
Deformability, 135, 137
Degradation of macromolecules, 122, 126, 128, 129
  of bioresorbable plastic materials, 136, 140—141
    rates, 134, 135
    sheep, quantitative evaluation, 133—134
  of polymers, see Enzymatic degradation
  rate
    for organic bioresorbable internal fixation devices, 121—123
    for poly(lactic acid), 125
  resistance to, 11
Degradation reactions, 10
Density gradient, see also Copolymer; Ultracentrifugation, 50
Density of cross-linking, 192
Deoxyribonucleic acid (DNA), 196
Depolymerization, see Polymerization
Deposition, 226
Design
  parameters, 10
  specifications, 15
Determination, for the determination of specific properties, see specific properties
Dextran, 206
Dialyzance, 151
Dialysis, 51, 144, 145
Dialyzed solutions, 51
Dialyzer
  coil, 145, 147
  hollow fiber, 145, 146, 148
  parallel plate, 145, 147
Dialyzing solution, 52
Diastereoisomers, lactide, 126
Diethylene glycol dimethacrylate, 196
Diffusing molecule, molecular size of, 193
Diffusion coefficient, 190—191, 193
Dihydropyran(s), 204
3,4-Dihydro-2H-pyran, 204
Diisocyanates, 198
Dilute solution, viscometry, 27
Dimethylaminoethyl methacrylate hydrogel, 225
Diols, polyurethane, 12
Dipyridamole, 227
Disk(s), 188
Disk viscometers, 31
Dispersed drug, 186—188
Dissolution process, 23
Dissolved drug, 186, 187
Dissolved macromolecules, size and shape parameters, 24
Distribution coefficient, 185, 193, 194
Ditazole, 227
DIVEMA, 207
Dogs, dysplasia in, 116
Domain binding, of counterions, 53
Drug(s)
  dispersed, 186—188
  dissolved, 186, 187
  release, 189
    controlled, 4
    from polymeric devices, 187
Drug carriers, 182
Dysplasia, in dogs, 116

## E

Ear, inserts into, 183
Ebulliometry, 31, 35—36
Ebullioscopic constant, 36
"Eddy-diffusion", see also Size exclusion chromatography, 46
Effective radius $R_e$, 25, 28
Einstein's equation, 28
Elastomers, 10
Electron microscopy, 240
Electrophoresis, 53
Enantiomeric composition, 125
Encapsulation, 247
  fibrous, 127
End group analysis, 31, 35
End group titrations, 29
Endocytosis, 206
Endotracheal tubes, 6
Entanglements, 69
Environmental resistance, 11
Environmental stress cracking, 86
Enzymatic degradation, see also Molecular weight distribution, 23, 46
Enzymatic inhibitors, action of, 222
Enzyme(s), see also Polyelectrolytes, 23, 202
Enzyme-deficiency diseases, 205
Eosinophile, polynuclear, 217
Epoxy resins, 5
Equilibrium constant, 52—53
Erodible devices, 184, 190
Erythrocytes, 217
Estradiol, 194
Estradiol diacetate, 190
Ethyl 2-cyanoacrylate monomers, 205
Ethylene, 230
Ethylene oxide sterilization, 204
Ethylene-propylene, 192
Ethylene-vinyl acetate, 189, 195
Ethylene-vinyl alcohol copolymers, 169
Ethylmethacrylate copolymers, 225
Extracorporeal blood purification, membrane separation, 144—149
Extraction, see also Fractionation, 47
Extrinsic pathway, for activation of blood coagulation, 220, 222
Extrusion, 15
Eye, conjunctival sac of, inserts into, 183

## F


Fabrication, of plastic materials, 14—15
Factor V, 93
Factor IXa, 223
Factor Xa, 93, 219, 223
Factor XII, 219
Factor XIIa
Factor XIII, 93
Factor XIIIa, 93
Fatigue, in composites, 11
Fatigue failure, 11
Fatigue life, 11
Federal German Government regulations, 15
Fertility control, 203
Fiber(s)
  PET, 75, 77—78
  polypropylene film, 16
Fibrin, 92
  fibrinogen, 92
  formation, 92—93
  insoluble, 93
  in medicine and surgery, 96
    Bioplast, 97—105
    fibrin film, 97
    fibrin foam, 97
    fibrinogen clotted *in situ*, 96—97
    fibrin powder, 97
  preparation, 93
  properties, 93—95
  structure property relationships, 95—96
Fibrinogen, 93, 217, 224, 226
  clotted *in situ*, 96—97
  structure, 92
Fibrin stabilizing factor, 93
Fibrosis, constrictive, 247
Fibrous capsule, 241, 247
Fibrous encapsulation, 127
Fibrous sheet, 240
Fick's first law, 185
Fillers, 4, 11, 194
Film(s), 114, 203
Flat sheet membranes, properties
  Cuprophan, 155
  Nephrophan, 156
Fleece, collagen, 114
Fleece-like collagen products, 114
Flex, 11
Flexibility, 11
Flow rate of circulating blood, 224
Fluorethylene, 193
Fluoropolymers, 5
Foam(s), 16
Foamy cells, 241, 242
Food and Drug Administration (FDA), 15
Foreign body giant cells, 244
Foreign body granulomas, 241
Fractionation, see also Cross-fractionation; Molecular weight distribution, 46—47, 51
Fracture(s), see also Bone fractures of polymers, 86—88
Fracture strength, 10
Free energy of mixing, 192
Free radial polymerization, 12
Free volume, 191


## G


Gamma irradiation, 11
Gastrointestinal (GI) tract, 183, 205
Gel(s), 228, 240
Gelatin, 112, 189
  articles, preparation of, 115—116
  cups, cross-linked, 116
  manufacture of, 115—116
  sponges, 112, 116
  structure of, 114
  tubes, 116
Gelatine, 116
Gel-filled bags, 247
Gel-filled prosthesis, 247
Gel particles, 24
Gel permeation chromatography, see also Size exclusion chromatography, 44, 46
Gentamycin, 196
Giant cell(s), 241, 242, 244, 247
Glass, 225
Glass transition temperature, 7, 66, 71
  of organic bioresorbable internal fixation devices, 121
Glassy polymers, 79, 83
  yielding of, 84
Glassy state, 67, 91
  polymers, 71—73
Glaucoma, 195
γ-Globulins, 224, 226
Glucosamine-HCl, 161
Glucose, 161
Glucuronic acid, 161
Glucuronolactone, 200
Glutardialdehyde, 228
Glutamic acid, copolymers of, 205
$-SO_2$ Glutamic acid, 232
Glutamine, 161
Glutaraldehyde, 225
$-SO_2$ Glycine, 232
Glycol dimethacrylate, 13
Glycolic acid (GA), 122, 203
Glycolide, 124, 203
Glycoprotein extracts, 225
Good solvents, 24
Grafting of hydrogel, 225
Graft polymers, see also Copolymers, light scattering; Copolymers, viscometry, 50
Granular matrix tablets, 188
Granuloma(s), 247
  foreign body, 241
Graphite-benzalkonium-heparin (GBH), 227

## H

Half-time escape rate ratios, of copolyether-ester membranes, 161
Hardening, see Orientation hardening; Work hardening
Hardness, 23
HDPE, 83, 84
creep, 81
creep modulus, 82
strain fracture, 88
stress relaxation data, 82
Heat-setting, 77
Hemodialysis, 145—148
hemofiltration, 148
in hepatic failure, 163
ultrafiltration, 148, 149
Hemodialysis membranes, compared to hemoperfusion membranes, 149
Hemofiltration, 148
Hemofiltration membranes, compared to hemodialysis membranes, 149
Hemoperfusion, 169
Heparin, see also Polyelectrolytes, 51, 196, 233
activity of, 228
blood coagulation and, 222—224
coating of surfaces with, 227—229
covalent attachment of, 169
ionic attachment of, 169
ionic binding of tocationic surfaces, 227
low molecular weight affinity for antithrombin III, 229
polysaccharide chain of, 232
solution properties of, 25
structure of, 51
surface-attached, 227
Heparin analogs, see Heparin
Heparin-coated surfaces, 229
Heparinized acrylonitrile copolymer membranes, 169
Heparinized membranes, 168—169
Heparinized polymers, 169, 207
Heparin-like activity, 232
Heparin-like materials, 229—233
Heparin-like properties, 230
Hepatic failure, 163
High molecular weight kininogen (HMWK), 217, 219, 226
Histamine, 195
Histidine, 161
Histiocytes, 241—243, 245
Histology, 121
Hollow fiber dialyzer, 145, 146, 148
Hollow fiber membrane(s), 153
Cuprophan, properties, 155
polyacrylonitrile, 163
poly(methyl methacrylate), 165
Homopolymers, 203
HPLC, 44
HRSD, 228
Human blood, composition of, 217
Human umbilical collagen, 225
Human plasma, see Plasma, human
Hydraulic conductivity coefficient, 150
Hydraulic permeability, 149
Hydrocephalus, 4
Hydrocephalus prosthesis, 12
Hydrocortisone, 195
Hydrodynamic size, 44
Hydrogel(s), 5, 13—14, 184, 197, 225—226
microporous, 207
release of drugs from, 189
Hydrogel systems, 197—202
Hydrogen bonds, 58
Hydrolytic degradation, 10
Hydrophilic drugs, 195
Hydrophilicity, 225, 230
Hydrophilic polymers, 13
Hydrophobic drugs, 195
Hydrophobic nondegradable devices, 194—197
Hydrophobic materials, 224—225
α-Hydroxy acids, 124
Hydroxyethylmethacrylate (HEMA), 13, 14, 197
2-Hydroxyethylmethacrylate (HEMA)-methacrylic hydrogel, 225
$-SO_2$ Hydroxyproline, 232

## I

"Ideal reaction", 247
Immunoglobulin(s), 189, 217
Immunologic reactions, 248
Impact properties, 13
Impact resistance, 11
Impact strength, 133
bioresorbable plastic materials, 137
Implant(s), 17, 129
polymers, 207
Implantation, of organic bioresorbable internal fixation devices, 121
Incontinence, 4
Indomethacin, 195
Inflatable prostheses, 241
Influence of regenerated cellulose membranes, 150
Inhomogeneity, of macromolecular materials, on molecular scale, 25—27
Injectable powders, 184
Inserts, into accessible body cavities, 183
Insoluble fibrin, 93
Insulin, 189, 196
Interface, 223
Interfacial polymerization, 200
Interferon, 207
Intermediates, preparation of, 11—13
Internal fixation, 120
Internal fixation devices, see Bioresorbable internal fixation devices
International standards, lack of, 15
Intestinal permeability, 208
Intramolecular properties of macromolecules, 23
"Intrinsic equilibrium constants", 53

Intrinsic pathway, for activation of blood coagulation, 218—221
"Intrinsic viscosity", 28, 50
Inulin, 149, 162
Ionic binding
heparin, to cationic surfaces, 227
polyelectrolytes in solution, 51
Ioplex, 231
Irradiation, 184
IR spectroscopy, see also Copolymer, 47
Isocyanate groups, 228

## J

Joint prosthesis, wear processes in, 9
Joint replacement surgery, 4

## K

Kallikrein, 219
Kidney, 145
Kinetics of clot formation, 224
Kininogen, high molecular weight, 217, 219, 226
Kupffer cells, 206

## L

Lactic acid (LA), 203
polymers and copolymers, degradation rates, 122
Lactide, 124, 203
Lactide diastereoisomers, 126
"Lag" effects, 186
Lamellar crystals, 73, 75, 79
Leucine, copolymers of, 205
L-Leucine, 161
Leucocytes, see White blood cells
Light scattering, see also Copolymer
determination of molecular weight, 37, 40
copolymers, 49—50
determination of radius of gyration, 25, 36—41
general, 26, 27, 38
measurement, 38
viscometry and, 27
Light scattering meters, 40
Liposomes, 206
Liquid phase, 67
Low molecular weight, 191
Low molecular weight compounds, 126
Low molecular weight heparin, affinity for antithrombin III, 229
Lubricants, 11
Lymphocyte(s), 217
Lysozyme, 196

## M

Macromolecule(s), 5
Macrophage(s), 206, 243
Magnetic microspheres, 206
Maleic anhydride, 202
Mammaplasties, augmentation, 240, 248
Mammary prosthesis, 4, 12, 247
silicone rubber, 240, 241, 245
Mandibular fracture(s), 121
repair, 137
Manufacturing process(es), 27, 129
effects of, 11—16
Mark-Houwink relationship, 28, 29, 50
Materials properties, 11
Matrices, cross-linked, 184
Maxwell viscoelastic model, 71
Measurement, for the measurement of specific properties, see specific properties
Mechanical properties, 9
bioresorbable plastic materials, 141
organic bioresorbable internal fixation devices, 122
polymers, 78-88
Medicine, fibrin in, 96—105
Medullary rods, 122
Megestrol acetate, 194
Melt, 62, 75, 184
crystallized polymers, 73
index, 13
of polymers, 69—71
Melting point, 23
poly(L-lactic acid), 121
Membrane(s), 184—186
blood compatibility, 152
design, approaches to, 152—153
evaluation, 151
properties, 151—152
role in biomaterials, 144
selection, for extracorporeal blood purification, 149—170
used in extracorporeal blood purification, permeability, 150—151
Membrane osmometer, see Osmometer
Membrane osmometry, see Osmometry
Membrane permeability factor, 150
Membrane separation processes, 144—149
Membrane systems, 153—154
Mestranol, 194
Methacrylate-based membranes, 165—166
Methacrylate hydrogel membranes, 166
Methacrylic acid, 14
Methacryloyl chloride, 228
$-SO_3$ Methionine, 232
Methyl acrylate, 196
Methyl 2-cyanoacrylate monomers, 205
Methyl methacrylate, 14, 196
Methyltricaprylammonium chloride (MTCAC), 227
Microcapsules, 184, 205
injectable, 203
Microchannels, 189, 194, 196
Microchemical investigations, 240
Microcrystalline collagen, 113, 114
Microencapsulation, 197

Microphase structures, 226
Microporous hydrogels, 207
Microstructure of polymers, 65—78
Middle molecules, 150
Mixing, free energy of, 192
$M_n$, see Number average molecular weight
Molecular orientation, 71
Molecular properties of macromolecules, see also Molecular weight, distribution; Polyelectrolytes; Shape of macromolecules; Size of macromolecules, 23, 33, 42
  determination of, from solution properties, 22, 23
  examples, 22, 23
  for spectroscopic methods, 47, 51
Molecular size, of diffusing molecule, 193
Molecular structure of macromolecules, see Polymers
Molecular weight, see also Colligative properties; Copolymers; Light scattering; Osmometry; Polyelectrolytes; Size exclusion chromatography; Ultracentrifugation; Viscometry, 13, 126, 204
  definitions, 23
  determination of number average, 27, 29, 31, 35, 48
  determination of weight average, 29, 38, 48—49
  determination of z-average, 27
  distribution of weight average, 27
  influence on crystalline polymers, 23
  log, 191
  of organic bioresorbable internal fixation devices, 122
  viscosity averaged, 29
Molecular weight distribution (m.w.d.), see also Copolymers; Fractionation; Polyelectrolytes; Size exclusion chromatogrpahy; Ultracentrifugation, 6, 13, 41
  bioresorbable plastic materials, 133
  definition, 23, 26
  determination of, 27, 41
  polymers, 64—65
  study of blood or tissue compatibility, 23
  study of polymer reaction, 47
  study of postpolymer reaction, 25, 26, 46
Monocyte, 217
Monofilament, 112
Monomer, residual, 126
Monomer molar concentration, ($C_m$), 52
Morphine sulfate, 189
Mouth, inserts into, 183
Multifilament, 113
m.w.d., see Molecular weight distribution
Myofibroblasts, 244, 245
Myoglobin, 162
Mylar, see Polyester

## N

Naloxone, 204
Naltrexone, 204
Nanocapsules, 184
Nanoparticles, 205
Narcotic addiction, 203
Narcotic antagonists, 203
Native collagen fibril, 113
Natural rubber, 6
Neck, 82, 83
Necrosis, 16
Negative charge, 230
Nephrophan membranes, 156
Network chain, 66—68
Neuroma(s), 241, 247
Nitrobenzene, 36
Nondegradable devices, hydrophobic, 194—197
Nonideality, 24
Nonideal solutions, 24
Nonthrombogenic behavior, 14
Norethisterone, 194
19-Norprogesterone, 194
Number average molecular weight ($M_n$), 35
  determination of, 31
Nylon, see Polyamide

## O

Ocusert Pilo, 195
Ophthalmic disorders, 4
Ophthalmic hydrogels, 13—14
Orbital floor, blow-out fractures of, 121
Orientation hardening, of polymers, 83
Oriented collagen fibril, 113
Oros, 200
Orthopedic applications of bioresorbable plastic material, 140
Osmometer, 32
Osmometry, 31—35
  determination of molecular weight, copolymers, 49
  general, 27
  vapor pressure, 36
Osmosis, 144, 145
Osmotic pumps, 201
Osteomyelitic bone cavities, 114
Osteosynthesis, 16, 120, 122, 133, 141
Osteotomy, 137
Ovalbumin, 162
Overall permeability, 150
Oxygen diffusion, 14

## P

Parallel plate dialyzer, 145, 147
δ-Parameters, 24
Particle scattering factor, see also Light scattering; Shape of macromolecules, 37, 39
Partition, 185
Partition coefficient, 185
Partition controlled release, 190
Partition membranes, 161

PEPT, tensile stress strain curve, 83
Peracetic acid treatment, 229
Permeability, 186
measurement, of collagen membranes, 160
of polymer membranes, 150—151
PET, see Polyethylene terephthalate
PETP, 72
molecular conformations, 74
PGA
fast-cured, 123
implants, 133
resorption, 132
Phagocytic cells, 206
Phagocytosis, 206, 243—245
"piggy-back", 206
Phagosomes, 244
Phase contact, 221
Phase separation, see also Fractionation, 46
Phenol, 36, 160
D,L-Phenylalanine, 161
Phosphate, 160
*p*-Phylene mustard, 205
"Piggy-back" phagocytosis, 206
Pigments, 11
Pilocarpine, 195
Pin(s), absorbable, 122
Pinocytosis, 206
PLA 100
boneplates, 133
implants, 133
mechanical and biological properties, factors affecting, 123, 126
screws, histological appearance, 139
PLA X GA.Y copolymers, 129
PLA X stereocopolymers, 127
Plasma, human, 217—218
Plasma levels, 182, 183
Plasma proteins, 207
Plasmin, 223
formation, 221, 222
Plasminogen activator(s), 227, 233
Plasticization, of cellulose membrane, 154
Plasticizer(s), 4, 10, 11, 194
Plastic materials
fabrication, 14—15
mechanical properties and biological performance, 10—11
structure, 5—6
strength and, 6—7
surface and, 7—10
Plate(s), 121, 122
Platelet, 217
activation, 218—220, 230
adhesion, 225, 229, 230, 232
inhibition of, 226
aggregation, 229, 230, 232
inhibition of, 226
aggregation inhibiting effect, 228
count, in ultrathin cellulose membranes, 158
heparin coated surface interactions, 229
Platelet aggregation inhibitor(s), 227, 233
Pluronic-F68, 207
PMMA, see Polymethylmethacrylate
Poiseuille's law, 30
Polyacetals, 5
Polyacetyl, 15
Polyacrylamide, 197, 198
Poly(acrylamide), 189
Polyacrylic acid, 51
Polyacrylonitrile, 162—163
Poly(alkyl 2-cyanoacrylates), 205
Polyamide (nylon), 5, 193, 200, 216
aromatic, 166
based membranes, 166
epoxide resin copolymer, 166
polymer, structure, 201
Polyamidoamine-styrene, 228
Poly(amino acids), 205
Polyanions, synthetic, 230
Poly(butyl methacrylate-costyrene), 45
Poly(n-butyl methacrylate), 45
Poly(caprolactone), 204
Poly(caprolactone)glycol, 12
Polycarbonate (Lexan), 193
based-membranes, 163, 165
glassy, 72
Polycarbonate-polyether block copolymer membrane, 165
Polydiastereoisomer type of copolymer, 124
Polydimethylsiloxane (PDMS) (Silastic), 58, 59, 193, 240, 247
yield stresses, 82
Poly(dioxanone), 204
Polydisperity ratio, 27
Polyelectrolyte(s), see also Inhomogeneity, 23, 51, 167, 207, 230
charge, 51
evidence of, 52
examples, 51
as biopolymers, 23
membranes, 167
molecular weight (distribution), 52
solution properties, 51, 52
solutions, 51
viscometry, 31, 52
Polyelectrolyte-complexes, 53, 231
Polyester (Dacron, Mylar), 7, 120, 193, 216, 225
fiber, 4, 7
uses, 5, 7
Polyether block copolymer membranes, 161
Polyethylene (PE), 25, 188, 193, 196
chain regularity, 62
cold flow (creep), in, 4
commercial, 63
crazing of, 83
environmental stress cracking agents, 86
microstructure, 82
molecular weight distribution, 65
silica in, 15
spherulites, 75, 76
temperature variation, 79
strain fracture, 88

stress concentrations, 86
structures, 192
substituted, 195—197
temperature variation, 80
tensile stress strain curve, 83
unit cell, 73, 74
uses, 58, 59
yielding, 84, 86
yield stresses, 82, 83
Young's modulus, 80
Polyethylene (low density LDPE), 5
Polyethylene (UHMWPE), 5
Poly(ethylene glycol), 195, 199
structure, 198
uses, 196, 198, 201, 207—208
Polyethyleneimine, 51
Poly(ethylene oxide), 198
Polyethyleneterephthalate (PET), 58, 59, 65
crystallization rate, 77
fibers, 75, 77—78
yield stresses, 82
temperature variation, 80
yielding, 86
Poly(ethyl methacrylate), 45
Poly(ethyl methacrylate-costyrene), 45
Poly(glutamic acid), 205, 206
Polyglycolic acid (PGA), 5
Poly(glycolic acid) (PGA), 122, 204
Poly(α-hydroxy acids), 124—127
Poly-α-hydroxy acid type polyester, 120
Polyhydroxyethyl methacrylate, 13
Poly(hydroxyethyl methacrylate), 192, 196
Poly(2-hydroxyethyl methacrylate) (HEMA), 189, 198
Poly(hydroxyethylmethacrylate-glycidyl-methacrylate), 228
1,4-*cis* Polyisoprene, 230
Polylactic acid, 5
Poly(lactic acid), 121, 122, 184, 204
degradation rate, 125
Poly(lactic/glycolic acid), 204
Poly(DL-lactic acid), 121
Poly(L-lactic acid), 12, 121—123, 203
Polylysine, 51
Polymer(s)
commonly used as biomaterials, 59
cross-linking of, 65, 66
current clinical use, listing of, 5
degradation rates, 122
mechanical properties, 78—88
membrane, see Membrane
microstructure, 65—78
molecular structure, 58—65
molecules
shape of, see Shape of Macomolecules
size of, see Size of macromolecules
solubility of, 23—25
structures of, 23
surfaces, cracking of, 10—11
thermoplastic, 184
used in surgery, see also Blood compatibility; Tissue compatibility, 23
Polymeric devices, drug release from, 187
Polymeric drugs, 182, 184
Polymerization, see also Molecular weight distribution, 11—14, 62
reaction, 46, 184
ring-opening, 124
step-growth, 124
study of, 46
Polymer tissue interactions, 207—208
Polymethacrylic acid, 230
Poly(methacrylic acid), 207
Poly(methylacrylate), 228
Polymethylacrylate-*N*-acrylhexamethylene-diamine, 228
Poly(methyl 2-cyanoacrylate), 192
Polymethylmethacrylate (PMMA), 58, 59, 62
craze formation, 87
environmental stress cracking agents, 86
strain fracture, 88
temperature variation, 80
yielding, 73, 84
yield stresses, 82
Poly(methyl methacrylate), 192, 196, 228
hollow fiber membrane, 165
Poly(α-methyl) styrenes, 33
Polynuclear neutrophile, 217
Poly(n-octyl methacrylate), 45
Poly(n-octyl methacrylate-costyrene), 45
Polyols, 198
Poly(orthocarbonates), 203
Poly(orthoesters), 203
Poly(oxyethylene), block copolymers of, 207
Poly(1,2 oxyethylene) glycol, 12
Polyoxymethylene, 59
Poly(oxypropylene), block copolymers of, 207
Polypeptide(s), 52, 53
membranes, 166, 167
Polypropylene (PP), 16, 62, 226
film, fibers from, 16
molecular conformations, 74
Polysaccharides, see also Polyelectrolytes, 23
Polystyrene (PS), 7, 45, 59, 62, 225
based membranes, 166
crazing, 84
resins, activity coefficients of, 232
Polystyrene copolymer, 193
Polystyrene-polyamine, 226
Polystyrene sulfonates, 230
Polysulfone membranes, 167, 168
Polytetrafluoroethylene (Teflon), 59, 216
Poly(tetramethylene adipate), glycol, 12
Polyurethanes, 7, 12, 59, 158, 225, 228
Poly(vinyl acetate), 196
Polyvinyl alcohol, 29
Poly(vinyl alcohol), 167—168
Poly(vinylbenzyltrimethyl ammonium) chloride, 231
Polyvinylchloride (PVC), 45, 59, 62
plasticized, 15
Poly(vinyl chloride) (PVC), 5, 6, 188, 196, 198

Poly(vinyl-N-pyrrolid-2-one), 198
Poly(vinylpyrrolidone), 166, 189
Pores, 188—189
Postpolymer reactions, see also Molecular weight distribution, 25, 46
Powder(s), injectable, 184
Powdered formulations, 203
Precipitation, see also Fractionation, 47
Prekallikrein, 217, 219
Prepolymers, preparation of, 11—13
Processability, control of, 11
Processing, 86, 126
Product
  design, faulty, 86
  properties, 59
Progestasert, 195
Progesterone, 193—194
L-Proline, 161
$-SO_3$ Proline, 232
Prostacyclin ($PGI_2$), 221
Prostaglandin ($E_2$), 182, 189, 199, 227
Prosthetic devices, 12, 182, 247
  removal, 248
  types of film in, 247
Protein(s), see also Polyelectrolytes, 23, 25, 52, 194
  adsorption, 226
  coatings, 224—226
  concentrations, in normal human plasma, 217
  kinetics of deposition, 226
  relative affinities of, 226
Pyran copolymer, 207
Pyrimethamine, 195, 204

## Q

Quality control, 13
Quinoxaline di-*N*-oxide, 200

## R

Radiation, 126
Radiolabeling, of organic bioresorbable internal fixation devices, 121
Radio opacifier, 13
Radius of gyration
  applicability to light scattering and, 41
  determination, 36—41
  of macromolecules in solution, 25
Raffinose, 162
Raffinose-$5H_2O$, 161
Random copolymers, see also Copolymers, light scattering; Copolymers, viscometry, 50
Raoult's law, 24
Rats, resorption of bioresorbable plastic materials in, 127—129
Rayleigh ratio, see also Light scattering, 38, 41
Reconstituted fibrous collagen products, 112
Reconstructive facial surgery, 137
Rectal route of administration, 205
Rectum, inserts into, 183
Refractive index increment, see also Light scattering, 41, 43, 45, 46, 49
Regenerated cellulose membranes, influence of, 150
Repeating units, 124
Resorbable composite material, 135
Resorption, 122
  of bioresorbable plastic materials, in sheep, 129—133
Retromammary prosthesis, 247
Rigidity, 11
Ring-opening polymerization, 124
Rods, 184
Role of membranes, in biomaterials, 144
Rotational isomers, 59—62
Rubber, see also Silicone rubber, 62, 65, 79, 192
  elastic properties, 67, 68
  natural, 6
  synthetic, 5
Rubber state, 66, 191
  of polymer, 66—69
Run number R, see also Copolymer, 47

## S

Saline, 247
Salt effects, 52
Sarcosine, 161
Screw(s), 121, 122
Sedimentation coefficient, 42, 43
Semicrystalline polymers, yielding, 84
Semicrystalline state of polymers, 72—75
  fibrous, 75—78
Sequence distribution, see also Copolymer, sequence distribution, 47
D,L-Serine, 161
Serum albumin, 51
Shape factor $\nu$, 28
Shape of macromolecules, insolution, see also Light scattering; Particle scattering factor; Radius of gyration; Shape factor $\nu$; Sedimentation coefficient; Viral coefficient; Viscometry, 22, 24
  determination of, 28, 38
    viral coefficients, 33
Sheep
  degredation of bioresorbable plastic materials in, quantitative measurement, 133—134
  resorption of bioresorbable plastic materials in, 129—133
Sheets, 122
Short term devices, 229
Si contamination, 247
Sieving coefficients, 149
SiK-$\alpha$-radiation, 245, 246
Silastic, 193, 195, 216
Silastomers, 240
Silica
  filler, 4

in polyethylene, 15
Silicon ($SiO_2$), 245
Silicone(s), 189, 190, 193—195
concentration, 247
droplets, 247
elastomer, 4
manufacture, 12
standards, 13
gel, 241, 247, 248
diffusion of, 4
membranes, diffusion rate of steroids accross, 194
oil, 247, 248
polymer, 240, 247, 248
Silicone prostheses, 245, 246
Silicone rubber (Silastic), 10, 216, 225, 226
decomposition of, 247
fibrous capsules surrounding, 242
mammary prosthesis, 241
phagocytized, 243—245
properties, 12
use, 5, 14, 194, 240
Siliconomas, 241, 242, 247
Sink conditions, 186
"Site binding", of counterions, 53
Size, macromolecules, in solution, determination of, see also Effective radius; Hydrodynamic size; Molecular weight; Radius of gyration; 22, 24,s3628
viral coefficients, 33
Size exclusion chromatography, 27, 44—47
determination of molecular weight (distribution), copolymer, 50
general, 46
polyelectrolytes, 52
Skin, 183
Skull fracture, repair, 137, 140
Slabs, diffusional release of solutes from, 187
$-SO_3^-$, 232
Sodium, 160
Sodium alginate, 189, 195
Sodium dextran sulfate, 231
Sodium iodide, 189
Sodium poly(styrene sulfonate), 231
Solubility, 193
polymers, 23—25
Solute(s)
diffusion across membrane, 185
diffusional release of, 187
Solute flux, 150
Solute transport characteristics, of cellulose membranes, 154
Solution properties of macromolecules, as tool for definition of molecular properties, see also Blood compatibility; Dissolution of macromolecules; Nonideality; Polyelectrolytes; Tissue compatibility, 22
"θ Solutions", 24
Solvents, 24, 25
θ Solvents, 24, 29
Sorbent membranes, 169—170
Soy bean trypsin inhibitor, 196
Spacer groups, 184
C-NMR Spectroscope, see also Copolymer, 47
Spectroscopic techniques, see also Copolymers, 47, 51
IR Spectroscopy, see also Copolymer, 47
Spheres, 203
diffusional release of solutes from, 187
Spherulites, 75, 76, 79
breakdown of microstructure, 83
Spherulitic semicrystalline state, of polymer, 73—75
Sponge(s)
collagen, 114
gelatin, 112, 116
Spray-drying, 184
Squalane, 45
"Square meter-hour" hypothesis, 150
Stabilizer(s), 4, 11
Standards, international, lack of, 15
Step-growth polymerization, 124
Stereocopolymers
biodegradation, 133
degradation, 129—131
enantiomeric composition, 25
PLA X, 127
stereosequence redistribution, 125
Stereoisomerism, 62, 63
Stereosequence redistribution, 125
Sterilization, 15, 126
Steroid(s), 194, 195
Steroidal contraceptives, 203
Stiffness, 23
Strand, 112
Strength, 11
Streptokinase, 227
Stress-corrosion crack, 10
Stress cracking, environmental, 86
Stress relaxation, 81, 82
Structural analysis, of organic bioresorbable internal fixation devices, 13
Structure, see also Molecular properties of macromolecules, 126
determination, 22
plastic materials, 5—6
strength and, 6—7
surface and, 7—10
polymers, 23
Styrene, 13
Substituted polyethylene, 195—197
Sucrose, 161, 162
Sulfadiazine, 204
Sulfamide groups, 231
Sulfonate, 231
Surface, 23
composition, 226
polymer, cracking of, 10—11
Surface chemistry, of plastic materials, 7—8
Surface grafting, 14
Surface morphology, of plastic materials, 8

Surface properties, effect on biological performance, 8—10
Surgical prosthesis, use of gelatin in, 114
Suture(s)
absorbable, 114
collagen, 112, 113
Swelling, 24, 184, 189, 192, 193
Synthesis, 120
Synthetic polyanions, 230
Synthetic polymers, 240
Synthetic rubbers, 5
Synthetic surfaces, 207

## T

Targeting, 205—207
Teeth, 200
Teflon, see Polytetrafluoroethylene
"θ-Temperature", 24
Tendon collagen fibril, 113
Tensile properties, 13, 16
Tensile strength, 23
bioresorbable plastic materials, 133, 134
measurement, 121
variations, 136
Tensile stress strain curve, 83
Testosterone, 194
Tetrahydropyranyl, 204
Thermal analysis differential, of organic bioresorbable internal fixation devices, 121
Thermal curing, 184
Thermal degradation of macromolecules, 126
Thermodynamic considerations, 6
Thermoplastic polymers, 184
Thiamine chloride, 161
Thiophosgene, 228
D Threonine, 161
$-SO_3$ Threonine, 232
Thrombin, 227, 232
Thrombogenicity, 225—226, 231
Thromboresistance, 225, 229
Thromboresistant surface, 227
Tissue(s), regeneration of, 114
Tissue collagen, 225
Tissue compatibility, and molecular properties of macromolecules, see also Molecular properties, 23
Tissue-polymer interactions, 207—208
Tissue reaction, 127
to poly(lactic acid), 121
Tortuosity, 189
Toughness, 11
Toxicity testing, 15
Toxic residues, 197
"*Trans*" (t) conformation, 60
Transdermal delivery, 205
Transdermal therapeutic system, 83
Transesterification reactions, 125
Tridocylmethylammonium chloride (TDMAC) treatment, 227, 230
Tripelenamine-HCl, 189
Tropocollagen, 113
Trypsin inhibitor, 196
Tubes, tubing, 114
gelatin, 116
gelatine, 116
small-diameter, 232
Tumorigenesis, 207
Turbidity, see also Light scattering, 37

## U

Ultracentrifugation, see also Copolymer, ultracentrifugation, 41—47, 50
copolymers, 75
determination of molecular weight (distribution), 42, 43
general, 26, 27
Ultracentrifuge, 41, 42, 50
Ultrafiltration, 144, 145
controlling of, with cellulose acetate membranes, 159
hemodialysis, 148, 149
Ultra high molecular weight polyethylene (UHMWPE), 4, 5, 9, 10
standards, 13
Ultrathin cellulose membranes, 157, 158
Umbilical collagen, 225
Urea, 160—162
Urease, 203
Urethane polymers, 200
Uric acid, 160, 161
Urokinase, 227
Uterus, inserts into, 183

## V

Vagina, inserts into, 183
Vaginal route of administration, 205
L-Valine, 161
Van der Waals' bonds, 58
Van der Waals' forces, 72, 79
Van't Hoff's law, 32—33
Vapor pressure measurement, 31
"Vapor pressure osmometry", 31, 36
Vascular prosthesis, 4, 7
Vessels, small diameter, for long term uses, 232
Vinyl acetate, 230
Vinyl alcohol, copolymerization, 69
Vinyl pyrrolidone, 14, 166
Vinyl sulfonates, ethylene with, 230
Vinyl coefficients
copolymers and, 49
determination, 33
nonideality of solutions and, 24
shape of macromolecules in solution and, 33
Viral series, see Viral coefficients
Viscoelastic model of polymers, 71, 80, 81
Viscometer(s), 30, 31

Viscometry, see also Shape factor $\nu$, 27—31
  determination of molecular weight, 29
  light scattering and, 27
  viscometric properties
    copolymer solutions, 50
    polyelectrolyte solutions, 31
Viscose process, 154
Viscosity(ies), see also Viscometry, 13, 82
  intrinsic, 28, 50
  of liquid, 27
  of polymer, 69
  specific, of organic bioresorbable internal fixation devices, 121
Viscosity-averaged molecular weight $M_v$, see also Molecular weight
Vitamin $B_{12}$, 149, 162
Volume flux, 150
von Willebrand factor, 218, 226


## W


Water, 36
  structure, organized, 225
Water-soluble drugs, 183
Water-swollen matrices, 183
Wear resistance, 11
Weight average molecular weight $M_w$, see also Molecular weight, 26
Wet gel membranes, 156—157
White blood cells, 217
  count, in ultrathin cellulose membranes, 158
Work-hardening of polymers, 83
Workshop processes, 11
Wound dressings, 205


## X


X-parameters, 24
X-ray diffraction, 121


## Y


Yielding, of polymers, 72, 73, 84, 87
Yield stresses, of polymers, 82
Y-irradiation, 196
Young's modulus, 23
  of polymers, 78—79, 82, 84
    of polyethylene, 80, 81


## Z


Z-average ($M_z$), see also Molecular weight, 27
Zimm plot, 40